163 Topics in Current Chemistry

Photoinduced Electron Transfer IV

Editor: J. Mattay

With contributions by
R. F. Khairutdinov, H. D. Roth, P. Suppan,
K. I. Zamaraev

With 81 Figures and 14 Tables

Springer-Verlag
Berlin Heidelberg GmbH

This series presents critical reviews of the present position and future trends in modern chemical research. It is addressed to all research and industrial chemists who wish to keep abreast of advances in their subject.

As a rule, contributions are specially commissioned. The editors and publishers will, however, always be pleased to be receive suggestions and supplementary information. Papers are accepted for "Topics in Current Chemistry" in English.

ISBN 978-3-662-14987-4 ISBN 978-3-540-46734-2 (eBook)
DOI 10.1007/978-3-540-46734-2

Library of Congress Catalog Card Number 74-644622

Originally published by Springer-Verlag Berlin Heidelberg New York in 1992
Softcover reprint of the hardcover 1st edition 1992

Typesetting: Th. Müntzer, Bad Langensalza;

51/3020-5 4 3 2 1 0 – Printed on acid-free paper

Guest Editor

Prof. Dr. *Jochen Mattay*
Organisch-Chemisches Institut,
Westfälische Wilhelms-Universität Münster,
Orléansring 23, D-4400 Münster

Editorial Board

Preface to the Series on Photoinduced Electron Transfer

The exchange of an electron from a donor molecule to an acceptor molecule belongs to the most fundamental processes in artificial and natural systems, although, at the primary stage, bonds are neither broken nor formed. However, the transfer of an electron determines the chemical fate of the molecular entities to a great extent. Nature has made use of this principle since the early beginnings of life by converting light energy into chemical energy via charge separation. In recent years man has learnt, e.g. from X-ray analyses performed by Huber, Michel and Deisenhofer, how elaborately the molecular entities are constructed within the supermolecular framework of proteins. The light energy is transferred along cascades of donor and acceptor substrates in order to prevent back electron transfer as an energy wasting step and chemical changes are thus induced in the desired manner.

Today we are still far away from a complete understanding of light-driven electron transfer processes in natural systems. It is not without reason that the Pimentel Report emphasizes the necessity of future efforts in this field, since to understand and "to replicate photosynthesis in the laboratory would clearly be a major triumph with dramatic implications". Despite the fact that we are at the very beginning of knowledge about these fundamental natural processes, we have made much progress in understanding electron transfer reactions in "simple" molecular systems. For example most recently a unified view of organic and inorganic reaction mechanisms has been discussed by Kochi. In this context photochemistry plays a crucial role not only for the reasons mentioned above, but also as a tool to achieve electron transfer reactions. The literature contains a host of examples, inorganic as well as organic, homogeneous as well as heterogeneous. Not surprisingly, most of them were published within the last decade, although early examples have been known since the beginning of photochemistry (cf. Roth's article in Vol. 156). A reason is certainly the rapid development of analytical methods, which makes possible the study of chemical processes at very short time ranges. Eberson in his monograph, printed by this publishing company two years ago, nicely pointed out that "electron transfer theories come in cycles". Though electron transfer has been known to inorganic chemists for a relatively long time, organic chemists have still to make up for missing concepts (cf. Eberson).

A major challenge for research in future, the "control of chemical reactions" as stated by the Pimentel Report, can be approached by various methods; light-driven processes are among the most important ones. Without interaction of the diverse scientific disciplines, recent progress in photochemistry, as well as future developments would scarcely be possible. This is particularily true for the study of electron transfer processes. Herein lies a challenge for science and economy and the special fascination of this topic – at least for the guest editor.

The scope of photochemistry and the knowledge about the fundamentals of photoinduced electron transfer reactions have tremendously broadened within the last decade, as have their applications. Therefore I deeply appreciate that the Springer-Verlag has shown interest in this important development and is introducing a series of volumes on new trends in this field. It is clear that not all aspects of this rapidly developing topic can be exhaustively compiled. I have therefore tried to select some papers which most representatively reflect the current state of research. Several important contributions might be considered missing by those readers who are currently involved in this field, however, these scientists are referred to other monographs and periodical review series which have been published recently. These volumes are meant to give an impression of this newly discovered reaction type, its potential and on the other hand to complement other series.

The guest editor deeply appreciates that well-known experts have decided to contribute to this series. Their effort was substantial and I am thankful to all of them. Finally, I wish to express my appreciation to Dr. Stumpe and his coworkers at the Springer-Verlag for helping me with all the problems which arose during the process of bringing the manuscript together.

Münster, December 1989 Jochen Mattay

Preface to Volume IV

The fourth volume of the PET (Photoinduced Electron Transfer) series is again devoted to some fundamental topics. All the chapters present comprehensive reviews. I am very grateful to the authors whose efforts have made this venture possible, especially, since two of them have already contributed to previous issues.

Photoinduced electron tunneling reactions have not yet been adequately reviewed in the literature although their use is widespread in photochemistry and photobiology. In the first chapter, this topic is discussed in depth covering organic and metal complex chromophores, bifunctional molecules, organized molecular assemblies as well as biological systems. The ability of electron tunneling at large distances may lead to unique opportunities in organizing photochemical conversions on the molecular level.

The second article deals with probably the most fascinating predictions of modern electron transfer theories i.e. the "Marcus Inverted Region (M.I.R.)". It was shown only one decade ago, nearly 20 years after the first formulation of the Marcus theory, that the M.I.R. does indeed exist: First for thermal charge shifts and later for charge recombination. Even a charge separation reaction was recently found to behave according to the Marcus theory. Nevertheless, many reactions do not follow the Marcus model and therefore the second contribution of this issue is mainly concerned with this question.

The final contribution is focussed on organic radical cations in a comprehensive and fundamental manner. It starts out with experimental methods of generation and characterization followed by a discussion of various types of electron transfer induced reactions. In the last section unusual structures of radical cations are described.

The articles of this volume again clearly demonstrate that "photoinduced electron transfer is one of the very few fundamental reaction types" (Roth) both for natural and artificial systems.

Münster, January 1992 Jochen Mattay

Table of Contents

Table of Contents of Volume 156

Table of Contents of Volume 158

Table of Contents of Volume 159

Photoinduced Electron Tunneling Reactions in Chemistry and Biology

K. I. Zamaraev and **R. F. Khairutdinov**

Institute of Catalysis of the Siberian Branch of the Russia Academy of Sciences, Prosp. Akad. Lavrentieva, 5, Novosibirsk 630090, Russia
Institute of Chemical Physics of the Russia Academy of Sciences, Kosygina str. 4, Moscow 117977, Russia

Table of Contents

A comprehensive review is given of the works on electron tunneling at large distances in photoinduced electron transfer reactions. Evidence is presented showing that electron tunneling reactions are rather widely spread in photochemistry and photobiology. The ability to participate in such reactions is inherent in both excited and ground states of various organic and inorganic molecules and radical ions, transition metal complexes and clusters, porphyrins, redox sites of proteins, etc.

The regularities of photoinduced electron tunneling are discussed in detail. The most outstanding feature of electron tunneling is demonstrated to be its ability to provide the occurrence of both primary and secondary reactions of PET between remote electron donor and electron acceptor sites, at distances sometimes as great as several tens of angstroms. It is shown to be responsible also for rather unusual kinetic regularities of certain PET reactions. Electron tunneling can also provide a new type of photoinduced electron transfer reactions, i.e. PET stimulated by illumination into the tunnel electron transfer band.

The rate of the tunnel PET reactions increases strongly when mediator centres are placed in the space between the donor and the acceptor. This effect is the more pronounced the smaller the gap is between the energy of the tunneling electron and the energy levels of the mediator.

The ability of electron tunneling to provide PET at large distances is shown to open up unique opportunities in organizing photochemical conversions on the molecular level. These opportunities are widely used in photobiology as well as in photochemistry of:

1) specially designed organic molecules containing photochemically active electron donor and acceptor sites which are oriented in a certain fashion and linked together by one or several molecular bridges, and

2) photochemically active organized molecular assemblies such as molecular multilayers, micelles, vesicles, etc.

1 Introduction

Until recently it seemed to be self-evident that chemical reactions are possible only for molecules which were in direct contact with each other. The reason for this viewpoint was a sharp decrease of the interaction energy between molecules with increasing intermolecular distance, so that at distances exceeding the sum of the radii of the reagents the interaction becomes too small to cause chemical transformations.

Over the last two decades, however, numerous data have been obtained that do not fit within this postulate of chemical kinetics which had previously seemed to be unshakeable. In several laboratories, inter- and intramolecular chemical and biochemical redox reactions have been observed between particles separated by distances considerably exceeding the sum of their van der Waals' radii, and amounting to several tens of angstroms. All these reactions have been found to proceed via one and the same mechanism: by tunnel electron transfer through a potential barrier separating the reagents. The barrier may be rather high (up to several electron volts), while the energy of interaction ensuring the electron transfer may be very low, far lower than that of thermal molecular motion. Reactions of this kind are usually referred to as reactions of long-range electron tunneling or simply electron tunneling reactions.

Concepts relating to the tunneling of particles through a potential barrier were introduced in pioneering works in physics immediately after the creation of quantum mechanics and were used to account for such phenomena as α-decay of atomic nuclei [1, 2], cold emission of electrons from metals [3] and the ionization of atoms in strong electric fields [4].

Tunneling of atomic nuclei in the course of an elementary act of chemical reaction was first considered theoretically in Refs. [5–8] soon after quantum mechanics had been created. It has been shown that nuclear tunneling may lead to unusually large isotope effects for reactions in which light atoms (hydrogen, deuterium, tritium) are transferred and to a decrease in the effective activation energy of chemical processes as the temperature decreases.

Convincing experimental evidence of nuclei tunneling in the course of chemical reactions has been obtained over the last two decades in studies of chemical reactions at low temperatures (below 100 K, as a rule) in condensed media. Nuclear tunneling in chemical reactions, however, usually occurs only over short (less than 1 Å) distances. Elementary estimations show that in chemical reactions practically no tunneling of nuclei can be observed at distances exceeding the sum of the van der Waals' radii of reacting molecules. This is due to the large mass of the atomic nuclei.

1.1 Tunneling in Dark Electron Transfer Reactions

The mass of the electron is small compared with that of the nuclei and so the electron can tunnel over much greater distances. For example, in solid state physics, it is well known that electron tunneling at great distances between the impurity centres of semiconductors plays an important role in providing their conductivity. In contrast, until recently, no strong experimental evidence had been reported in favour of intermolecular electron tunneling at the great distances in systems that are important for chemistry and biology. It should be noted that attempts have repeatedly been made to estimate theoretically the distances to which an electron can tunnel in the course of chemical reactions. However, due to a rather uncertain choice of parameters needed for the calculation, such estimates are not considered to have been sufficiently reliable.

A great role in substantiating the importance of electron tunneling reactions was played by Ref. [9], where the characteristic time, $\tau_{1/2}$, of electron transfer from the heme site of the cytochrome molecule to the chlorophyll molecule in a bacterium was shown to be constant within the temperature range of 130 to 4.2 K. The temperature independence of $\tau_{1/2}$ permitted one to reject a diffusion mechanism for the process. However, it was still impossible to exclude the possibility of the reaction proceeding via direct contact between the active sites of the reacting molecules.

To our knowledge, the first process in which at the moment of reaction, the reagents were proved by direct experiment to be widely separated was the reaction between electron, e_{tr}^-, trapped in an alkaline vitreous solution and the anion-radical O^- studied in Ref. [10]. By measuring the width of the e_{tr}^- electron paramagnetic resonance (EPR) line, it was possible to prove that, for most pairs of reagents, the distance could not be less than 14 Å. Further analysis of the e_{tr}^- EPR line shape showed this distance to be actually longer and in fact to exceed 20 Å. The coincidence of kinetic curves for this reaction at 77 and 4.2 K allowed one to reject unambiguously the control of the reaction rate by thermal diffusion, which is known to be slowed down dramatically as the temperature decreases. As proved by subsequent and more detailed kinetic research in the temperature range 4.2–93 K, the main channel of the reaction between e_{tr}^- and O^- is non-activated electron tunneling over a great distance and, at higher temperatures, activated electron tunneling over a great distance. At still higher temperatures, the process consists of two stages:

1) the approach of the reacting particles via diffusion up to the distance R_D at which the characteristic time of tunneling becomes, within an order of magnitude, equal to the time of a diffusion jump, and

2) the subsequent electron tunneling to the distance R_D.

For electron tunneling reactions in vitreous solutions the change of the reagent concentration with time has a rather unusual (logarithmic) character. Therefore, the quantitative investigation of the kinetics of such reactions has to be done over a wide interval of time, t, (e.g. for the above-mentioned reaction of e_{tr}^- with O^-, this range amounted to five orders of magnitude). The combination of the measurements made using the pulse radiolysis technique with those made for a very long t, made it possible to carry out, for the reaction $e_{tr}^- + Cu(en)_2^{2+} \rightarrow Cu(en)_2^+$ (where en stands for ethylenediamine), a quantitative study of the electron tunneling kinetics within a range of time as broad as 13 orders of magnitude, from 10^{-7} to 10^6 s under the conditions of controlled spatial distribution of the reagents [11].

Up til now many kinetic investigations of electron tunneling over large distances have been made. They include processes not only of such "exotic" particles as e_{tr}^-, but also conventional chemical compounds. The range of these reactions is rather wide. It includes electron transfer from organic anion-radicals to organic molecules [12, 13], reactions between compounds of metals of variable valence [14], transfer from inorganic anion-radicals to compounds of metals [15], in systems modelling the chain of electron transfer in biological objects [16], between electron donor and electron acceptor centers on the surface of heterogeneous catalysts [17], intramolecular long-range electron transfer between donor and acceptor fragments separated by an inert bridge [18].

The results of experimental research have also stimulated the appearance of theoretical papers devoted to the analysis of an elementary act of electron tunneling reactions in terms of the theory of non-radiative electron transitions in condensed media and to the derivation of the kinetic equations of long-range electron transfer processes [19–30].

The comprehensive review of experimental and theoretical works on long-range electron tunneling in chemistry and biology can be found in Ref. [31].

1.2 Kinetic Peculiarities of Electron Tunneling Reactions

Two main conclusions about the kinetics of electron tunneling follow from theoretical works on electron tunneling. One of them is that the dependence of the tunneling probability, W, on distance R between the donor and the acceptor can be well approximated by the exponential function

$$W = \nu_e \exp(-2R/a_e), \tag{1}$$

where ν_e is the frequency factor, which depends on the nuclear motions in reagents and a medium and may reach 9 value up to $10^{20}\,s^{-1}$, and $a_e \sim 1\,Å$ is a parameter depending on the overlapping of electron wave functions of the donor and the acceptor. The other is that the kinetics of the processes of long range electron transfer depends on the mutual spatial distribution of the donors and acceptors, and in vitreous matrices is described by unusual (logarithmic in time) kinetic equations that deviate dramatically from the usual first and second order kinetic equations.

Describing electron tunneling reactions in vitreous matrices it is convenient to use such a characteristic as the tunneling distance, R_t,

$$R_t = (a_e/2) \ln \nu_e t, \tag{2}$$

i.e. the distance at which the condition $W(R_t)\,t = 1$ is fulfilled. Because of a very strong dependence of W on R for the pairs of donors and acceptors with the distance between them smaller than R_t the probability of the reaction occurring by time t is close to 1, while, for the pairs with $R > R_t$, it is practically equal to zero.

As is seen from Eqs. (1) and (2), it makes sence to choose parameters ν_e and a_e as the basic kinetic characteristics for electron tunneling reactions. Knowing them, it is easily possible to find, with the help of Eq. (1), the probabilities W(R) of electron tunneling for various distances R between the reagents or to calculate with the help of Eq. (2), the distance R_t to which the electron will tunnel within time t.

At present numerous data on the values of ν_e and a_e for various pairs of reagents are available [31]. According to these data ν_e can vary over a very broad range around a typical value $\nu_e \simeq 10^{15}\,s^{-1}$, reaching $10^{20}\,s^{-1}$ as the upper extreme value. The typical values of a_e lie within the interval 0.5–2.5 Å, though occasionally even larger values of a_e were observed. These data can be used to estimate the maximum distances R_t at which electrons can tunnel during PET reactions. In order to this one has to substitute into Eq. (2) the maximum values $\nu_e = 10^{20}\,s^{-1}$ and $a_e = 2.5$ Å and an appropriate value of time t. For primary processes of electron transfer from an excited state of a photosensitizer to an electron acceptor or from an electron donor to an excited state of a photosensitizer, the life time τ of the excited state of the photosensitizer should be used as time t. For secondary dark processes their characteristic times τ_s should be used as t. Table 1 presents the maximum tunneling distances R_t^{max} for primary and secondary processes of PET reactions calculated using the values $\nu_e = 10^{20}\,s^{-1}$, $a_e = 2.5$ Å for values of τ from 10^{-9} s to 1 s and τ_s from 10^{-9} s to 10^6 s.

Table 1 also presents the tunneling distances R_t^{typ} calculated for most typical values of $\nu_e = 10^{15}\,s^{-1}$ and $a_e = 1$ Å for the same times t. As seen from Table 1, electron tunneling can indeed provide PET reactions between rather remote partners.

The ability of electron tunneling to provide PET at large and various distances when put together with a rather sharp exponential dependence of the tunneling probability on the distance can result in a rather unusual character of the reaction kinetics. The details of these kinetics depend substantially on the character of the spatial distribution of the reagents as well as on how mobile they are. We shall discuss this problem here only very briefly, just to provide better understanding of the data presented below on electron tunneling in PET. For a more detailed discussion see Chapter 4 of Ref. [31].

Consider first the kinetics of electron tunneling between immobile reagents. In this case the character of the spatial distribution of the reagents has the strongest influence on the reaction kinetics. From the practical point of view two types of

Table 1. Tunneling distances R_t^{max} and R_t^{typ} for primary and secondary processes of PET reactions calculated using Eq. (2) with $\nu_e = 10^{20}\,s^{-1}$, $a_e = 2.5$ Å (R_t^{max}) and $\nu_e = 10^{15}\,s^{-1}$, $a_e = 1$ Å (R_t^{typ}) for various values of time t. For primary processes the values $t = \tau = 10^{-9} - 1$ s can be taken as typical while for secondary processes all the values $t = \tau_s$ are possible

t, s	10^{-9}	10^{-6}	10^{-3}	1	10^3	10^6
R_t^{max}, Å	32	40	49	57	66	75
R_t^{typ}, Å	7	10	14	17	21	24

spatial distribution are most important: the pairwise distribution and the non-pairwise random distribution.

For pairwise distribution the kinetics of electron tunneling has the simplest form for the rectangular distribution function, i.e. when immobile reagents are located in isolated pairs with the distribution over the distance R of form

$$f(R) = \begin{cases} (R_2 - R_1)^{-1} & \text{for} \quad R_2 \geq R \geq R_1 \\ 0 & \text{for} \quad R > R_2 \quad \text{and} \quad R < R_1 \end{cases} \tag{3}$$

In this case for the observation times t, satisfying the condition $R_1 \ll \frac{a_e}{2} \ln \nu_e t \ll R_2$, the concentration of pairs at a certain instant of time t will be connected with their concentration at the initial instant by the relationship [10]

$$\frac{n(t)}{n(0)} = \frac{R_2 - \frac{a_e}{2} \ln \nu_e t}{R_2 - R_1} \tag{4}$$

Equation (4) has a simple physical meaning. Actually, Eq. (4) can be represented in the form

$$\frac{n(t)}{n(0)} = \frac{R_2 - R_t}{R_2 - R_1} \tag{5}$$

where R_t is given by Eq. (2). Since for pairs with $R < R_t$ the probability of decay by time t is close to 1, while for pairs with $R > R_t$ it is practically equal to zero, both the right- and the left-hand sides of Eq. (5) represent the fraction of pairs which avoided decay by the time t.

For other than rectangular types of pairwise spatial distribution of the reagents, the kinetics may deviate somewhat from Eq. (4). Note, however, that in most cases this deviation is not expected to be too large since, due to a very sharp exponential dependence of the tunneling probability on R, the kinetics of electron tunneling is not that sensitive to the exact character of a pair-wise distribution.

If monomolecular decay of the donor particles with the rate constant k is possible along with tunneling decay by the reaction with the acceptor particles, the following expression describes the change of the concentration of the donor particles with time

$$\frac{n(t)}{n(0)} = \left[\frac{R_2 - \frac{a_e}{2} \ln \nu_e t}{R_2 - R_1} \right] \exp(-kt) \tag{6}$$

This equation should be used, e.g. to describe the kinetics of the excited donor fluorescence decay via two independent channels, i.e. spontaneous intramolecular deactivation and electron transfer to an acceptor, provided that the donor and the acceptor are distributed in pairs with rectangular distribution over the distances R.

For the random distribution of immobile reagents, when concentration of the acceptor N exceeds the initial concentration of the donor n(0) considerably ($N \gg n(0)$), the current concentration of the donor n(t) will be related to n(0) by the equation

$$n(t)/n(0) = \exp(-(\pi a_e^3 N/6) \ln^3 \nu_e t) \tag{7a}$$

or

$$n(t)/n(0) = \exp(-\tfrac{4}{3}\pi R_t^3 N) \tag{7b}$$

This equation has also a simple physical meaning. Actually $\frac{4}{3}\pi R_t^3$ is the volume of sphere around the acceptor such that if a donor-particle gets into it, this will result in its decay during time t with a probability practically equal to unity. The term $\exp(-\frac{4}{3}\pi R_t^3 N)$ represents the probability for the donor particle not to get into this sphere around an arbitrary acceptor particle, i.e. the probability that a particle of the donor will not decay by time t as a result of electron tunneling.

If monomolecular decay of the donor particles with the rate constant k is possible along with tunneling decay, the following expression for the change of the concentration of the donor particle with time can be obtained [26a, 32]

$$\frac{n(t)}{n(0)} = \exp\left(-kt - \frac{\pi a_e^3}{6} N \ln^3 \nu_e t\right) \tag{8}$$

Such a situation occurs, for example, for electron tunneling reactions involving excited donor particles when these particles can disappear not only as a result of the electron tunneling reaction, but also due to spontaneous loss of excitation.

Figure 1 illustrates the excellent agreement between the experimental data on electron tunneling and Eq. (7a) for the decay of e_{tr}^- by the reaction with $Cu(en)_2^{2+}$ (en represents here ethylenediamine) in a water-alkaline (10 M NaOH) vitreous matrix at 77 K. The random character of $Cu(en)_2^{2+}$ spatial distribution was controlled in this experiment by special measurements. In Fig. 1 the solid lines represent theoretical curves calculated by means of Eq. (7a) and the optimal values of the parameters $\nu_e = 10^{15.2}\ s^{-1}$ and $a_e = 1.83$ Å selected so as to fit best all the four experimental curves simultaneously. Equation (7a) is seen to describe quite well the reaction kinetics over 13 orders of magnitude variation of time and 1 order of magnitude variation of acceptor concentration.

Consider now the kinetics of electron tunneling between mobile reagents. In this case the reagents can decay via both direct electron tunneling and via the preliminary approach to shorter distances and then tunneling. As a result, on the kinetic curve of electron tunneling reaction, two characteristic

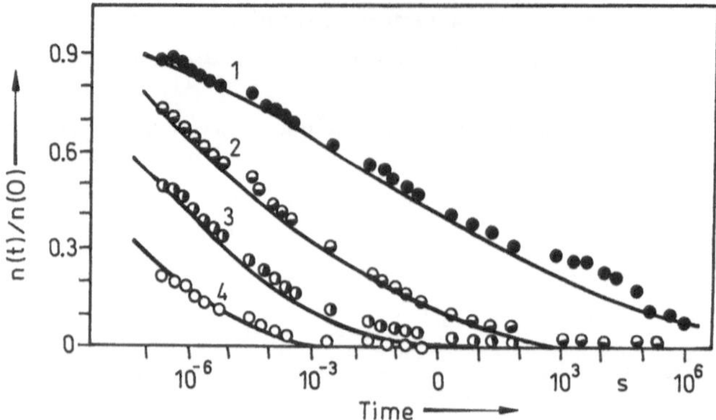

Fig. 1. Kinetics of the decay [11] of e_{tr}^- at 77 K by reaction with $Cu(en)_2^{2+}$ in vitreous 10 M aqueous NaOH solutions at $Cu(en)_2^{2+}$ concentrations of (*1*) – 10^{-2}, (*2*) – 2.5×10^{-2}, (*3*) – 5×10^{-2}, (*4*) – 10^{-1} M. The *points* represent the experimentally observed values and the *solid lines* are cuves drawn from calculation using Eq. (7a) and the parameters $\nu_e = 10^{15.2}\ s^{-1}$ and $a_e = 1.83$ Å

segments can be singled out. For observation time $t \ll \tau_D$ ($\tau_D = a_e^2/D$, where D is the diffusion coefficient, τ_D is the time of reagents shifting to distances of the order of a_e) the reaction between those reagents is ovserved which are located initially at distances $R < R_D = \frac{a_e}{2} \ln \nu_e \tau_D$. For these particles the rate of tunneling reaction exceeds that of diffusion jumps ($W(R) > \tau_D^{-1}$), and they succed in reacting via electron tunneling before performing a diffusion jump. At $t > \tau_D$ the decay is observed of those reagents which are located initially at distances $R > R_D$. At such distances the electron tunneling reaction is a slower process than the diffusion jumps ($W(R) < \tau_D^{-1}$). Such particles therefore first approach via diffusion to the distance $R = R_D$ and only then do they react via the electron tunneling mechanism. Under these conditions reaction is controlled by diffusion and is described by a conventional kinetic equation with the rate constant equal to $4\pi R_D D N$ (for the first order kinetics) and $4\pi R_D D$ (for the second order kinetics).

For the most important practical case of random reagents distribution and $N \gg n$, the kinetics of electron tunneling reaction at $t \ll \tau_D$ (fast tunneling) is described by Eq. (7a) and at $t \gg \tau_D$ (fast diffusion) by the following equation [26j, 27a, b]

$$\frac{n(t)}{n(0)} = \exp\left(-\frac{4}{3}\pi R_D^3 N - 4\pi R_D D N t\right) \tag{9}$$

Figure 2 presents kinetical data on the decay of e_{tr}^- in water-alkaline (6 M NaOH) glass via reaction with CrO_4^{2-} at 178 K, i.e. at a temperature at which defreezing of diffusion has occured. The experimental data of Ref. [33] are compared with the results of calculation of the kinetics by means of Eq. (7a) and Eq. (9) [34]. The description by Eq. (7a) with the values $\nu_e = 10^{10.4}\ s^{-1}$ and

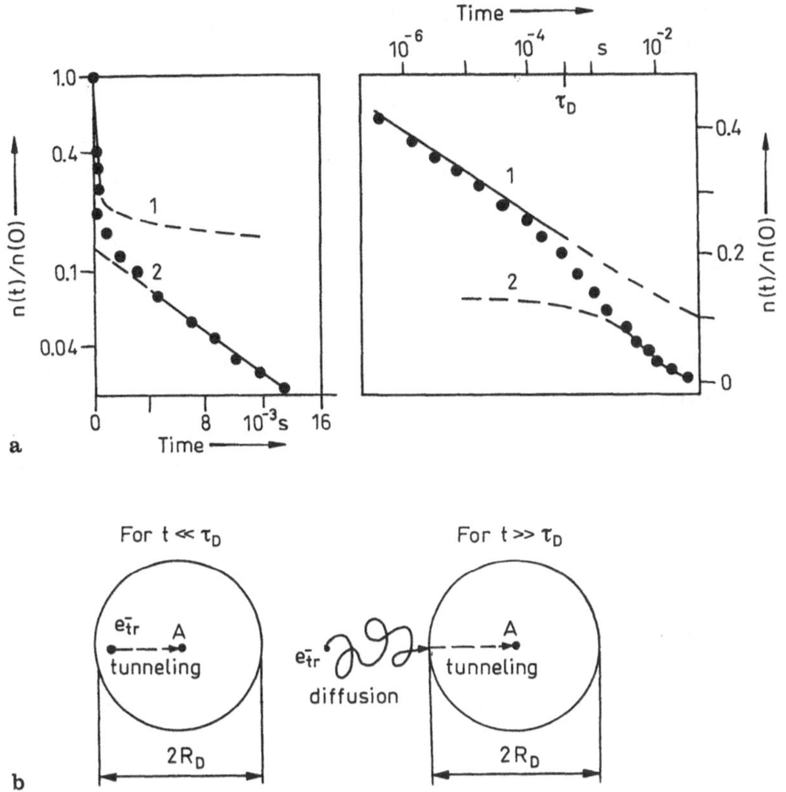

Fig. 2. (a) – Kinetics of the decay of e^-_{tr} at 178 K by reaction with CrO_4^{2-} in vitreous 6 M aqueous NaOH solutions containing 0.025 M CrO_4^{2-}. The *points* represent the experimentally observed values [33]; the *lines* are the curves drawn from calculations using Eq. (7a) with $\nu_e = 10^{10.4}\,s^{-1}$ and $a_e = 2.9$ Å *(curve 1)* and Eq. (9) with $R_D = 31$ Å and $D = 2.2 \times 10^{-12}\,cm^2\,s^{-1}$ *(curve 2)*
(b) – The schematic picture of electron transfer for short $t \ll \tau_D$ and long $t \gg \tau_D$ times

$a_e = 2.9$ Å is seen (curve 1) of the initial segment of the kinetic curve, and of the terminal segment – by Eq. (9) with the values $D = 2.2 \times 10^{-12}\,cm^2/s$ and $R_D = 31$ Å (curve 2).

1.3 Tunneling in Electron Transfer Reactions of Excited Molecules

It is well known that excitation of molecules to higher electronic states leads to a decrease in their ionization potential [35]. Therefore it could be expected that excitation of donors will increase the probability of electron tunneling from these donors to acceptors so that at sufficiently high acceptor concentrations (that is at short enough distances between donor and acceptor particles) the sub-barrier electron transfer (tunneling) from a donor to acceptor will compete with the over-barrier transfer.

Electron tunneling from electron-excited donor molecules to acceptors may display itself, for example, in quenching the luminescence of excited molecules. The electron tunneling from the excited donor to an acceptor as one of the possible mechanisms of luminescence quenching was first mentioned in Ref. [36]. In this work quenching of the fluorescence of pyrene by some aromatic amines and cyano compounds in aqueous solutions at 300 K has been studied. To our knowledge, direct experimental evidence for the real occurrence of this mechanism in practice was first reported independently in Refs. [37] and [38]. The first work deals with electron phototransfer reactions between In^+ ions and the latter-with electron transfer from excited organic molecules to CCl_4 acceptors.

Since that time there have been a lot of publications on photoinduced electron tunneling reactions in various fields of chemistry and biology. In the present review an attempt is made to consider comprehensively both theoretical and experimental data which have been obtained to date on photoinduced electron tunneling reactions and to discuss the role played by these reactions in different areas of chemistry and biology.

2 Intermolecular Electron Tunneling in PET Reactions with Non-Biological Molecules

As already noted above, the first experimental evidence of electron tunneling from excited donor particles to acceptors was obtained independently in Refs. [37] and [38].

2.1 Reactions of Excited Inorganic Ions in Crystals

On illuminating KCl crystals containing additives of In^+ ions in concentrations of 10^{-4} to 10^{-2} M, electron transfer from the excited ion of indium to the indium ion in the ground state was observed [37]

$$*In^+ + In^+ \longrightarrow In^{2+} + In \tag{10}$$

At low temperatures (7–100 K), the probability, W, of this reaction has been found [37] to be independent of temperature while at $T > 100$ K, the reaction rate was observed to increase with increasing temperature. The presence of a temperature-independent region for the dependence of W on T and the unusually sharp dependence of the probability of photoionization of In^+ centres on their concentration (i.e. on the distance between them) have been acounted for by electron tunneling from $*In^+$ to In^+. The probability of tunneling has been found to decrease by *e* times at a distance of one lattice constant. With the exponential dependence of W on the distance R between the $*In^+$ and In^+ particles of the form (1), such a decrease of W with increasing R corresponds to the value of $a_e = 12.6$ Å. Such an unusually high value of the parameter a_e for the reaction in question is not surprising and is due both to the low value of the electron-binding energy in $*In^+$ ($I_d = 0.1$ eV) and to the fact that this reaction occurs in a crystal rather than in an amorphous vitreous matrix.

2.2 Reactions of Excited Organic Molecules in Vitreous Solutions

The other pioneering work on electron tunneling in reaction of excited particles dealt with the photoionization of organic molecules in vitreous matrices [38]. The studies made in this work were based on the following idea.

Owing to a relatively high (compared with molecules in the ground electronic state) probability of electron tunneling for excited molecues, this process, at sufficiently short distances between the excited molecules and the particles of electron acceptors, can compete with the ordinary over-barrier electron transfer

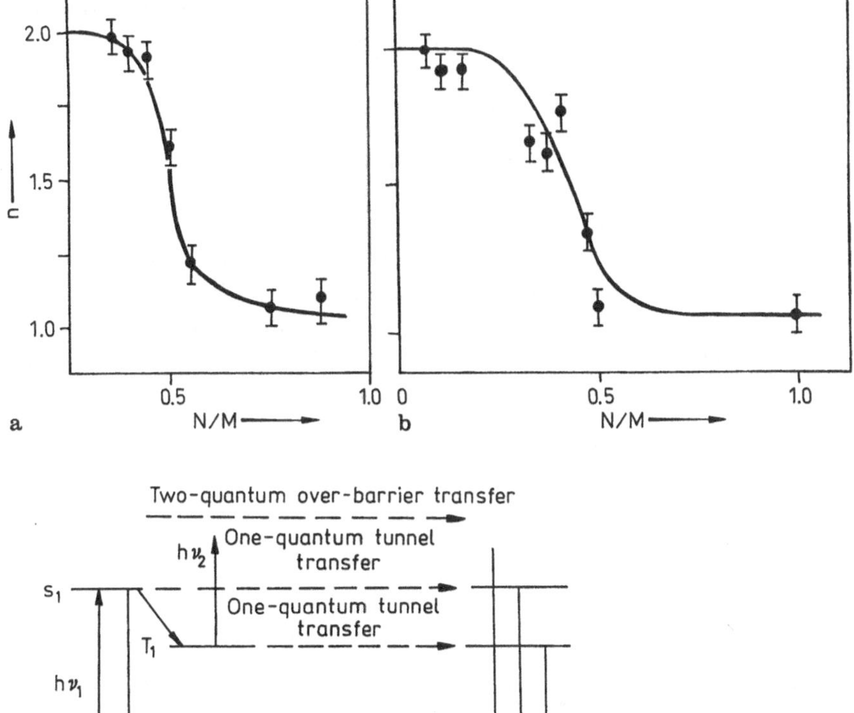

Fig. 3a–c. The effect [38] of the concentration of CCl_4 on the order, *n*, of the electron phototransfer reactions from **(a)** naphthalene and **(b)** diphenylamine to CCl_4 with respect to light intensity and **(c)** schematic representation of the two quantum over barrier and the one-quantum tunnel electron phototransfer from a donor to an acceptor

(see the scheme in Fig. 3). In practice this effect was expected to manifest itself in the transition, as the concentration of acceptor rises, from the usual two-quantum photoionization of organic donor molecules D

$$^{S_0}D \xrightarrow{h\nu_1} {}^{S_1}D \longrightarrow {}^{T_1}D \xrightarrow{h\nu_2, B} D^+ + B^- \qquad (11)$$

to the one-quantum one (see the same scheme in Fig. 3)

$$\begin{array}{ccccc} ^{S_0}D & \xrightarrow{h\nu_1} & ^{S_1}D & \longrightarrow & ^{T_1}D \\ & & \searrow_B & & \swarrow_B \\ & & & D^+ + B^- & \end{array} \qquad (12)$$

and hence, could be recorded by means of studying the dependence of the rate of electron phototransfer on the intensity of light.

The transition from the two-quantum to the one-quantum mechanism was indeed discovered [38] in studying the reactions of ionizing naphthalene (Nh) and diphenylamine (DPA) in the presence of CCl_4 *in vitreous alcohol* (CH_3OH) *matrices at 77 K*

$$Nh(DPA) + CCl_4 \xrightarrow{h\nu} Nh^+(DPA^+) + CCl_4^- \qquad (13)$$

The order, n, of the reaction of the CCl_4^- radical formation with respect to the light intensity ($W \sim J^n_{exc}$, where W is the reaction rate and J_{exc} is the light intensity) has been found to be close to 2 at low concentrations of CCl_4 and tend to 1 at high ones (see Fig. 3). In other words, to photoionize a molecule of Nh or DPA, two quanta of light have to be spent at low concentrations of CCl_4 when the

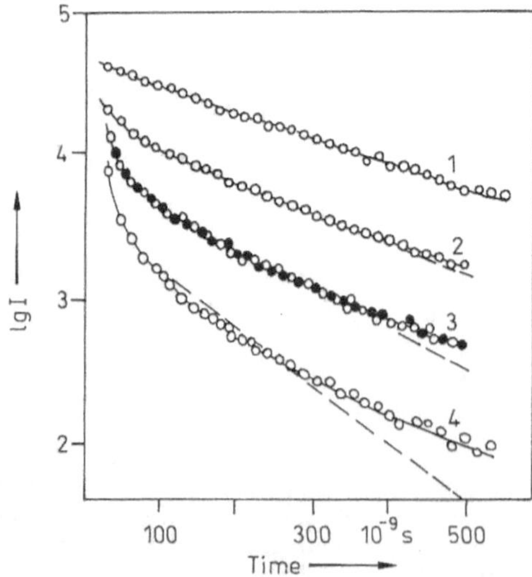

Fig. 4. The kinetics [39] of the decay of naphthalene fluorescence in the presence of CCl_4 in ethanol. Concentration of CCl_4: (*1*) – 0, (2) – 2, (*3*) – 2.5, (*4*) – 3 M. The *points* indicate the experimental data; the *broken lines* were calculations using Eq. (14); the *solid lines* were calculations using Eq. (15) ○, T = 77 K; ●, T = 140 K. I is the fluorescence intensity in arbitrary units

distance between the excited donor molecule and CCl_4 is large, while only one quantum is needed at high concentrations of CCl_4 when this distance is small.

The nature of the excited state responsible for the one-quantum photoionization of donor molecules in the presence of CCl_4 has been established by studying the kinetics of quenching the luminescence of the singlet and the triplet excited states of Nh molecules in the presence of CCl_4 [39]. The curves of the fluorescence quenching are presented in Fig. 4. As is seen from the figure, at high acceptor concentrations ($N \gtrsim 1$ M) the intensity of the fluorescence of napthalene decreases non-exponentially with time.

The decay kinetics of excited electron donor molecules (the intensity of fluorescence is proportional to the concentration of excited molecules at any given time) can be interpreted in two ways. First, one may try to approximate it with the sum of two exponents, one of which refers to the decay of the fluorescence of free donor molecules and the other to that of the complex between the donor and the acceptor. This interpretation is similar to the description of the two-exponential decay of the fluorescence observed in the presence of two compounds containing heavy atoms [40]

$$\frac{n(t)}{n(t_0)} = (1-\alpha)\left\{\exp\left[-k_0(t-t_0)\right] + \frac{\alpha}{1-\alpha}\exp\left[-k_1(t-t_0)\right]\right\} \quad (14)$$

where k_0 and k_1 are the rate constants for the decay of the excited free donor molecule and of the complex, respectively, while α is the fraction of the donor molecules that have formed the complex.

The other interpretation is based on the idea of electron tunneling from excited molecules. In this instance there occurs an overlap of the exponential decay at a rate constant k_0 which refers to spontaneous deactivation of the Nh molecules from the excited to the ground state and of the logarithmic kinetics characteristic of electron tunneling reactions (cf. Chap. 4, Sect. 2, Ref. [31])

$$\frac{n(t)}{n(t_0)} = \exp\left[-k_0(t-t_0) - \frac{\pi a_e^3}{6} N(\ln^3 \nu_e t - \ln^3 \nu_e t_0)\right] \quad (15)$$

This equation follows directly from Eq. (8). Comparison of experimental data with the results of calculations using Eqs. (14) and (15) has shown that the experimental data can be better described by a model implying electron tunneling from the first singlet excited state of naphthalene, ^{S_1}Nh.

Owing to a relatively short time interval within which the kinetics of tunneling decay could be observed, the main tunneling parameters ν_e and a_e could not be separately estimated with sufficient accuracy. From the kinetic curves, however, one could readily obtain the value of $\beta = (\pi a_e^3/2)\ln^2 \nu_e t_0$, which is a combination of these parameters. Indeed, if $t/t_0 < 10^4$ (this condition is satisfied for the process discussed) Eq. (15) is transformed to (see also Chap. 4, Sect. 2 of Ref. [31])

$$\frac{n(t)}{n(t_0)} = \exp\left[-k_0(t-t_0) - \beta N \ln(t/t_0)\right] \quad (16)$$

An analysis of the curves in Fig. 4 carried out by using Eq. (16) has made it possible to find the value $\beta = (0.205 \pm 0.010)\ M^{-1}$ at $t_0 = 10^{-8}$ s.

An attempt was made to determine the feasibility of electron tunneling from the triplet states of Nh and DPA to CCl_4 by studying the kinetics of the decay of naphthalene and diphenylamine phosphorescence in alcohol glasses. This has been found to be exponential with virtually the same rate constant, k_0, over the whole range of CCl_4 concentrations studied (0–3.5 M). This means that the tunneling mechanism cannot be responsible for the electron transfer from the triplet excited states of the Nh and DPA molecules to CCl_4 in alcohol matrices.

Thus, the one-quantum phototransfer of the electron from *Nh and *DPA to CCl_4 via the tunneling mechanism occurs from the first singlet excited state of Nh and DPA molecules rather than from the first triplet excited state. Hence, the photoinduced electron transfer from naphthalene and diphenylamine to CCl_4 is performed via the reactions

$$^{S_1}Nh + CCl_4 \longrightarrow Nh^+ + CCl_4^- \tag{17}$$

$$^{S_1}DPA + CCl_4 \longrightarrow DPA^+ + CCl_4^- \tag{18}$$

Note that in Ref. [41] the growth was observed of the intensity of the Nh phosphorescence on adding CCl_4. This fact indicates the existence of the reverse electron transfer from CCl_4^- to Nh^+ to form a Nh molecule in the triplet excited state.

According to the theory [19–24], the probability of electron tunneling must essentially depend on the motion of atom nuclei in reacting particles. This effect has been experimentally observed [42] for reaction (17). The rate of this reaction has been found to change drastically on deuterating naphthalene.

As seen from Fig. 5, the non-exponential drop in the intensity of fluorescence for Nh-d_8 is faster than that for Nh, i.e. the rate of reaction (17) increases upon

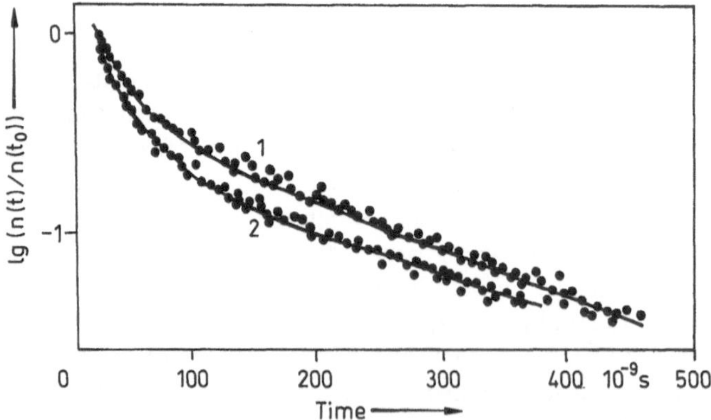

Fig. 5. The kinetics [42] of the decay of (*1*) Nh and (*2*) Nh-d_8 fluorescence in vitreous ethanol in the presence of 2.4 M CCl_4. The *points* indicate experimental data; the *lines* were calculated using Eq. (16). T = 77 K

naphthalene deuteration. Thus, for this reaction, an inverse kinetic isotope effect is observed as distinct from the process of spontaneous deactivation, for which a normal isotopic effect ($\tau_d > \tau_h$) is observed. The mean value of the parameter β obtained for Nh-d_8 from experiments with various concentrations of CCl_4 proved to be $\bar{\beta} = (0.24 \pm 0.01)\,M^{-1}$ at $t_0 = 10^{-8}$ s. As the effect of the nuclear motion on W(R) must be reflected more in the value of ν_e than in that of a_e it seems natural to connect the difference observed between the value of β for Nh and that for Nh-d_8 with the change in the parameter ν_e. At the value of $a_e \simeq 1$ Å typical of tunneling reactions, the difference observed in the values of β corresponds to an approximately 2.5-fold increase of ν_e upon naphthalene deuteration. With an increase in temperature from 77 to 140 K, the parameter β remained virtually unchanged, although the time, τ, for spontaneous deactivation was markedly reduced. Thus, tunneling reaction (17) proceeds via a non-activated mechanism.

To study the influence of the motion of the nuclei of the medium molecules on the kinetics of reaction (17), the values of β were measured in a number of solvents: C_2H_5OH, CD_3OD, CH_3OH, and toluene. In all cases these values were found to be the same. The data obtained show that, in the dissipation of energy released in the electron tunneling from the excited molecules of Nh and Nh-d_8 to CCl_4, the principal part is played by excitation of vibrations in the reacting molecules while the high-frequency vibrations of the molecules of the medium appear to be less important.

The fluorescence of N,N′-diethylaniline, N,N,N′,N′-tetramethyl-p-phenylenediamine (TMPD) and tetrakis(dimethyl-amino)ethylene has been reported to be quenched by electron acceptors [43] in vitreous trans-1,5-decalindiol. The efficiency of quenching the fluorescence of aromatic molecules in this matrix remains unchanged when the temperature is raised from room temperature to 363 K. The efficiency of quenching has been measured by the classic static method using the Perrin equation. The Perrin quenching radii, R_q, have been used as a quantitative measure of the quenching efficiency. For the quenching processes caused by electron tunneling the radius R_q coincides with the tunneling distance R_t. The values of R_q have been found to fall within the range 7.9–15.3 Å and to correlate with the free energy changes in the reactions of electron transfer from the excited donor molecule to the acceptor, $-\Delta G°$, estimated from the formula

$$-\Delta G° = E(S_1) - E(D/D^+) + E(B/B^-) + \frac{e^2}{\varepsilon_s R_q} \tag{19}$$

where $E(S_1)$ is the excitation energy of the donor singlet S_1, $E(D/D^+)$ is the donor oxidation potential, $E(B/B^-)$ is the acceptor reduction potential, and $e^2/\varepsilon_s R_q$ is the Coulomb term taking account of the electrostatic interaction. The values of $-\Delta G°$ calculated in this way are expected to be somewhat different from the real values since, in calculating them, use was made of the values of redox potentials in other media rather than in vitreous decalindiol. On the whole, however, the values of $-\Delta G°$ can be expected to reflect the trend and scale of changes in the Gibbs energy of electron transfer for the series of compounds under discussion. The graph of the dependence of R_q on $-\Delta G°$ for the processes studied in Ref.

[43] is depicted in Fig. 6. In accordance with the theory [19–24, 44], the efficiency of electron transfer (characterized by the radius R_q) is seen first to increase with increasing values of $-\Delta G^\circ$, but then to approach a plateau rather than to fall with further increase of $-\Delta G^\circ$ as is expected from the theory (see below). Such unexpected behaviour of R_q at high values of $-\Delta G^\circ$ can be accounted for, for example, by excitation of quantum vibrational degrees of freedom or by the appearance of yet another channel of the process, i.e. tunneling to the electronically excited levels of the acceptor.

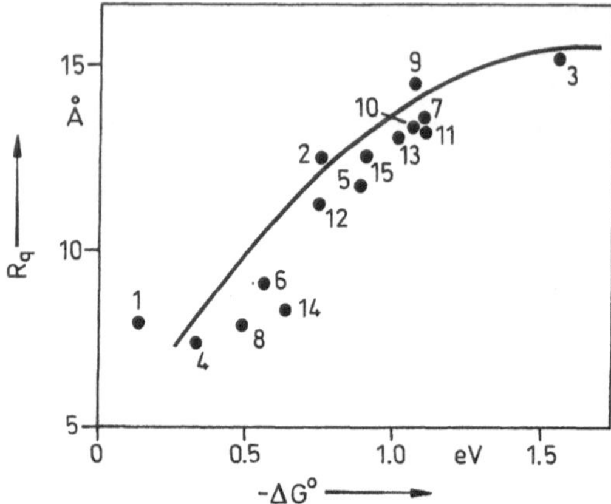

Fig. 6. Perrin quenching radii, R_q, [43] vs. variations of the free energy, $-\Delta G^\circ$, of electron transfer from the excited donor molecule to the acceptor molecule for donor-acceptor pairs in vitreous *trans*-1.5-decalindiol. 1, Rubrene + *N,N'*-diethylaniline (DEA); 2, rubrene + *N,N,N',N'*-tetramethyl-*p*-phenylenediamine (TMPD); 3, rubrene + tetrakis (dimethylamino)-ethylene; 4, tetracene + DEA; 5, tetracene + TMPD; 6, 9, 10-dinaphthylanthracene + DEA; 7, 9, 10-dinaphthylanthracene + TMPD; 8, perylene + DEA; 9, perylene + TMPD; 10, 9-methylanthracene + TMPD; 11, 9, 10-diphenylanthracene + TMPD; 12, coronene + TMPD; 13, benzo[ghi] perylene + TMPD; 14, fluoranthene + DEA; 15, acridine + DEA

Quenching the luminescence of TMPD by phthalic anhydride (PA) and of pyromellitic dianhydride (PMA) by hexamethyl triindan (HMTI) in vitreous MTHF at 77 K has been discussed in Ref. [45]. The donors were the excited molecules of *TMPD or molecules of HMTI in the ground state while PA molecules in the ground state or the excited molecules of *PMA served as the electron acceptor. Quenching of both the fluorescence and the phosphorescence of excited molecules was observed. The authors deny the possibility of quenching the luminescence of *TMPD and *PMA via the mechanism of energy transfer since the luminescence has been quenched by molecules of PA and HMTI possessing higher energies of excited states than those possessed by TMPD and PMA molecules. The estimations made have shown that the non-luminescent donor-acceptor complexes, if formed before the solutions have been frozen, cannot explain the observed luminescence quenching either. The authors, therefore, assume the luminescence to be quenched by electron transfer

$$*TMPD + PA \longrightarrow TMPD^+ + PA^- \tag{20}$$

$$HMTI + *PMA \longrightarrow HMTI^+ + PMA^- \tag{21}$$

Just as for the reaction of Nh with CCl_4 [39], no ion formation could be detected as a result of luminescence quenching. This has been explained by subsequent fast recombination of the ions formed. According to the estimation of Ref. [45] based on the values for the added quencher concentrations, the distances of electron transfer from the triplet excited state of a TMPD molecule to PA and from HMTI to *PMA are equal to about 25 Å.

Electron transfer kinetics from the triplet excited state of TMPD to PA in polystyrene has been monitored by phosphorescence emission decay in Ref. [46]. The rate constant has been found to be invariant over the temperature interval 77–143 K. Parameters a_e and ν_e calculated from the phosphorescence decay using Eq. (8) were found to be $a_e = 3.46$ Å and $\nu_e = 10^4\ s^{-1}$.

Electron tunneling from the first singlet excited state of pentacene to duroquinone in sucrose octaacetate glass was discovered [47] by observing the time dependence of fluorescence emission. The tunneling distance, $R_t = 14.3$ Å during the lifetime, τ, of the pentacene excited state and the parameter $a_e = 0.7$ Å were extracted from the time-dependent curves by using a kinetic equation analogous to Eq. (8). From these values of a_e and $R_t = (a_e/2) \ln \nu_e\tau$, using the lifetime $\tau = 9.6 \times 10^{-3}$ s for the pentacene excited state, one can obtain $\nu_e = 10^{16.9}\ s^{-1}$. The influence of the angular dependence of the electron transfer rate on the fluorescence decay kinetics was analyzed. For this purpose, theoretical calculations based on models assuming electron transfer between two *p* orbitals as well as from a *p* orbital to an *s* orbital were carried out. They demonstrated that only for a short time (less than a few hundred picoseconds) and high acceptor concentrations does the time dependence of the fluorescence intensity for the ange-dependent theory deviate from that for the angle-independent theory. For longer durations ($>10^{-9}$ s) the angle-independent theory with the angle-averaged electron tunneling parameters may be used for the description of experimental data.

2.3 Reactions of Excited Metal Complexes in Solid Matrices and Liquid Solutions

Electron transfer from the excited states of Fe(II) to the H_3O^+ cation in vitreous aqueous solutions of H_2SO_4 which results in the formation of Fe(III) and of H atoms has been studied in Refs. [48, 49]. The quantum yield of the formation of Fe(III) in 5.5 M H_2SO_4 at 77 K has been found to be only two times smaller than at room temperature. Photo-oxidation of Fe(II) is also observed at 4.2 K. The actual, very weak dependence of the efficiency of Fe(II) photo-oxidation on temperature indicates the tunneling mechanism of this process [48, 49]. A detailed theoretical analysis of the mechanism of electron transfer from the excited ions Fe(II) to H_3O^+ in solutions was made in Ref. [50] in terms of the theory of radiationless transitions. In this work a simple way is suggested for an a priori estimation of the maximum possible distance, R_{max}, of tunneling between a donor and an acceptor in solid matrices. This method is based on taking into account the dependence of the energy, E(R), of donor cation and acceptor anion formation on the distance, R, of electron transfer. This dependence is due to the Coulomb interaction of ions in a solid matrix

$$E(R) = E_\infty - \frac{e^2}{\varepsilon_S R} \tag{22}$$

where E_∞ is the energy of formation of ions when they are at infinite separation and ε_s is the dielectric permeability (in the case of a solid solution the authors of Ref. [50] used the optical dielectric constant as ε_s). The value of E_∞ was calculated by the authors from the values of the redox potentials of the reagents in the same, but liquid, matrix with corrections for the non-equilibrium character of solvation of electron phototransfer products in a vitreous matrix. These corrections were calculated by formulae essentially similar to the Born formula.

As an example, in Fig. 7 a dependence of E(R) is cited which has been calculated [50] for electron phototransfer from the excited state of the Fe(II) ion to H_3O^+. Using the energy conservation law one can estimate the maximum distance of tunneling, R_{max}, from the formula $E_{min} = E(R_{max})$, where E_{min} is the minimum energy of a light quantum capable of inducing an electron transfer. This formula implies that tunneling takes place from the lowest vibrational level of the donor. It can be seen from the Fig. that the maximum distance of tunneling is 16 Å. The real distance of tunneling may be shorter (but not longer) than R_{max}, e.g. because the lifetime, τ, of the donor's excited state may be too short for the electron to tunnel to the distance R_{max} [$R_{max} > R_t(\tau)$]. Note also, that the use of the optical dielectric constant in the expression for E(R) appears to result in overestimating the contribution of Coulomb interaction to E(R) and hence in overestimating the value of R_{max}.

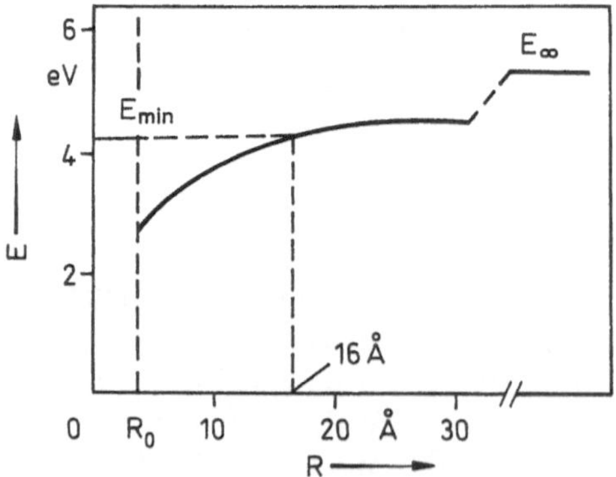

Fig. 7. The interaction energy of Fe(III) and H formed in non-equilibrium solvation states upon photolysis of Fe(II), vs the distance between these particles [50]. The energy of the initial state $Fe(II)_{aq} + H_3O^+$ is taken to be zero. R_0 is the radius of the $Fe(II)_{aq}$ ion. E_∞ – interaction energy at infinitely large distance, E_{min} – the minimum energy of a light quantum capable of inducing an electron transfer

*Electron transfer quenching of the charge transfer excited state of tris (3,4,7,8-tetramethylphenanthroline)ruthenium(II) *Ru(II) (Me_4phen)$_3$ by methylviologen (N,N'-dimethyl-4,4'-bipyridinium, MV^{2+}) in glycerol* cooled to $T \lesssim 250$ K was studied in refs. 51 and 52. The viscosity of glycerol at $T \lesssim 250$ K is $\eta > 10^2$ Pa s. The distance to which the reagents can migrate via translational diffusion during $\tau = 7 \times 10^{-6}$ s (*Ru(II) (Me_4phen)$_3$ excited state lifetime) in the absence of acceptors is smaller than 0.8 Å. Thus the distance between the donor and the acceptor was actually fixed in those experiments on the time scale of the excited state lifetime. Both static and dynamic quenching measurements were made. The "Perrin critical distance", R_q, was found to be 17.3 Å. The emission decay curve in the absence of quencher was exponential, while the addition of 0.4 M MV^{2+} resulted in non-exponential decay at $t \leq 10^{-6}$ s. The value of a_e obtained in [52] from the best fit of the emission decay curves by the appropriate kinetic equation was found to be 1.4 ± 0.3 Å. From this value and $R_q = R_t = (a_e/2) \ln \nu_e\tau = 17.3$ Å, using $\tau = 7 \times 10^{-6}$ s one can obtain $\nu_e = 10^{15.8 \pm 3}\ s^{-1}$ [31].

Electron transfer from excited (2,2'-bipyridine)ruthenium(II) Ru(II) (bpy)$_3$, to MV^{2+} as well as the back reaction between MV^+ and Ru(III) (bpy)$_3$ were studied in Ref. [53] using cellophane, a regenerated form of cellulose, as a matrix. In cellophane the reacting species are trapped by shrinkage of the structure upon dehydration at room temperature. Ion pair formation is negligible in such systems, even at rather high concentrations of the reacting species, so that complications in the kinetics that may sometimes be associated with frozen systems at high concentrations of ionic reagents are eliminated.

The build-up of reduced methylviologen (monitored at 400 nm) and the decay of Ru(II) (bpy)$_3$ luminescence (monitored at 610 nm) are shown in Fig. 8.

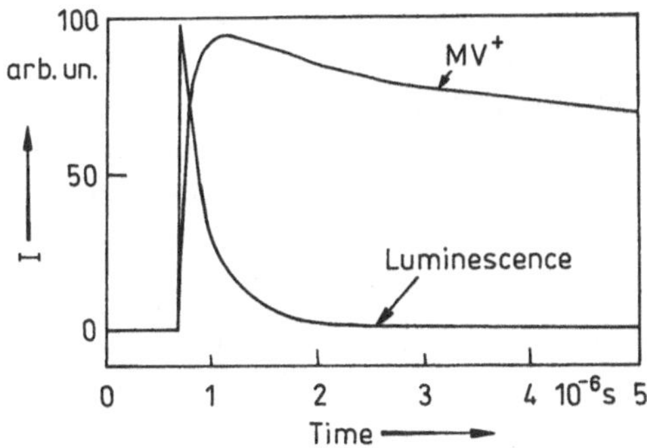

Fig. 8. Reduced methylviologen build-up and Ru(II) (bpy)$_3$ luminescence decay both obtained by using the sample containing 0.01 M Ru(II) (bpy)$_3$ and 0.04 M MV^{2+}. T = 295 K. From Ref. [53]

Due to the complicated kinetics for both processes no attempt was made in Ref. [53] to treat the data quantitatively. It was estimated, however, that the back electron transfer reaction is slower by about 3 orders of magnitude than that of the forward electron transfer. At the same time, the free energy change for the forward reaction ($\Delta G^\circ = -0.4$ eV) is smaller than for the back electron transfer ($\Delta G^\circ = -1.7$ eV). This decrease of the reaction rate at large exothermicity $J = -\Delta G^\circ$ was attributed [53] to the decrease of the Franck-Condon factors with increasing J in the situation when J exceeds the reorganization energy E_r.

Quantitative investigations of the photoinduced electron transfer from excited $^*Ru(II)(bpy)_3$ to MV^{2+} were made in Ref. [54], in which the effect of temperature has been studied by steady state and pulse photolysis techniques. The parameters ν_e and a_e were found in Ref. [54] by fitting the experimental data on kinetics of the excited $Ru(II)(bpy)_3$ decay with the kinetic equation of the Eq. (8) type. It was found that a_e did not depend on temperature and was equal to 4.2 ± 0.2 Å. The frequency factor ν_e decreased about four orders of magnitude with decreasing the temperature down to 77 K, but the Arrhenius plot for W was not linear, as is shown in Fig. 9.

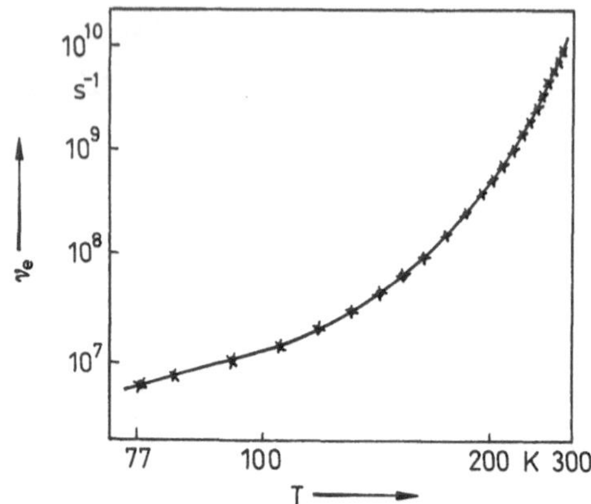

Fig. 9. Temperature dependence of the frequency factor ν_e for the electron transfer from excited $Ru(II)(bpy)_3$ to MV^{2+} in cellophane matrix at $Ru(II)(bpy)_3$ concentration of 0.01 M and MV^{2+} concentration of 0.1 M [54]

Electron transfer was interpreted in Ref. [54] in terms of the nonradiative decay process [20–24, 44]. For an up-to-date review of theoretical works on electron transfer see the relevant chapter in this volume (R. A. Marcus – Recent developments in fundamental concepts of PET in biological systems).

A complete quantum mechanical description of electron transfer provides the following expression for the rate with a single, averaged effective internal vibration coupled to the electron transfer [23]:

$$W = \frac{2\pi}{\hbar} |V(r)|^2 (4\pi E_r k_B T)^{-1/2} \sum_{m=0}^{\infty} (e^{-s} S^m/m!) \exp\left[-\frac{(E_r + \Delta G^\circ - m\hbar\omega)^2}{4E_r k_B T}\right] \quad (23)$$

where $S = E_i/\hbar\omega$, E_i is the part of the reorganization energy arising from vibrations with $\hbar\omega \gg k_B T$, $V(r)$ is an electron coupling energy, k_B is the Boltzmann constant, T is the temperature, E_r is the part of the reorganization energy arising from vibrations with $\hbar\omega \ll k_B T$, ΔG° is the Gibbs free energy change in the electron transfer reaction. In the high temperature limit ($\hbar\omega \ll kT$) Eq. (23) reduces to the classical Marcus formula [23, 44]:

$$W = \frac{2\pi}{\hbar} |V(r)|^2 (4\pi E_r k_B T)^{-1/2} \exp\left[-\frac{(E_r + \Delta G^\circ)^2}{4E_r k_B T}\right] \quad (24)$$

Thus as $-\Delta G^\circ$ increases, W increases at $-G^\circ < E_r$, reaches a maxium at $-\Delta G^\circ = E_r$ and then decreases at $-\Delta G^\circ > E_r$. The latter region is known as the Marcus inverted region [23, 44].

The best fit between experimental data on long-range electron transfer from $*Ru(II)(bpy)_3$ to MV^{2+} and Eq. (23) for the rate constant of tunnel electron transfer was obtained in Ref. [54] with the following parameters: $\Delta G^\circ = -0.1$ eV, $\hbar\langle\omega\rangle = 0.044$ eV, $S = 24$, $V = 19\ cm^{-1}$. Note that the value of the electron coupling energy obtained is significantly higher than the values estimated in Ref. [23]. Superexchange type interactions may be one of the explanations of such high value of the coupling energy in the system under consideration [54].

Electron tunneling reactions were also studied [55] in a rigid polymer (polycarbonate) matrix by following *the reductive quenching of a series of excited homologues,* $*Ru(II)(LL)_3$*, of* $Ru(II)(bpy)_3$ *by a series of aromatic amines, D*

$$*Ru(II)(LL)_3 + D \rightarrow Ru(I)(LL)_3 + D^+ \tag{25}$$

The tunneling distance, R_q, for various reactions of this type were obtained by fitting the static quenching of $*Ru(II)(LL)_3$ by various concentrations of D, to the Perrin equation. R_q was found to depend on reaction exothermicity and to reach 20 Å for the reaction between $*Ru(II)(ester)_3$ and TMPD(ester = 4,4'-diisopropyl ester 2,2'-bipyridine).

The decay curve of $Ru(II)(ester)_3$ luminescence in the presence of TMPD is shown in Fig. 10. Superimposed on the experimental decay curve is a theoretical fit of the appropriate kinetic equation similar to Eq. (8) with $a_e = (2 \pm 0.04)$ Å and $\nu_e = 10^{13.8 \pm 1.3}\ s^{-1}$.

The effect of long-range electron transfer on chemiluminescence of the hexanuclear cluster ion $Mo_6Cl_{14}^{2-}$ *in the presence of three series of structurally and electronically related organic compounds* (aromatic amine radical cations A^+, nitroaromatic

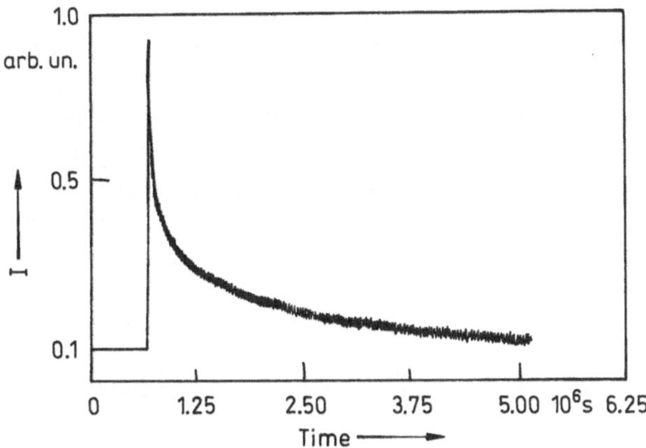

Fig. 10. Luminescence decay of $*Ru(II)$ (4,4'-diisopropylester 2,2'-bipyridine)$_3$ in the presence of 0.0728 M TMPD in polycarbonate. Superimposed on the experimental decay curve is a theoretical fit to Eq. (8). The theoretical and experimental curves coincide completely. From Ref. [55]

radical anions D^- and pyridinium radicals D) in liquid CH_3CN solutions at room temperature has been investigated [56]. The formation of the electronically excited $^*Mo_6Cl_{14}^{2-}$ ion was due to the electron transfer reactions involving electrochemically generated $Mo_6Cl_{14}^{3-}$ or $Mo_6Cl_{14}^-$ species:

$$Mo_6Cl_{14}^{3-} + A^+ \rightarrow Mo_6Cl_{14}^{2-} + A \quad (26)$$

$$Mo_6Cl_{14}^{3-} + A^+ \rightarrow {}^*Mo_6Cl_{14}^{2-} + A \quad (27)$$

or

$$Mo_6Cl_{14}^- + D^-(D) \rightarrow Mo_6Cl_{14}^{2-} + D(D^+) \quad (28)$$

$$Mo_6Cl_{14}^- + D^-(D) \rightarrow {}^*Mo_6Cl_{14}^{2-} + D(D^+) \quad (29)$$

The electron transfer chemistry of $Mo_6Cl_{14}^-$ and $Mo_6Cl_{14}^{3-}$ ions can be described in terms of two competing reaction channels: a highly exergonic electron transfer pathway yielding ground-state products, and less exergonic electron transfer leading to the formation of electronically excited $^*Mo_6Cl_{14}^{2-}$ ion. The dependence of the luminescence efficiency, φ, on the free energy of reactions (27) and (29) was similar for all three series of reagents A^+, D^- and D. In agreement with the theory, for reactions with positive ΔG very small values of φ were observed ($\varphi < 10^{-6}$). Over a narrow free energy range around $\Delta G = 0$, φ rapidly increases with decreasing ΔG, asymptotically approaching a plateau value $\varphi_p \lesssim 0.1$ at $\Delta G \rightarrow -2$ eV. Non-zero values of φ_p clearly indicate that reaction pathway (27) and (29) leading to the formation of excited species, are kinetically competitive with reactions (26) and (28) leading to the formation of the species in the ground states, even though the latter are favoured thermodynamically. Electron transfer distances for the processes (26)–(29) are estimated to be from 11 to 18 Å [56].

In Ref. [57], *electron transfer to methylviologen from a series of Ru(II) (bpy)$_3$ compounds with several hydrocarbon chains of various length attached to each bpy ligand* is reported.

The quenching of the luminescence of such "hedgehog" complexes with isolating spacers of increasing thickness around core, upon collision with methylviologen molecules was found to occur via electron transfer over the distances of 15 Å and more. It was shown that the quenching efficiency k_q evaluated using Stern-Volmer equation [58] decreased exponentially with increasing length of the hydrocarbon chain up to $n = 7$.

2.4 Electron Tunneling in Photoinduced Decay of Trapped Electrons

As is known from the literature [59a], illumination of frozen solutions containing trapped electrons in the e_{tr}^- absorption band results in a decrease in the concentration of e_{tr}^-. It is also well known that, for water-alkaline glasses, with such illumination, a transition of an electron from a trap into the conduction band occurs [59b]. In order to explain these facts a model was proposed in Ref. [60] according to which excitation of an electron into the conduction band by light absorption results initially only in a change of the site of the electron location. However, during this light-induced migration the electron may come to be located so close to an acceptor that within the time preceding the next act of absorbing a photon there occurs a tunneling recombination between e_{tr}^- and this acceptor. Since light absorption involves the transition of an electron to the conduction band, it is reasonable to assume that, as a result, the electron becomes localized far enough from the trap from which it was disloged by the photon.

If illumination is started long enough after the end of irradiation that produced e_{tr}^-, then the kinetics of a tunneling reaction in the presence of light-induced electron diffusion and under the condition of random spatial distribution of acceptor particles, provided $n \ll N$ and $\tau \ll t$, can be described by the expression [60]:

$$\frac{n(t)}{n(0)} = \exp\left(-\tfrac{4}{3}\pi R_\tau^3 N t/\tau\right) \tag{30}$$

Here n and N are the concentrations of e_{tr}^- and the acceptors, respectively, t is the time elapsed from the moment the light is switched on, τ^{-1} is the probability of the transition of an electron to a quasi-free (mobile) state per unit time under the action of light, $R_\tau = (a_e/2) \ln \nu_e \tau$ is the distance of electron tunneling from a trap to an acceptor within the time τ.

Equation (30) has a simple physical meaning. Indeed, the right hand side of Eq. (30) is the probability that the electron will not get into the volume $NV_\tau = (4/3)\,\pi R_\tau^3 N$ as a result of light-induced jumpwise migration through the sample, and t/τ is the number of jumps made within the time t. As follows from the definition of R_τ, the entry of the electron into a volume V_τ around an acceptor leads, within the time τ of its residence in this volume, to its tunnel decay on an acceptor with a probability virtually equal to unity.

In order to check the proposed model of e_{tr}^- photobleaching, in Ref. [34] the kinetics of e_{tr}^- photobleaching in the presence of acceptor additives in vitreous

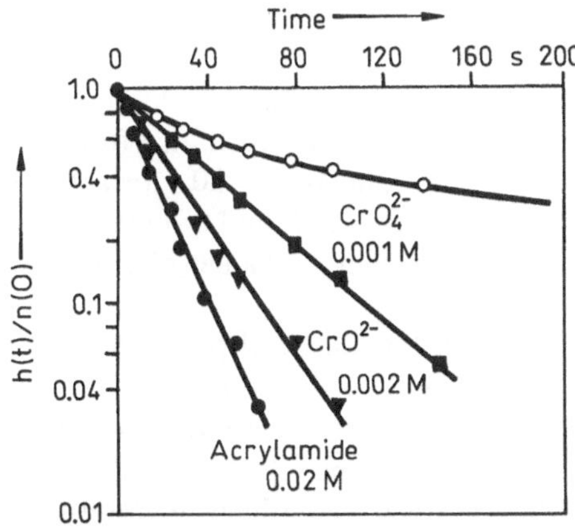

Fig. 11. Kinetic curves for the photobleaching [34] of e_{tr}^- in a vitreous 10 M aqueous NaOH solution irradiated at 77 K in the absence (○) and in the presence *(solid symbols)* of acceptor additives, in the coordinates of Eq. (30)

water-alkaline and water-ethylene glycol matrices at 77 K was studied. Typical curves for e_{tr}^- photobleaching are presented in Fig. 11. The addition of acceptors is shown to result in an essential increase in the rate of e_{tr}^- photobleaching, the kinetics of e_{tr}^- decay in the presence of additives being described by an exponential law in accordance with Eq. (30).

The analysis of the curves of e_{tr}^- photobleaching in the presence of a number of acceptors by means of Eq. (30) made it possible to determine the ratios V_τ/τ (Table 2). From Eq. (30) these ratios are seen to serve as effective rate constants for photobleaching processes. Knowing the values of ν_e and a_e from the experiments on the dark decay of e_{tr}^- by reactions with the same acceptors (see Ref. [31]), one can further determine from these ratios the time τ. Since τ is independent of the kind of acceptor, upon determining τ in a given matrix for one of the acceptors, one can calculate the values of V_τ for other acceptors from the photobleaching curves provided the bleaching conditions are identical.

Table 2. Effective rate constants V_τ/τ of the light-induced decay of trapped electrons and the radii, R_τ, of the capture of e_{tr}^- by acceptors calculated from the kinetics of e_{tr}^- photobleaching by means of Eq. (30). T = 77 K [34]

Acceptor	Matrix	V_τ/τ $M^{-1}\,s^{-1}$	R_τ, Å
CrO_4^{2-}	10 M NaOH + H_2O	6.4	37
$Fe(CN)_5NO^{2-}$	10 M NaOH + H_2O	2.7	28
$Fe(CN)_6^{3-}$	10 M NaOH + H_2O	1.6	23
Acrylamide	10 M NaOH + H_2O	1.3	22
NO_3^-	10 M NaOH + H_2O	1.1	21
CrO_4^{2-}	Ethylene glycol + H_2O (1:1 by volume)	1.2	
Acrylamide	Ethylene glycol + H_2O (1:1 by volume)	0.53	
NO_3^-	Ethylene glycol + H_2O (1:1 by volume)	0.3	

Table 2 lists the values of the effective rate constants of the light-induced e_{tr}^- decay, V_τ/τ, and of the radii of capture of e_{tr}^- by acceptors, R_τ, calculated from the e_{tr}^- photobleaching curves by means of Eq. (30). The value of $\tau = 20$ s for a 10 M aqueous solution of NaOH needed for such calculations, was found from the curve of e_{tr}^- photobleaching in the presence of CrO_4^{2-} and from the values of ν_e and a_e for this acceptor given in Ref. [31].

Thus, the data on the kinetics of decay of e_{tr}^- in the course of photobleaching in matrices containing acceptor additives are well described by a model implying a migration of electrons by a jumpwise mechanism via excitation to the conduction band from traps located far from acceptor particles to those located close to them with the subsequent capture of electrons by these particles via a tunnel mechanism.

2.5 Summary

Let us summarize the results of research on intermolecular electron tunneling in reactions of excited particles. At the present time long-range electron tunneling has been observed in a large variety of reactions of these particles. These reactions include those of electron transfer with the participation of both organic and inorganic compounds in vitreous solutions and crystals. This leads us to believe that electron tunneling can play an essential part in photochemistry. At the same time, many important features of electron tunneling reactions of excited particles have not yet been cleared up. For example, it is not yet quite clear which factors favour and which hinder tunneling. Discussed above are numerous reactions where long-range electron tunneling occurs. At the same time, numerous examples are known when reactions of excited state quenching via electron transfer have been observed to occur in liquids via collisions and have not been detected in vitreous solutions where no collisions are possible between excited particles and quencher molecules. Thus, for example, in Ref. [41] the effect of additives of the quencher CCl_4 on the luminescence kinetics of the $Ru(bpy)_3^{2+}$ in alcohol matrices was studied at 77 K. At this temperature the lifetime of the excited state of this complex is long enough (more than 10^{-6} s), nevertheless the acceptor additives in concentrations of above 3 M do not affect the kinetics of quenching.

3 Electron Tunneling in Reactions Involving Porphyrin Molecules

For a better insight into the mechanism of the primary stages of photosynthesis it is of interest to study electron transfer reactions with the participation of chlorophyll and its synthetic analogs. As far back as 1948, a reversible reaction was discovered – the photoreduction of chlorophyll, Chl, in solutions containing ascorbic acid, AH, (the Krasnovskii reaction) [61]

$$\mathrm{Chl} + \mathrm{AH} \xrightarrow{h\nu} \mathrm{Chl}^- + \mathrm{AH}^+$$

The electron transfer proceeded with the participation of the triplet excited state of Chl. The discovery of the reversible reaction of chlorophyll photoreduction served as a stimulus for starting systematic research on photochemical redox reactions of chlorophyll and its synthetic analogs, i.e. various metalloporphyrins (MP). For a review of the physical and chemical properties of MP, see, for example, Ref. [62].

An important role in the development of the views on the significance of electron tunneling in reactions involving electron-excited molecules of MP has been played by the work of Carapellucci and Mauzerall [63], in which the quenching of the phosphorescence of negatively charged zinc uroporphyrin by different ions in liquid solutions was studied. The intensity of quenching has been found to be only slightly dependent on the charge of the quencher and virtually independent of its redox potential. At the same time, the radius of quenching calculated by the Debye formula [64], depends on the charge of the acceptor and at the zero ion strength of the solution changes from 15 Å for positively charged *N*-benzylnicotinamide to 30 Å for $Fe(CN)_6^{3-}$. Since the sum of the donor and acceptor radii did not exceed 11 Å, the authors suggested a tunnel mechanism of electron transfer from an excited molecule of zinc uroporphyrin to an acceptor.

If account is taken of the relatively small ionization potential and of the long lifetime of triplet excited states of porphyrins, then such distances of tunneling appear to be reasonable. However, due to the high diffusional mobility of reagents, for electron transfer with the participation of metalloporphyrins in liquid solutions it is usually rather difficult to reject the possibility of an occurrence of the reaction with direct contact of reagents and to prove unambiguously that it proceeds via electron tunneling at large distances. Far better opportunities of doing so are offered by the studies of photochemical processes in vitreous matrices at low temperatures under conditions ruling out any diffusion of particles. And indeed, vitreous solutions of metalloporphyrins containing additives of acceptors and donors of electrons proved convenient model systems for studying the role of long-range electron tunneling in processes of charge separation.

Porphyrins of zinc and magnesium are very convenient objects for modelling the processes in reaction centres of biological photosystems. This is due to the fact that the atom of zinc, like that of magnesium, forms complexes with the poryhyrins in which the metal atom donates two electrons to the ligand. The compound formed can be regarded as a compound of a porphyrin dianion with a metal dication. This complex is readily ionized by the action of light. In addition, singlet excited states of zinc and magnesium porphyrins have rather high probabilities of interconversion into triplet states (0.6–0.7 for MgP and 0.9–1.0 for ZnP) and comparatively long lifetimes of those states. This ensures high values of the quantum yield of charge separation of photochemical reactions with the participation of magnesium and zinc porphyrins.

3.1 Electron Transfer Reactions with the Participation of Poryphyrins in a Singlet Excited State

The possibility of photo-oxidizing zinc and magnesium porphyrins in vitreous matrices in the presence of the efficient acceptor of electrons $C(NO_2)_4$ was first reported in Ref. [65]. The authors observed the formation of NO_2 free radical and the cation-radical of metalloporphyrin, MP^+, upon illumination of solutions in the MP absorption band.

The first detailed research on the mechanism of PET with the participation of MP in vitreous matrices appears to be that detailed in Ref. [66], which deals with the processes of charge separation in vitreous solutions of zinc and magnesium porphyrins in ethanol containing some addition of CCl_4 at 77 K. Illumination of the solutions in the Soret band or in long-wavelength of MP absorption results in a one-quantum ionization of MP and formation of MP^+ cation-radicals and

CCl_4^- anion-radicals recorded according to their characteristic EPR and optical spectra [67a, b].

$$MP \xrightarrow{h\nu} {}^*MP \xrightarrow{CCl_4} MP^+ + CCl_4^- \qquad (31)$$

In Ref. [66] the electron transfer to the acceptor is shown to proceed from singlet excited states of MP.

The kinetics of low-temperature ($T = 77$ K) electron tunneling from singlet excited states of zinc and magnesium porphyrins to CCl_4 has been studied [16a]. In this work vitreous solutions of zinc tetraphenylporphin, ZnTPP, and magnesium Etio-1-porphyrin, MgEtio-1, in ethyl alcohol containing different amounts of CCl_4 (1–2 M) have been investigated. Metalloporphyrins were excited with laser pulses with a wave-length $\lambda_{exc} = 530$ nm and a duration of 10^{-11} s. A non-exponential part has been found to appear on the curve of fluorescence decay of metalloporphyrin in the presence of CCl_4. Such a shape of the fluorescence decay kinetics of metalloporphyrins can be accounted for, for example, by the appearance, in the presence of CCl_4, of an additional channel for fluorescence quenching via electron tunneling from ${}^{S_1}MP$ to CCl_4, just as in the case of quenching excited states of simpler organic molecules in the presence of CCl_4 (described in Sect. 2.2).

Upon excitation of the system with a series of saturating laser pulses (i.e. pulses of light causing a transition of practically all the molecules of MP to an excited state), the initial non-exponential part of the decay curve of MP fluorescence disappears. These phenomena have been explained [16a] by a change in the spatial distribution of CCl_4 molecules near molecules of MP under the action of the laser pulses. This change is due to the fact that the first particles to be reduced as a result of electron tunneling from singlet excited metalloporphyrin molecules to CCl_4 are those CCl_4 molecules which are located closest to the donor *MP. In this case, with the increasing number of pulses due to the irreversible decomposition of part of the CCl_4^- anion-radicals into CCl_3 and Cl^-, the number of *MP ... CCl_4 pairs with short distances between *MP and CCl_4 particles decreases. Under these conditions the contribution of electron tunneling to the decay of the luminescence of *MP decreases monotonously with the increase of the number of laser pulses and finally reaches zero. Note that the initial non-exponential part in luminescence decay curves will disappear completely after the charges in all the MP ... CCl_4 pairs with distances between MP and CCl_4 shorter than $R_\tau = (a_e/2) \ln \nu_e \tau$ have been separated irreversibly under the action of light. R_τ represents the distance of electron tunneling from the excited state of metalloporphyrin, *MP, to the acceptor within the characteristic time, τ, of *MP fluorescence decay.

The effect of nuclear motions on the efficiency of electron tunneling from the excited singlet states of porphyrin molecules to acceptors in vitreous matrixes has been studied in Refs. [68a, b]. The rate of electron tunneling from excited zinc and magnesium porphyrins to CCl_4 or $CHCl_3$ or from excited dimers of these pigments to their monomers has been found to increase drastically upon deuteration of the donors and of the molecules of the medium (C_2H_5OH, CH_3OH). For the reverse processes of the charge recombination between CCl_4^- or $CHCl_3^-$ and MP^+ deuteration of the molecules of the medium as well as deuteration of

$CHCl_3^-$ donor also increased the rate of tunneling. However, in both cases the rate of tunneling has been found to be independent of the deuteration of the acceptor particles ($CHCl_3$ for the direct process and MP^+ for the reverse processes). Analysis of the kinetics of electron tunneling in matrices with the solvent molecules deuterated at different positions (CD_3CH_2OH, CD_3CH_2OD, CH_3CD_2OH, C_2H_5OD, C_2D_5OD) suggests that deformational vibrations of the solvent molecules are coupled to electron tunneling more strongly than the valence vibrations.

3.2 Spontaneous and Photostimulated Recombination of Photoseparated Charges

A detailed study has been made [69] of the spontaneous recombination of MP^+ and CCl_4^- particles formed upon illumination at 77 K of vitreous solutions of zinc and magnesium porphyrins containing CCl_4

$$MP^+ + CCl_4^- \longrightarrow MP + CCl_4 \tag{32}$$

This process is the reverse of the stage

$$^*MP + CCl_4 \longrightarrow MP^+ + CCl_4^- \tag{33}$$

of Eq. (31) except, that Eq. (32) results in the formation of an MP particle in the ground state rather than of an electron-excited *MP particle.

It has been found [69] that, at 77 K in vitreous ethanol, methanol and MTHF, the process kinetics demonstrates a close to linear dependence of the concentration of the $MP^+ \dots CCl_4^-$ pairs on the logarithm of the observation time, t, which is characteristic of electron tunneling reactions. The process rate increases dramatically upon illuminating the solutions in the absorption bands of MP^+ ($\lambda_{max} \simeq 400$ and 700 nm) or CCl_4^- ($\lambda_{max} \simeq 365$ nm). The mechanisms of the process acceleration upon illumination in the absorption bands of MP^+ and CCl_4^- have proved to be essentially different.

The increase in the rate of recombination upon illumination in the CCl_4^- absorption band is due to the CCl_4^- photoionization with an electron being transferred to a continuous spectrum and subsequently captured by an arbitrary MP^+ particle. As seen from Fig. 12, in this case the process kinetics (curve 1) is described by the seond-order equation

$$\frac{n(t)}{n(0)} = [1 + n(0)\,kt]^{-1} \tag{34}$$

where n(0) and n(t) are the concentrations of the $MP^+ \dots CCl_4^-$ pairs at the initial instant of illumination and at a time t after the start of illumination and k is the process rate constant. The observation of such a kinetic law, characteristic of chemical reactions under the conditions of a complete defreezing of reagent mobility, proves that one of the reagents taking part in process (32) (i.e. the electron

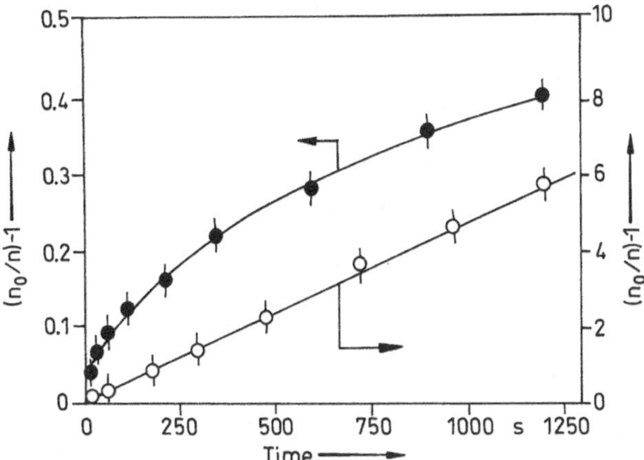

Fig. 12. Kinetics [69] of the photostimulated recombination of ZnP^+ and CCl_4^- upon illumination of vitreous solutions in the absorption bands of (*1*) CCl_4^- and (*2*) ZnP^+

initially localized on CCl_4^-) becomes mobile as a result of photoionization. Also, the observation of the kinetic law of the second, rather than the first, order proves conclusively that upon photoionization of CCl_4^- the electron acquires the ability to recombine with an arbitrary MP^+ cation rather than with the parent cation.

The character of the charge recombination kinetics in the presence of illumination in the MP^+ absorption band remains the same as in the absence of illumination. A considerable (more than 10^3-fold) increase, however, is observed in the recombination rate of the $MP^+ \ldots CCl_4^-$ pairs. Thus, the rate of electron tunneling from CCl_4^- to MP^+ increases essentially upon photoexcitation of MP^+. A detailed analysis of the photostimulation of tunneling recombination has led the authors of Ref. [70] to the conclusion that the acceleration of this process upon photoexcitation of MP^+ is due to the transfer of electron excitation energy from $^*MP^+$ to the tunneling electron (not to CCl_4^-). This brings about a virtual decrease in the height of the barrier to tunneling and a significant rise in the value of the parameter a_e.

3.3 Detection of Anisotropy of Electron Tunneling from CCl_4^- to MP^+

In Refs. [71 a] and [71 b], a substantial change of the EPR spectra of the MP^+-CCl_4^- radical pair was detected upon the change of the polarization of the light producing this pair and the rotation of the sample (cylindrical tube) containing these radical pairs in the cavity of the EPR spectrometer around the axis of the cylinder.

As a typical example, Fig. 13 shows the EPR spectra of the MP^+-CCl_4^- radical pair (MP^+ is the cation-radical of zinc mesotetra-α,α,α,α-*o*-pivalamidophenylporphin) formed upon illumination with linearly polarized light in vitreous alcohol solution. The vitreous solution, containing MP and CCl_4, was illuminated in cylindrical tubes with light whose vector $\vec{E}$ was parallel to the axis $\vec{y}$ or to the $\vec{z}$

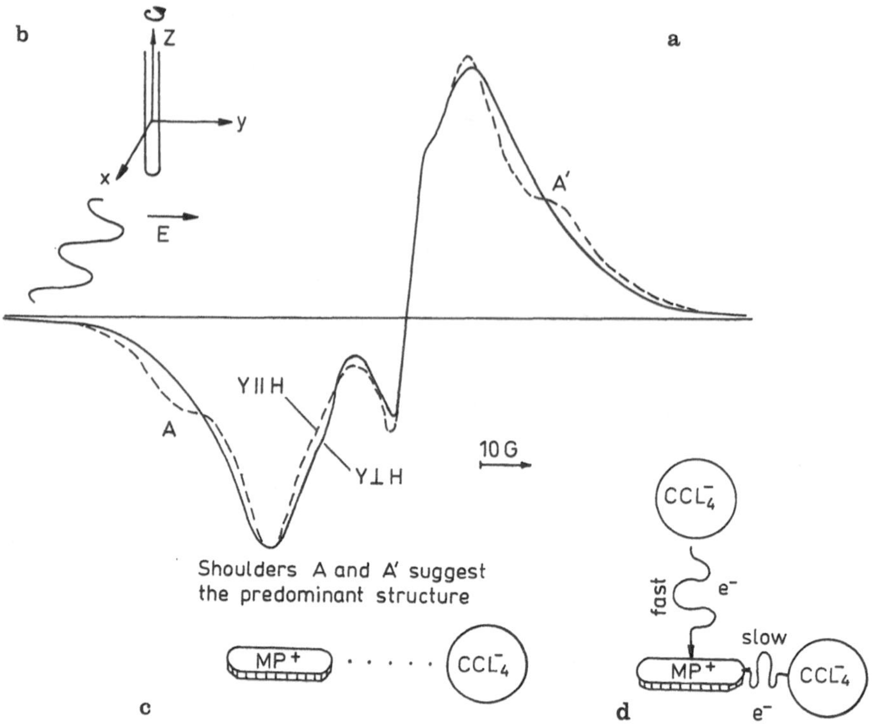

Fig. 13. (**a**) – EPR spectra of the MP^+-CCl_4^- radical pairs formed upon illumination by polarized light of a vitreous ethanol solution containing 10^{-4} M zinc *meso*-tetra-α,α,α,α-*o*-pivalamidophenylporphin and 1 M CCl_4 for two orientations of the sample (cylindrical tube) in the magnetic field, $\vec{H}$, of he EPR spectrometer. The *broken line* corresponds to the orientation $\vec{y} \parallel \vec{H}$, and the *solid line* to the orientation $\vec{y} \perp \vec{H}$. $\vec{z}$ is the axis of the cylindrical tube; $\vec{y}$ is the axis which is parallel to vector $\vec{E}$ of the polarized light; and $\vec{H}$ is the direction of the magnetic field of the EPR spectrometer. Data taken from Refs. [71a, b]. (**b**) – Scheme of the illumination of the samples by polarized light. (**c**) – Schematic picture of the predominant spatial positions of MP^+ and CCl_4^- in the sample. (**d**) – Scheme illustrating why MP^+-CCl_4^--pairs with structure (c) are predominantly accumulated

axis (see the notations of the axes in Fig. 13). For $\vec{E} \parallel \vec{y}$, the EPR spectrum was found to change and for $\vec{E} \parallel \vec{z}$ not to change upon rotation of the tube around the $\vec{z}$ axis. As shown in Ref. [71a], the appearance of the shoulders A an A′ in the EPR spectrum of the MP^+-CCl_4^- pair suggests that anion radicals CCl_4^- are located mainly in the equatorial plane of MP^+. Such a location of CCl_4^- particles that have avoided recombination up to the moment of recording the EPR spectrum, indicates that the probability of electron tunneling from CCl_4^- to MP^+ depends on the mutual orientation of the MP^+ and CCl_4^- particles. Indeed, one can expect electron tunneling to be most efficient in the direction of the π orbitals of the porypyrin ring. In this case, the pairs in which CCl_4^- particles are located in the vicinity of the symmetry axis of the porphyrin ring, are expected to decay faster than those located in the vicinity of the equatorial plane of the ring. As a result, these are mainly the latter pairs that survive up to the moment of recording the EPR spectrum.

The observed dependence of the rate of electron tunneling on the orientation of the reacting particles suggests that orientation of porphyrin rings can perhaps serve as an instrument regulating the rate of electron transport during photosynthesis.

3.4 Electron Transfer Reactions with the Participation of Porphyrins in a Triplet-Excited State

Electron tunneling from triplet excited molecules of MP can be observed in the case of photo-oxidizing copper porphyrins in the presence of tetranitromethane in vitreous matrices. Copper porphyrins, CuP, are convenient subjects for studying electron tunneling reactions with the participation of triplet excited states of the porphyrin ring since the presence of the paramagnetic copper(II) atom in CuP molecules provides an efficient interconversion of the singlet state of the porphyrin ring, which is originally formed upon the excitation of MP compounds with light quanta, into a triplet excited state and thus prevents CuP from deactivation via fluorescence.

As observed in Ref. [72], the addition of tetranitromethane, $C(NO_2)_4$, electron acceptor to a solution of CuP in ethanol causes a decrease in the quantum yield of phosphorescence of CuP at 77 K and the appearance in the optical and EPR spectra of signals which are characteristic of CuP^+, NO_2, and $C(NO_2)_3^-$ particles. The formation of these particles points to PET from CuP to $C(NO_2)_4$. The decay curves of CuP phosphorescence in vitreous solutions containing $C(NO_2)_4$ in low concentrations are of an exponential character. At sufficiently high concentrations of $C(NO_2)_4$ (0.3–0.5 M), however, these curves deviate from the simple exponential form. The appearance of non-exponential parts of the decay curves has been accounted for by electron tunneling from the triplet excited state of CuP particles to molecules of $C(NO_2)_4$.

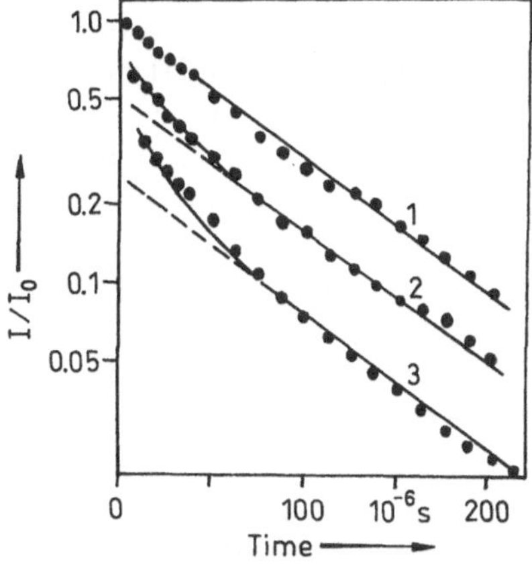

Fig. 14. The kinetics [72] of the phosphorescence decay of tetramethyl ether of copper hemato-IX porphyrin (CuP) in the presence of various concentrations of $C(NO_2)_4$: (*1*) – 0, (*2*) – 0.3, and (*3*) – 0.6. M The *points* represent the experiment and the *solid lines* reflect the calculation using Eq. (8)

The phosphorescence decay kinetics of the triplet excited states of CuP molecules (Fig. 14) is adequately described by Eq. (16). Using this equation one can obtain the values of the parameter $\beta = (\pi a_e^3/2) \ln^2 \nu_e\tau$ from the initial non-exponential part of the phosphorescence decay curves and the values of $\tau = 1/k$, i.e. the characteristic time of phosphorescence decay, from the final exponential part. Then the data on the dependence of the quantum yield of CuP phosphorescence on the concentration of $C(NO_2)_4$ have been used to estimate the effective radii of electron tunneling from triplet excited copper porphyrins to $C(NO_2)_4$ within the time τ: $R_\tau = (a_e/2) \ln \nu_e\tau$ (Table 3). In doing so, the quenching of CuP luminescence by electron abstraction was assumed to be the only process leading to a decrease in the quantum yield of CuP phosphorescence in the presence of $C(NO_2)_4$. From Table 3 an electron is seen to tunnel, within the lifetime of triplet excited states $\tau \simeq 10^{-4}$ s, from *CuP particles to $C(NO_2)_4$ molecules over the distance $R_\tau \simeq 11$ Å. Further, the parameter ν_e and a_e for different porphyrins were estimated from the values of β, R_τ, and τ. These values are also cited in Table 3.

Table 3. Parameters R_τ, a_e and ν_e for electron tunneling from electron-excited molecules of copper porphyrins to $C(NO_2)_4$ in vitreous ethanol at 77 K [72]

Parameter	Copper tetra-phenylporphin	Copper tetra-phenyl (pival-amide) porphin	Tetramethyl-ether of copper hemato-IX porphyrin
R_τ, Å	11.0 ± 0.5	10.8 ± 0.5	9.8 ± 0.5
a_e, Å	2.2 ± 0.4	1.8 ± 0.4	2.0 ± 0.4
ν_e, s^{-1}	10^8 ± 1.3	10^9 ± 1.5	10^7 ± 1.2

3.5 Mediator Assisted PET between Porphyrin Molecules

The role of mediator molecules in donor-acceptor electron transfer processes is an item of considerable recent interest [73–81]. A lot of research has been done on intermediate acceptors in the electron transfer in photosynthesis and theoretical studies bases on the superexchange interaction have been carried out [76–82]. In Refs. [83, 84], electron transfer in the presence of ordered mediator molecules with arbitrary energy levels in one-dimensional case [83] as well as electron transfer in the presence of one resonant mediator [84] were considered.

In a number of works, it was supposed that in biological systems, due to certain structures of the molecules of the medium with vacant electron levels, tunnel reactions are more effective than in non-ordered media. But until recently, no direct experimental evidences supporting this assumption had been presented.

The first work in which such experimental evidences were presented seems to be [85]. In this work, light-induced two-quanta electron transfer from one porphyrin molecule to another was studied in triethylamine at 77 K. The reaction was shown to proceed via the tunneling mechanism according to the scheme:

$$P \cdot [N(C_2H_5)_3]_2 \xrightarrow{h\nu_1} P^- \cdot [N(C_2H_5)_3]_2^+ , \tag{35}$$

$$P^- \cdot [N(C_2H_5)_3]_2^+ + P \cdot [N(C_2H_5)_3]_2 \xrightarrow{h\nu_2} P \cdot [N(C_2H_5)_3]_2^+ + P^- \cdot [N(C_2H_5)_3]_2 \tag{36}$$

Here $P \cdot [N(C_2H_5)_3]_2$ denotes a complex between porphyrin molecule and triethylamine. The stability of the long lived radicals $P^- \cdot [N(C_2H_5)_3]_2$ seems to be due to their remoteness from the "hole" $P \cdot [N(C_2H_5)_3]_2^+$.

The values of $\nu_e = 10\ s^{-1}$ and $a_e = 5$ Å were found from accumulation curves of $P^- \cdot [N(C_2H_5)_3]_2$ anion-radicals. The low value of the frequency factor ν_e was explained by low stationary concentration of the excited complexes $*(P^- \cdot [N(C_2H_5)_3]_2^+)$. As shown in Ref. [85] this concentration is contained as a multiplier in the expression for ν_e.

The efficiency, F, of the tunnel electron transfer from the excited complexes $*(P^- \cdot [N(C_2H_5)_3]_2^+)$ to porphyrin molecules of $P \cdot [N(C_2H_5)_3]_2$ complexes sharply increased when CCl_4 or $CHCl_3$ were added to the solution (see Fig. 15). The values of F were calculated by comparison of the number of $P^- \cdot [N(C_2H_5)_3]_2$ anion-radicals formed in the absence of CCl_4 or $CHCl_3$ with the total amounts of $P^- \cdot [N(C_2H_5)_3]_2$ and CCl_4^- or $CHCl_3^-$ anion-radicals generated during the same time in the samples containing additives. CCl_4^- and $CHCl_3^-$ anion-radicals were produced by means of the dark tunnel electron transfer from $P^- \cdot [N(C_2H_5)_3]_2$ to CCl_4 or $CHCl_3$. As can be seen from Fig. 15, almost 10^2-fold increase of the efficiency factor was observed at 1 M concentration of CCl_4. Addition of 1 M $CHCl_3$ increased the efficiency factor by the factor of 40.

Similar results were obtained for the ethanol solutions of covalently bonded diethylamine-porphyrin molecules, $P-N(C_2H_5)_2$.

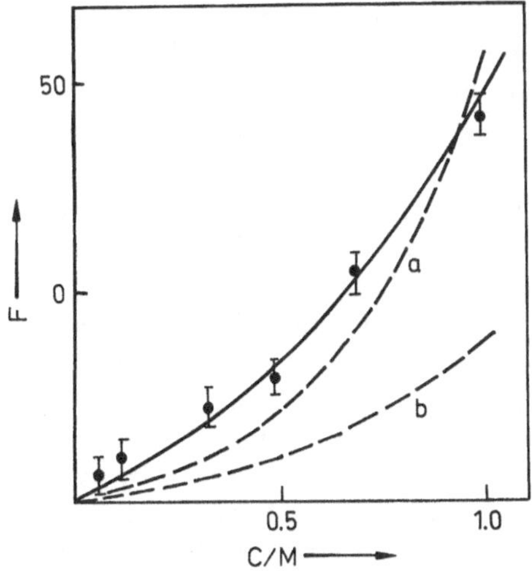

Fig. 15. The dependence of the efficiency F of electron tunneling in reaction (36) on the concentration of CCl_4 mediator. The *solid line* is the experimental curve; *dashed lines* are theoretical curves calculated for $\Delta = 0.5$ eV (*a*) and $\Delta = 0.7$ eV (*b*) [85]

A theoretical explanation of this phenomenon based on the assumption that CCl_4 or $CHCl_3$ molecules serve as mediators for the reaction (36) was offered in Ref. [85]. A simple formula was obtained for the dependence of the tunneling probability W(k, R) on the number, k, of mediator particles taking part in electron transfer and on the average difference Δ between the energy of a tunneling electron and the energies of the mediator levels:

$$W(k, R) = 2\nu_e[(1 + V/\Delta)^k - 1/2] \exp(-2R/a_e) \tag{37}$$

where $V \simeq 1$ eV is the energy of exchange interaction between porphyrin molecules being in direct contact with each other.

Numerical calculations of the dependences of the efficiency of tunnel electron transfer in triethylamine solutions of P on the concentration of mediator molecules calculated in Ref. [85] using Eq. (37), for various values of parameter Δ are presented in Fig. 15. From comparison of these results with the experimental data, it is seen that the best fit corresponds to $\Delta = 0.5$ eV. So one can suppose that the nearest vacant level of CCl_4 in triethylamine is 0.5 eV distant from the energy of the tunneling electron.

4 Intramolecular Electron Transfer in Non-Porphyrin Bridge Organic Molecules

During the last few years measurements of intramolecular electron transfer in bifunctional bridge organic molecules D-L-A containing two electron donor/acceptor groups linked by chains of chemical bonds (bridge L) have provided further insights into the effect of the distance, mutual orientation of D and A, reaction energetics, as well as of the structure of the bridges L, reagents D and A and medium on the rate of tunneling [86–110]. In the case of rigid bridges, owing to the absence of scatter over the distances between the fragments D and A, one can expect intramolecular electron tunneling in D-L-A systems to obey conventional first-order kinetics in both vitreous and liquid solutions. The process rate, however, will decrease exponentially with increasing bridge length. For flexible bridges, when scatter of the distances does exist, one can expect electron tunneling in D-L-A molecules to obey logarithmic kinetics at least in vitreous solutions.

The majority of publications on intramolecular long-range electron transfer in bridge molecules deals with dark secondary processes of radiation induced or photochemically induced electron transfer. Intramolecular electron transfer arises in this case from electron implantation to the donor fragments of the molecule.

Thus, e.g. as in the typical investigation in Ref. [110], *long-range intramolecular electron transfer has been detected for rigid bifunctional steroid molecules* dissolved in THF. The structural formulae of these compounds are represented in Fig. 16. The donor acceptor pairs I–V presented in this figure are separated by various numbers, n, of similar chemical bonds, the so-called "n-bond" system. The anion-radicals of molecules I–V were obtained via reactions of these molecules

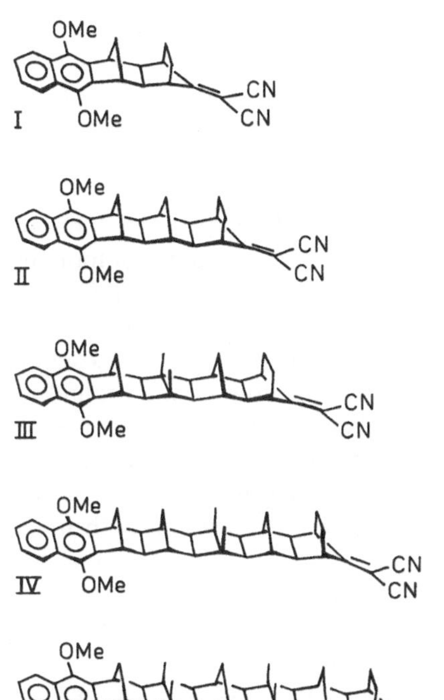

Fig. 16. Structure of bifunctional molecules with the rigid bridges used in Refs. [108–114] for studying intramolecular electron transfer

with solvated electrons generated by a short pulse ($\tau \simeq 3 \times 10^{-11}$ s) of an electron accelerator. The electrons were initially trapped by a 1,4-dimethoxynaphthalene group, and then intramolecular electron transfer took place from the 1,4-dimethoxynaphthalene fragment to the more electron-attracting dicyanovinyl group located at the opposite end of the bifunctional molecules. In so doing, a thermodynamically more advantageous state of the anion-radical was achieved. The rates of intramolecular electron transfer for $n = 4, 6, 8, 10, 12$ were found to exceed 1×10^9 s^{-1} for all five compounds.

These very fast electron transfers through as many as 12 saturated carbon-carbon bonds are consistent with fast electron transfer observed for photoinduced charge separation in the same compounds [108, 109]. In these works, the time-resolved microwave conductivity (TRMC) technique was applied to the investigation of charge separation in the donor-acceptor pairs shown in Fig. 16. This technique is based on the fact that a large change in dipole moment occurs on photoinduced electron transfer; this in turn causes a significant change in the solution conductivity which is monitored by a change in the dielectric loss. The latter can be measured quantitatively by a change in the microwave power reflected by a cavity containing the sample.

The TRMC measurements show that the light induced charge separation

$$\text{D-L-A} \xrightarrow{h\nu} \text{D}^*\text{-L-A} \longrightarrow \text{D}^+\text{-L-A}^- \tag{38}$$

occurs on a sub-nanosecond time scale for all five D-L-A molecules. The lifetime of the charge transfer state D^+-L-A^- towards recombination of D^+-L-$A^- \rightarrow$ D-L-A in benzene, τ_r increases dramatically from 10^{-9} s for compound I to 6×10^{-9} s, 3.2×10^{-8} s, 3.6×10^{-7} s, and 7.4×10^{-7} s for compounds II, III, IV and V, respectively. The efficiency of charge separation for all five compounds was found to be close to unity.

Using the obtained values of τ_r, the distances, R, separating the donor (dimethoxynaphthalene) and the acceptor (dicyanoethylene) groups in compounds I–V, and supposing an exponential dependence of τ_r on R, $\tau_r = \nu_e^{-1} \exp(2R/a_e)$, one can estimate $a_e = (1.0 \pm 0.3)$ Å and $\nu_e = 10^{12 \pm 2}$ s^{-1} for the recombination reaction. The values of these parameters for the forward charge separation reaction calculated from fluorescence lifetime of the photoexcited donors, were found to be notably bigger, $a_e = 1.92$ Å and $\nu_e = 10^{15}$ s^{-1} [111–113].

In Ref. [114], new optical electron transfer bands were observed in the visible-near-infrared absorption spectra of the anions of compounds II, III, IV and V. The maxima of these spectra were shifted to a shorter wavelength upon increasing solvent polarity. Their intensities decreased rapidly as the length of the bridge increased (by a factor of ~4 for each additional two bonds between dimethoxynaphthalene and dicyanovinyl groups). These bands seem to result from a light absorption process in which the photon removes an electron from one group and transfers it across the bridge to the other group (the so called "tunneling charge transfer bands"). The values of electron matrix elements for electron tunneling transition, V, (see Eq. (23)) calculated from optical spectra, were found to be 0.16, 0.06 and 0.03 eV across 4-, 6- and 8-bond bridges.

The works on electron transfer in steroid bridge molecules discussed above represent an example of a rather recent study on PET in D-L-A systems.

The earliest work on intramolecular PET involved *studies of charge transfer (CT) absorption bands in D-L-A complexes [117]*. The authors of this work observed distinct intramolecular CT bands in linked D-L-A compounds, which consisted of a series of methyl- and methoxy-benzene derivatives as donors and the 4-cyano-pyridinium cation as the acceptor, separated by one or two methylene groups.

In Ref. [118] the series of D-L-A compounds were studied in which the length of the saturated hydrocarbon bridge L was varied from zero to three carbon atoms. These compounds contained anthracene as the acceptor A and *N,N'*-dimethylaniline as the donor D. A new long-wavelength fluorescence was observed under anthracene excitation. This emission was found to shift markedly to longer wavelengths with increasing solvent polarity, indicating that the emitting state had a large dipole moment. According to Ref. [118], the emission originated from a CT state (A^--L-D^+) formed upon electron transfer occured from D to A*.

Similar observations were made in Refs. [86, 119] with molecules *containing methoxybenzene chromophore as D and dicyanoethylene acceptor fragment as A linked with hydrocarbon saturated bridge and in Refs. [115, 116] with D-L_2-A molecules containing acceptors linked to the donors via two bridges*, so that the acceptor is held in a relatively rigid, "capped" position over the donor.

5 Electron Tunneling in Bridge Porphyrin-Quinone Compounds

The study of electron tunneling in bridge molecules with porphyrin and quinone D and A fragments is important for elucidating the details of the initial stages of photosynthesis. Numerous studies of PET in such molecules have been carried out. The length of the bridging fragment L was up to 15–20 Å. One can point out three main directions of research on bridge porphyrin-quinone molecules: (1) study of the processes of charge separation in porphyrin-quinone molecules with a flexible bridge; (2) study of the processes of charge separation in compounds with a rigid bridge; (3) study of charge separation in covalently linked triads containing porphyrin covalently linked with quinone and an electron donor. Research and synthesis of the simplest porphyrin-quinone compounds, molecules with flexible bonding, was first carried out by the authors of Refs. [120, 121]. Transition from the systems with flexible bonding to porphyrin-quinone compounds with a rigid bridge has made it possible to study in more detail the dynamics of charge separation between prophyrin and quinone without the complicating influence of conformational motions that are characteristic of the systems with flexible bonding. The synthesis of covalently linked triads turned out to be the step towards an even more adequate modelling of reaction sites in photosynthesizing organisms. Let us first consider the processes of charge separation in systems with flexible bonding.

5.1 Processes of Charge Separation in Porphyrin-Quinone Compounds with Flexible Bonding

Kong and Loach [120, 121] seem to be the first to synthesize porphyrin-quinone compounds of the P-L-Q type. The structure of the compounds they obtained ($n = 2, 3$) is shown in Fig. 17. A strong quenching of the fluorescence of the porphyrin fragment by the quinone fragment was discovered in these compounds.

The suggestion that the quenching of the porphyrin fragment fluorescence by quinone is due to intramolecular electron transfer was first made in Refs. [122] and [123]. The Soret band of P-L-Q was found [123] to be notably broader than that of the free porphyrin, P, and the quantum yield of P fluorescence was observed to decrease by a factor of more than 10^3. Both these facts were explained by electron transfer from the *P fragment to the Q fragment. The hypothesis about electron transfer from *P to Q has since been repeatedly used to account for the nonexponential character of P fluorescence decay as well as for the dependence of the efficiency of fluorescence quenching on the length of the bridge L, the nature of the porphyrin fragment and the nature of the solvent [124–126]. Note that the data reported [127, 128] on the dependence of the efficiency of the porphyrin fluorescence quenching in homogeneous solutions of non-bridged P and Q species on their redox properties also agree with the mechanism of quenching via the electron transfer process of the $^*P + Q \rightarrow P^+ + Q^-$ type.

The formation of the ion-radical pair P^+-L-Q^- upon illumination of the P-L-Q species in vitreous solutions at $T = 90–160$ K was first demonstrated [129] using

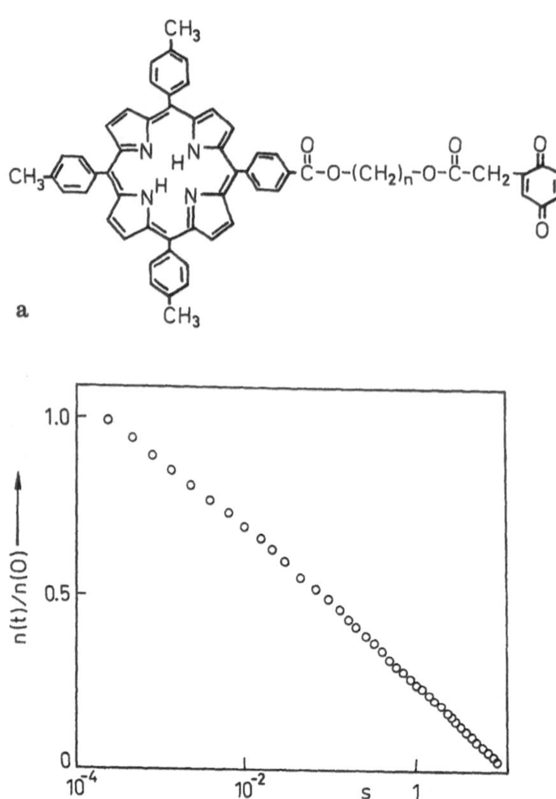

Fig. 17. (**a**) – Structure of molecules P-L-Q and (**b**) – the decay kinetics for P^+-L-Q^- ion-radical pairs with $n = 3$ at T = 160 K in a vitreous mixture of methanol + chloroform (98:2). The data are taken from Ref. [129]

the EPR method. The P-L-Q compound of the kind depicted in Fig. 17 ($n = 3$) in a methanol-chloroform mixture (98:2) was studied in this work. Along with the generation of the ion-radical pair, a decrease by a factor of 10 was observed in the intensity of the porphyrin fragment luminescence. After the light was switched off, a certain number of pairs were observed to decay due to the back transfer of the electron from Q^- to P^+. At 160 K the ion-radical pairs decay almost completely. The kinetics of the decay of P^+-L-Q^- ion-radical pairs within a range of observation times of about 4 orders of magnitude is characterized by a law of variation of P^+-L-Q^- concentration with time which is close to logarithmic rather than exponential (see Fig. 17). Note also that, at $T < 160$ K and observation times $t < 10^3$ s, part of P^+-L-Q^- pairs are kinetically stable. The number of stable P^+-L-Q^- pairs grows with decreasing temperature. The temperature dependence of the portion of P^+-L-Q^- pairs that had recombined by a definite time points to the activated character of recombination [31]. The logarithmic character of the process kinetics can be accounted for by different conformational state of the hydrocarbon chain binding the porphyrin with the quinone, which leads to different distances between P^+ and Q^- fragments and hence to different recombination rate constants for various P^+-L-Q^- ion-radical pairs.

The formation of P^+-L-Q^- ion-radical pairs upon illumination of the P-L-Q compounds of the kind depicted in Fig. 17 and of their zinc complexes within a notably wider range of temperatures (up to 300 K) was detected by the EPR method [124, 130]. The quantum yield of charge separation and the time of charge recombination at 300 K amounts to 5×10^{-3} and 10^{-3} s, respectively. In accordance with the modern models of electron transfer in condensed media the efficiency of charge separation in P-L-Q grows with increasing solvent polarity [124].

An analysis of the EPR spectra of the P^+-L-Q^- ion-radical pairs [131] made it possible to estimate the energy of magnetic dipole-dipole interaction between the paramagnetic fragments P^+ and Q^- in these particles and to calculate from this energy the mean distance $R_{PQ} = 10$–12 Å between P and Q fragments in the molecule of the kind, depicted in Fig. 17a ($n = 2$) in which the dicarboxyl bridge

$$-\underset{\substack{\| \\ O}}{C}-O-(CH_2)_n-O-\underset{\substack{\| \\ O}}{C}-CH_2-\,,$$

is replaced by the diamide bridge,

$$-\overset{\substack{O \\ \|}}{C}-\overset{\substack{H \\ |}}{N}-(CH_2)_n-\overset{\substack{H \\ |}}{N}-\overset{\substack{O \\ \|}}{C}-CH_2-$$

The existence of various conformations for molecules depicted in Fig. 17a at $n = 2, 3, 4$ was established [132] by observing the strong broadening of the Soret band and of fluorescence bands of these P-L-Q compounds. The highest efficiency of charge separation was observed at $n = 3$ [131, 132]. Comparison of these data with the results of Refs. [133–136], according to which the maximum efficiency of charge separation for a number of other donor-acceptor complexes of the D-$(CH_2)_n$-A kind is also reached at $n = 3$, showed that, in complexes with flexible bonding, the $-(CH_3)_3$-bridge is the optimal one for electron transfer. A bridge of this length appears to ensure both the proximity of donor and acceptor fragments and a sufficient conformational mobility, making it possible to reach the optimal position and orientation of the donor and the acceptor fragments within the lifetime of the excited donor fragment.

Two-step electron transfer along the gradient of the redox potential of accpetor groups was detected [100, 137] upon illumination of solutions of P-$(CH_2)_4$-Q_1-$(CH_2)_4$-Q_2 where Q_1 was benzoquinone and Q_2 trichlorobenzoquinone. Charge separation processes in these systems can be explained in terms of the scheme

$$\begin{array}{ccc} \text{P-L-}Q_1\text{-L-}Q_2 & \overset{h\nu}{\rightleftarrows} & P^+\text{-L-}Q_1^-\text{-L-}Q_2 \\ & \searrow \quad \nearrow & \\ & P^+\text{-L-}Q_1\text{-L-}Q_2^- & \end{array} \tag{39}$$

The characteristic time, $\tau < 10^{-11}$ s, of the two-step charge separation resulting in the formation of the P^+-L-Q_1-L-Q_2^- particles is far shorter than that, $\tau > 5 \times 10^{-10}$ s, of the one-step charge separation in a molecule even with a shorter bridge, P-$(CH_2)_6$-Q_2. These data agree with theoretical estimates [137, 138] according to which multi-step process of electron transfer along the gradient of the redox potential of the acceptor groups ensures a higher efficiency of charge separation than a one-step process with the same final change in the redox potential.

5.2 Charge Separation Processes in Porphyrin-Quinone Compounds with Several Flexible Bridges

The high conformational mobility of porphyrin-quinone compounds with flexible bonding makes it difficult to elucidate in sufficient detail the mechanism of electron transfer between porphyrin and quinone fragments. Far greater possibilities for determining the role of mutual orientation of P and Q and the distance between them are offered by P-Q compounds in which the P and Q fragments are linked by several bridges. A P-Q compound of this kind was first prepared by Gannesh and Sanders [139] (the synthesized compound is schematically depicted in Fig. 18). An analysis of the spectra of P-Q has shown that, for this compound the quinone fragment is, on the average, located perpendicularly to the porphyrin fragment which creates rather unfavourable conditions for electron tunneling from P to Q [139]. This results in a rather weak photochemical activity of the compound in question towards formation of P^+-Q^- species [140, 141].

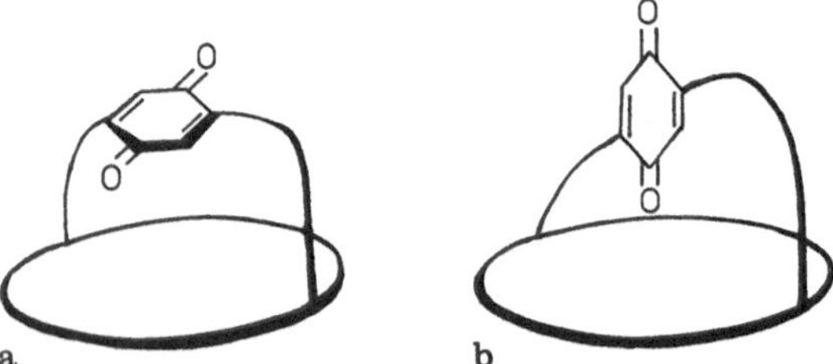

Fig. 18. Structure of quinone-capped porphyrin complexes [139]. Two extreme conformations in which the quinone fragment is (*a*)-parallel and (*b*)-perpendicular to the porphyrin are shown

In Ref. 142, a detailed analysis of the forward and reverse electron transfer rates for capped P-L_2-Q in a variety of solvents was given. The results show that forward electron transfer is in the normal region $E_r > -\Delta G°$ and charge recombination is in the Marcus inverted region, $E_r < -\Delta G°$.

In the case of co-facial quinone-capped porphyrins (P and Q are linked by four tetraamidophenoxy bridges and are located at a distance of about 10 Å from each other), the quantum yield of charge separation is much bigger and reaches 30% for short distances between P and Q [143, 144]. Luminescence quenching via electron transfer from *P to Q is observed for both singlet- and triplet-excited states of the porphyrin fragment of P-Q. The appearance of the additional channel for luminescence decay via electron transfer manifests itself in the biphase character of P-Q luminescence decay kinetics.

A decrease in the length of the bridge, and an orientation of P and Q fragments in co-facial quinone-capped porphyrins which is favourable for electron tunneling, results in a sharp rise of the efficiency of electron phototransfer from P to Q [144].

Despite the presence of four bridging groups the quinone is not held rigidly above the plane of the porphyrin. Two channels of electron transfer from the singlet excited state of the porphyrin were found for this compound [145]. They were ascribed to slowly equilibrating "introverted" and "extroverted" conformers in which the estimated interplanar porphyrin.quinone separation is respectively, 6.5 and 8.5 Å. The faster of the two rate constants is independent of the temperature over the range 80–300 K.

An interesting porphyrin-bis-quinone Q-L-P-L-Q molecule with two quinone fragments one of which was located above the porphyrin plane and the other below this plane, was synthesized in Refs. [146, 147]. As determined by X-ray technique, the two quinones are coplanar with the porphyrin and are symmetrically displaced at a distance 3.4 Å from it. This molecule shows a fluorescence decay profile in a liquid solution that is best fited by a distribution of lifetimes rather than a single lifetime, as expected if the quinones are capable of changing their position relative to the porphyrin.

5.3 Processes of Charge Separation in Porphyrin-Quinone Compounds with a Rigid Bridge

Another example of compounds with the fixed mutual location of porphyrin and quinone are the porphyrin-quinone compounds with a rigid bridge. Charge photoseparation in P-L-Q molecules in which L is the triptycene bridge, P is tetraphenylporphin, TPP, or its zinc complex, and Q is benzoquinone, BQ, naphthoquinone, NQ, or anthraquinone, AQ, has been studied [148]. The distance

Table 4. Quantum yields, φ of the fluorescence of the porphyrin fragment and free energy changes, $-\Delta G°$, for the electron transfer reaction from the *P fragment to the Q fragment of P-L-Q molecules in methylene chloride solution [148]

Compound	φ	$-\Delta G°$, eV
TPP	0.14	
TPP-AQ	0.119	0.02
TPP-NQ	0.001	0.29
TPP-BQ	<0.001	0.47
ZnTPP	0.023	
ZnTPP-AQ	0.002	0.43
ZnTPP-NQ	<0.001	0.70
ZnTPP-BQ	<0.001	0.88

between the centres of P and Q fragments in these compounds amounted to 10, 10.5 and 11 Å for TPP-L-BQ, TPP-L-NQ and TPP-L-AQ, respectively.

The efficiency of luminescence quenching (see Table 4) was found to correlate with the change in the free energy, $-\Delta G^\circ$, of the electron transfer reaction estimated according to the formula (see Sect. 2.2)

$$-\Delta G^\circ = E(S_1) - E(P/P^+) + E(Q/Q^-) \tag{40}$$

where $E(S_1)$ is the excitation energy of the donor singlet state S_1, $E(P/P^+)$ is the porphyrin oxidation potential, and $E(Q/Q^-)$ is the quinone reduction potential. For the sake of simplicity, the Coulomb term is omitted from the equation since it is the same for all three compounds.

When the light is switched off, the recombination process P^+-L-Q^- → P-L-Q is observed. The characteristic time of charge recombination is $\simeq 10^{-10}$ s. As distinct from the direct process of charge separation, whose efficiency increases with an increasing value of $-\Delta G^\circ$, the probability of charge recombination decreases on increasing the value of $-\Delta G^\circ$ [149]. This decrease appears to be due to the fact that, for the charge recombination processes in the systems under study, the exothermicity, J, is so large that it exceeds the reorganization energy, E_r, In this case, according to Eq. (24), a decrease must indeed be observed for the probability of electron tunneling with increasing $-\Delta G^\circ$.

The charge photoseparation in porphyrin-quinone compounds with a rigid bicyclo[2.2.2]octyl bridge, ensuring a distance between the centres of P and Q of about 16 Å, has been studied [103]. The rate constant of intramolecular electron transfer from *P to Q was found to depend on the dielectric properties of the medium and reached $3.3 \times 10^7\ s^{-1}$ for a solution of P-L-Q in propionitrile.

To determine the parameters of electron tunneling from *P to Q, the kinetics of *P fluorescence decay was studied for P-L-Q compounds where the rigid bridge was made up of one or two bicyclo[2.2.2]octyl fragments [101, 150] (for the structural formulae of the compounds see Fig. 19). Typical curves of fluorescence decay are presented in Fig. 20. The transition from a bridge containing two bicyclo[2.2.2]octyl fragments to a shorter one containing only one such fragment is shown to result

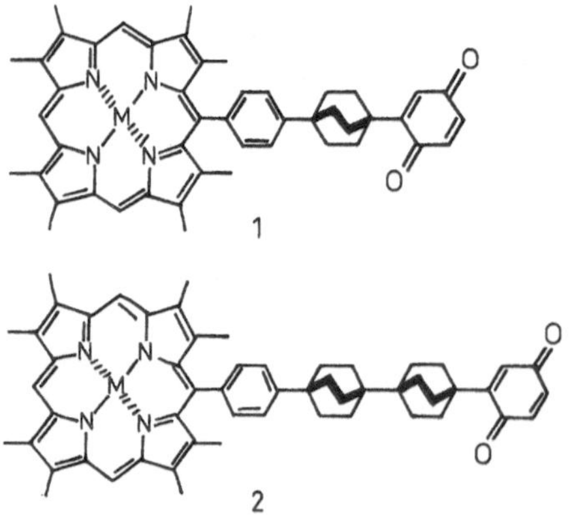

Fig. 19. Structure of porphyrin-quinone compounds with a rigid bicyclo[2.2.2]octyl bridge

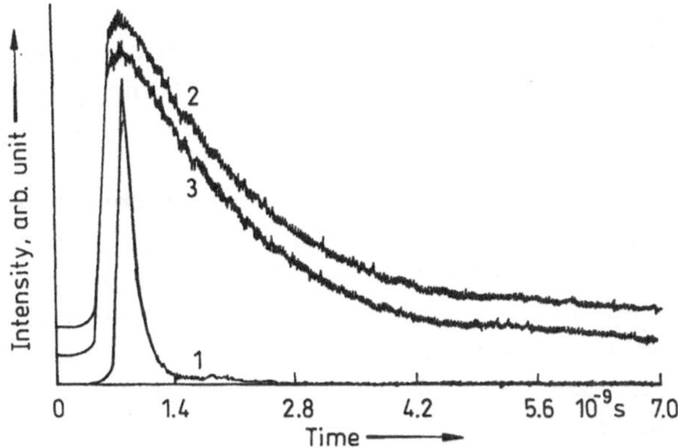

Fig. 20. Fluorescence decay curves of zinc *meso*-phenyl-octaalkyl porphyrin bound with benzoquinone by (*1*) – one or (*2*) – two bicyclo[2.2.2]octyls. (*3*) – Fluorescence decay curve for zinc *meso*-phenyl-octaalkyl porphyrin. T = 298 K. The data are taken from Ref. [150]

in a sharp decrease of the time of fluorescence decay. The rate constant of electron transfer from *P to Q calculated from the data of Fig. 20 proved to be $k_t^{(2)} \lesssim 10^7\ s^{-1}$ and $k_t^{(1)} = 10^{10}\ s^{-1}$ for P-L-Q with a long and short bridge, respectively.

Further, using Eq. (1) for the dependence of the probability of electron tunneling 1from *P to Q, one can easily calculate $a_e \lesssim 1.4$ Å and $\nu_e \gtrsim 10^{19}\ s^{-1}$. But, if instead of the intercentral distance, the shortest distance between P and Q is used as R, the value $\nu_e \gtrsim 10^{16}\ s^{-1}$ is obtained.

At T = 77 K in MTHF, the kinetics of fluorescence decay of P-L-Q with a bridge containing one bicyclo[2.2.2]octyl is of a non-exponential character. The effect can be explained by the coexistence in the frozen solution of several rotational conformations of the P-L-Q mole (rotation of the porphyrin fragment around the σ bond in its meso position is meant here). The characteristic time of the fluorescence decay for the predominant portion of the *P-L-Q particles at 77 K, $\tau \simeq 1.1 \times 10^{-10}$ s, virtually coincides with the value of $\tau = 1/k_t^{(t)}$ at 298 K, i.e. the rate of tunneling from *P to Q is independent of temperature. The exponential character of the fluorescence decay curve at 298 K indicates that, at this temperature, the rate of rotation exceeds $k_t^{(1)}$.

The measurements of the rate constants for the forward reaction from the porphyrin excited singlet state and for the reverse electron transfer reaction for the series of zinc porphyrin-anthraquinone molecules was made in Ref. [151]. The number of saturated carbon atoms between the porphyrin and the anthraquinone was varied from one to three. It was found that for the solutions of these compounds in butyronitrile both rates decreased exponentially with the edge-to-edge distance, with $\nu_e = 10^{12.9}\ s^{-1}$ and $a_e = 0.95$ Å for the forward process and $\nu_e = 10^{11.5}\ s^{-1}$ and $a_e = 1.25$ Å for the reverse one.

Another mode of rigid linking of P and Q has been suggested [152]. In this work porphyrin covalently linked with cyclodextrin, PC, was reported to have been synthesized. This molecule takes advantage of the ability of cyclodextrins to complex quinones into their central cavity. P^+-C-Q^- ion-radical pairs were found to be generated upon illumination of a solution of PC-Q in a 50:50 mixture of glycerol and dimethyl sulphoxide in long-wave bands of P. Efficiency of charge separation strongly depends upon the reduction potential of Q.

Some other covalently bound porphyrin-acceptor complexes such as porphyrin-viologen [153–158] and pyromellitimide-bridged porphyrins [159–161] have been synthesized and studied. As in the case of P-Q complexes, strong fluorescence quenching and ion-radical pair formation were observed in these systems under irradiation of complexes in porphyrin absorption bands.

Synthesis of PP-L-A molecules consisting of bisporphyrin PP linked to a pyromellitimide rather than quinone acceptor A was also reported in [162]. For the cofacial bisporphyrin strong quenching of fluorescence was found, while for the side-by-side bisporphyrin relatively weak quenching was observed. Fluorescence quenching data are supported by the direct ps laser studies of PET in PP-L-A molecules with cofacial and side-by-side bisporphyrin. These results show that the proximity of PP and Q is not sufficient for high efficiency of PET. Other factors, such as appropriate geometry of PP play an important role for efficient PET. Note that cofacial bisporphyrin models the "special pair" electron donor in the reaction centre of photosynthesis.

5.4 Charge Separation in Covalently Linked Triads

The next important step in modelling the process of charge separation during photosynthesis was the synthesis and investigation of molecules of the D-P-Q type containing covalently linked porphyrin, P, quinone, Q, and donor, D, fragments. The structural formula of one of these compounds is represented in Fig. 21a. Excitation of D-P-Q with a light pulse results in fast ($\tau < 10^{-10}$ s) formation of an D^+-P-Q^- ion-radical pair whose lifetime is equal to a few microseconds [163–165]. Charge separation proceeds via initial electron phototransfer from P to Q. Also observed is the subsequent electron transfer from D to the porphyrin cation-radical P^+. A large distance between the D^+ and Q^- fragments ensures a more than 10^6-fold increase in the lifetime of D^+-P-Q^- over the time of charge recombination in a similar P^+-Q^- compound containing no donor fragment. The quantum yield of D^+-P-Q^- formation is quite large and amounts to 0.25 for $n = 1$.

For the series of the compounds with $n = 1$–4 the lifetime of D^+-P-Q^- was found to decrease exponentially with the distance between Q^- and D^+ with $a_e = 3.3$ Å [166]. This value considerably exceeds the value of $a_e \simeq 1.25$ Å found for P^+-L-Q^- [151] and may arise from the amide group insertion into the linking bridge in contrast to hydrocarbon spacers in the latter molecules. In Ref. [167], triad molecule with $n = 1$ has shown the capability of facilitating light-induced charge separation across a bilayer lipid membrane.

Charge photoseparation in the covalently linked D-P-Q triad with rigid bridges (see Fig. 21b) has been studied [105]. The distance between the donor and acceptor

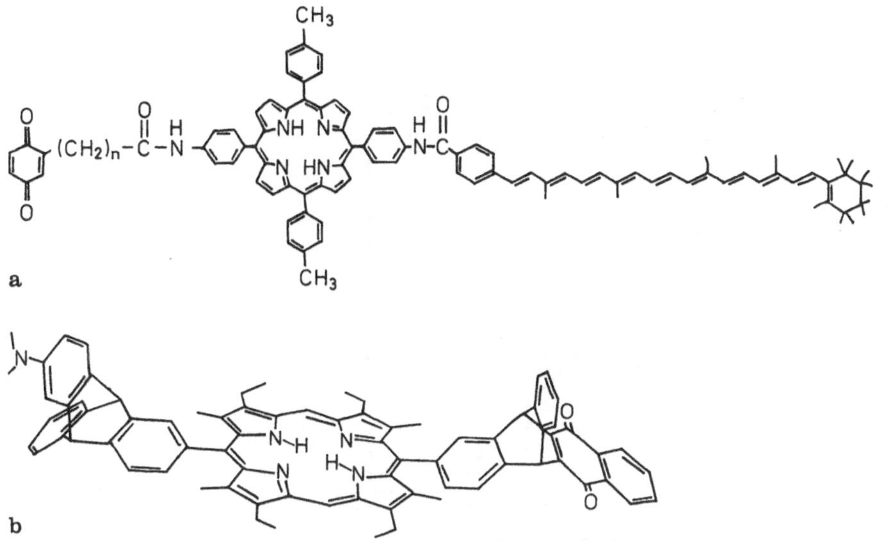

Fig. 21. Structure of covalently linked triads used in (a) – Refs. [163] and [164], and (b) – Ref. [105] for studying the sensitized electron phototransfer from the donor fragment to the acceptor fragment

fragments of this triad amounts to 25 Å. As for the compound depicted in Fig. 21a, the two-step mechanism of charge separation ensures a high rate of the direct process ($\tau < 10^{-10}$ s), a long lifetime ($\tau > 10^{-6}$ s) of the D^+-P-Q^- particles, and a large quantum yield of their formation ($\varphi \simeq 0.7$). Thus, the results reported in [105, 163, 164] corroborate experimentally the correctness of the theoretical model [138, 168] of electron transfer in photosynthesis according to which a step-wise character of the process is expected to ensure a high efficiency of charge separation in the reaction centres of photosynthesis.

Triads which are still more close to the natural photosynthetic system were studied in Ref. [169]. The primary electron donor in these compounds is pyropheophorbide *a*, a derivative of Chl *a*. The initial acceptor is naphthoquionone, a close homologue of the menaquinone contained in bacterial photosynthesis reaction centres [170]. The donor is the carotenoid. In addition to sequential long-range electron transfer, these triads exhibit efficient intramolecular energy transfer from the carotenoid to the pyropheophorbide *a* moiety (singlet excitation energy transfer in the forward direction and triplet energy transfer in the opposite direction).

5.5 Modelling of the Intermediate Step of the Charge Separation Process During Photosynthesis with Porphyrin-Quinone Systems

A simple system for modelling the intermediate step of the charge separation process during photosynthesis (the stage of electron transfer from the reduced pheophytin, i.e. chlorophyll deprived of the Mg atom, to quinone) has been

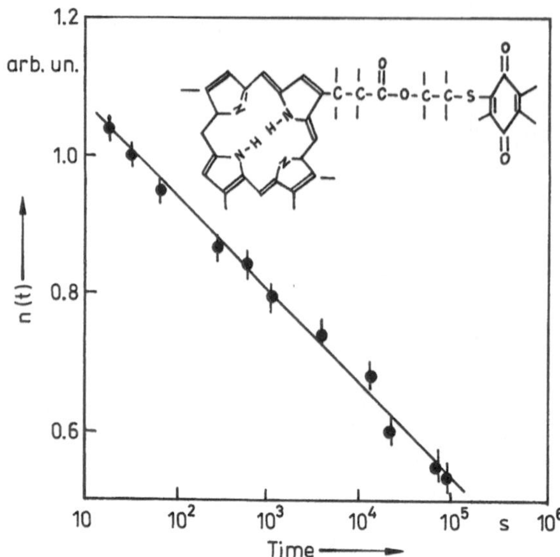

Fig. 22. The kinetics [171, 172] of the dark intramolecular electron transfer in a porphyrin-quinone compounds (P^--L-Q → P-L-Q^-) at T = 77 K. The structure of the compound is given in the upper part of the Fig.

advanced and studied [171, 172]. In these works the charge photoseparation process was studied in solutions of P-L-Q compounds in electron donor, Et_3N, solvent at 77 K. The structure of one of the P-L-Q compounds studied is given in Fig. 22. We will consider briefly the main results of Refs. [171, 172].

Illumination of vitrified solutions of P-L-Q in Et_3N at 77 K results in the formation of P^--L-Q and P-L-Q^- ion-radicals according to the scheme

$$\begin{array}{l} (\text{P-L-Q})_1 \xrightleftharpoons{h\nu_1,\ (\text{Et}_3\text{N})} \text{Et}_3\text{N}^+ + (\text{P}^-\text{-L-Q})_1 \\ \qquad\qquad\qquad\qquad\qquad \downarrow h\nu_2,\ (\text{P-L-Q})_2 \\ \qquad\qquad\qquad (\text{P-L-Q})_1 + (\text{P}^-\text{-L-Q})_2 \\ \qquad\qquad\qquad\qquad\qquad \downarrow \\ \qquad\qquad\qquad\qquad\quad (\text{P-L-Q}^-)_2 \end{array} \tag{41}$$

Upon absorbing the first quantum of light, the porphyrin fragment of P-L-Q is reduced by the nearest solvent molecule, Et_3N^+ and P^--L-Q particles being formed. Further, along with charge recombination, there also proceeds an electron phototransfer from the initially formed P^--L-Q particles to other (P-L-Q) molecules, which are more distant from Et_3N^+ cation formed in the first step of the reaction. In the above reaction scheme, the P-L-Q molecules participating in the initial act of charge separation and located closer to Et_3N^+ cations, are denoted as $(\text{P-L-Q})_1$, and those which are more distant from Et_3N^+ cations are denoted as $(\text{P-L-Q})_2$. An electron captured by the $(\text{P-L-Q})_2$ molecule is first localized on the porphyrin fragment, followed by a spontaneous intramolecular transfer of this electron from the porphyrin fragment to the quinone fragment. The kinetics of this process is represented in Fig. 22. The intramolecular electron transfer (P^--L-Q → P-L-Q^-) is seen to be well described by a linear dependence of the concentration of P^--L-Q particles on the logarithm of the observation time. Just as above the logarithmic character of the process kinetics can be accounted for by the coexistence of several different conformations of the bridge, L, which links the porphyrin with the quinone. This results in a scatter of the rate constants of electron transfer for various P^--L-Q particles.

The results of Refs. [171, 172] show that it is indeed possible to use solutions of porphyrin-quinone compounds in electron donor solvents for modelling the stage of electron transfer from pheophytin to quinone during photosynthesis. Further research on these relatively simple model systems may provide a still deeper insight into the mechanism of this stage of photosynthesis.

6 Photoseparation of Charges in Organized Molecular Assemblies

The tunneling mechanism of electron transfer provides a high quantum yield and a high efficiency of conversion of light energy into chemical energy during photosynthesis. The spatial isolation of oxidized and reduced particles formed at primary stages of photosynthesis and the impossibility of direct collisions between them make it possible to prevent harmful reverse processes of recombination of photoseparated charges. In creating artificial organized molecular systems for converting light energy into chemical energy, the problem of preventing rapid recombination of charges separated by light becomes rather serious and possible ways of solving it have been discussed [173–178]. The majority of them is based on the idea of creating (by analogy with photosynthesis) obstacles to direct collisions between very reactive primary particles formed under the action of light. The concrete ways of realizing this idea, however, are rather varied. Thus, suggested now as organized molecular assemblies providing electron phototransfer

$$D + A + h\nu \longrightarrow D^+ + A^-$$

betwen the spatially separated particles D and A, but hindering the reverse recombination reaction in the absence of light

$$D^+ + A^- \longrightarrow D + A$$

are molecular monolayers, flat lipid bilayers between two water solutions, lipid vesicles, micelles, microemulsions, and laminated systems consisting of several different molecular monolayers. In most cases, the particles D and A between which there occurs electron phototransfer are located on the opposite sides of the interface created by the organized molecular structure. In this case direct or stepwise (with the participation of intermediate particles, photocatalysts or electron carriers specially introduced into the separating layers as additives) electron tunneling between spatially separated donor and acceptor particles may be one of the principal mechanisms of electron phototransfer. Data corroborating the possibility of electron tunneling from electron excited donor to acceptor particles separated from each other by an organized layer of inert molecules seem to have been first obtained by the author of Ref. [178].

6.1 Electron Tunneling in Molecular Layers

The phototransfer of electrons between the donor, octadecylsubstituted cyanine, and the acceptor, *N*,*N'*-dioctadecyl-4,4'-bipyridine (dioctadecyl viologen) cation, separated by a monolayer of molecules of saturated fatty acids has been observed and investigated [178]. The monolayer thickness was varied by increasing the number of CH_2 fragments in the acid molecule. A schematic representation of this structure is shown in Fig. 23. The excitation of donor particles, D, with light has been found to result in an electron transfer from *D to the acceptor. The formation of A^- particles was recorded by their characteristic optical and EPR spectra [179]. Along with the formation of A^- particles, the quenching of the fluorescence of cyanine dye donor molecules was observed. The absence of an overlap of the spectrum of *D luminescence and the absorption spectrum of A particles has made it possible to conclude that the quenching of fluorescence is mostly due to electron transfer from *D to A rather than to energy transfer. The

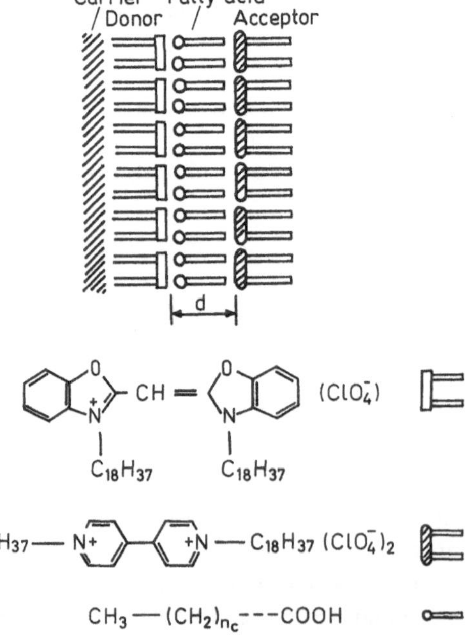

Fig. 23. Schematic representation of a multilayer molecular structure consisting of donor and acceptor particles separated by a layer of fatty acid molecules, on a glass carrier. The structure of the compounds is given at the bottom of the Fig.

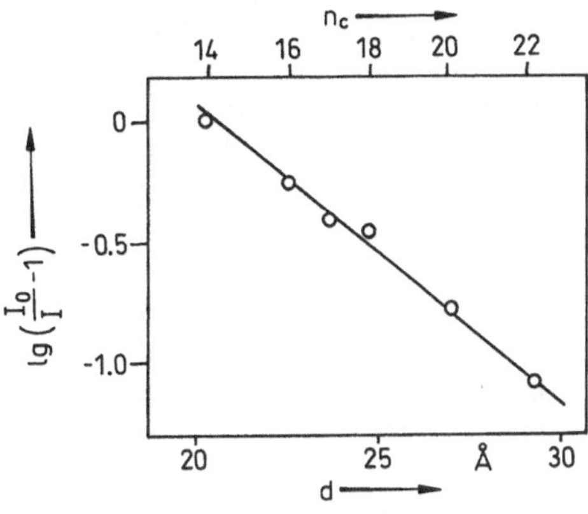

Fig. 24. Fluorescence intensity, I of donor particles vs the thickness of the isolating layer, d, in the multilayer molecular structure depicted in Fig. 23. I_0 is the fluorescence intensity of donor particles in the absence of acceptor particles; n_c is the number of $-CH_2$-fragments in a fatty acid molecule. From Ref. [178].

stationary intensity of fluorescence I, is exponentially dependent on the thickness of the isolating layer, d, (see Fig. 24):

$$I_0/I - 1 = B \exp(-2d/a_e) \tag{42}$$

where I_0 is the stationary intensity of fluorescence in the absence of a layer of acceptor molecules, B and a_e are the parameters.

The exponential dependence of the efficiency of fluorescence quenching on the distance between a donor and an acceptor may be explained by the tunneling mechanism of electron transfer from a singlet-excited molecule of the donor to the acceptor. Indeed, in the case of stationary excitation of donor particles, the value of I is determined by the stationary concentration n_{st}^* of the excited donor particles: $I = An_{st}^*$ where A is a constant. The value of n_{st}^* is, in turn, inversely proportional to the rate constant, k, of deactivation of excited particles: $n_{st}^* = nJ_{exc}\varphi/k$, where J_{exc} is the intensity of the exciting light, φ is the quantum yield of excited molecules, and n is the concentration of nonexcited donor molecules. Thus, $I = AnJ_{exc}\varphi/k$. hence, one can easily obtain

$$k_q = \tau^{-1}(I_0/I - 1) \tag{43}$$

where k_q is the rate constant of donor luminescence quenching by reaction with the acceptor and τ is the lifetime of the excited state of the donor in the absence of the acceptor. From Eqs. (42) and (43) it follows that for the system discussed

$$k_q = B\tau^{-1} \exp(-2d/a_e) \tag{44}$$

i.e. the rate constant of donor luminescence quenching by electron transfer to acceptor particles is exponentially dependent on the distance to the acceptor. It is precisely this dependence which would be expected in the case of a tunneling mechanism of electron transfer from excited donor molecules to acceptor molecules. Studies of molecular models suggest that the thickness of the isolating layer in the structures studied in Ref. [178] and depicted in Fig. 23, is 20–30 Å.

From the data of Fig. 21, and with the help of Eq. (42), one can readily obtain the values of the parameter a_e : a_e = 6.8 Å. The found value of a_e essentially exceeds its typical values observed for electron tunneling processes in vitreous matrices. The reasons for such a high value of a_e in the process under discussion have yet to be found. But one can point out several reasons which, in principle, might give rise to such a high value of a_e. First of all, it may be due to the low value of the effective mass of a tunneling electron, e.g. due to the regularity of the structure of molecular layers. Also, it is not excluded that, in the molecular structures depicted in Fig. 23, the donor and the acceptor molecules oscillate along the direction of electron transfer penetrating, in so doing, the isolating layer to a certain distance. In this case, the distance of electron tunneling, R_t, must be less than the thickness of the isolating layer, d, and its actual variation with increasing number of $-CH_2$-fragments in an acid molecule may perhaps differ from the change in the layer thickness expected on the basis of the study on molecular models.

The quenching of the fluorescence of donor particles *D and the formation of reduced acceptor particles A^- in photochemical reactions followed by a slow decay of particles A^- have also been observed in a number of other donor-acceptor molecular layers of a similar structure [180, 181]. The data obtained have also been explained by electron tunneling from the excited donor particles to the acceptor particles over distances exceeding 20 Å.

The results of the investigations [178, 180, 181] point to the theoretical possibility of using monolayers for efficient charge separation in artificial systems for light energy utilization. Such systems must contain a donor D and an acceptor A, between which electron transfer in the absence of light is an endoergic process, as well as a photocatalyst PC whose role can be played either by the D or A molecules themselves or by special substances introduced additionally into the system. The absorption of light by the photocatalyst must result in an electron transfer from D to A. In this case, the time of the inverse process, recombination of D^+ and A^-, must be longer than that needed to utilize the stored energy in the processes of further conversions of the D^+ and A^- particles to stable energy-rich chemical products and for returning the whole system $D^+ + PC + A^-$ to its initial $D + PC + A$ state. An attempt to use monolayers on the basis of fatty acids for constructing such systems has been made [182].

In this reference the appearance of a light-induced potential difference between two electrodes separated by a specially organized molecular multilayer is reported. A barium electrode and a semitransparent aluminium electrode, which have substantially different electronic work functions $J^{Al} > J^{Ba}$, have been used in these studies. The two electrodes were separated by a multilayer system consisting of a layer of isolating molecules covered by a layer of dye molecules capable of losing an electron on absorbing light, then by another layer of isolating molecules and, lastly, by a layer of acceptor molecules. The isolating monolayers consisted of molecules of the fatty acid CH_3-$(CH_2)_{18}$-COOH. The layer of dye molecules contained molecules of octadecyl-substituted cyanine dye. The acceptor layer included molecules of dioctadecyl viologen. The essential difference between electronic work functions for the aluminium and the barium electrodes stimulated electron transfer from the aluminium electrode towards the barium electrode, which is assisted by the excited molecules of the dye. Such a directed electron transfer led to the appearance, under the action of light, of a potential difference, ΔV. That the induced potential difference occurred precisely upon exciting the dye molecules was evidenced by the coincidence of the spectrum of the photopotential action (i.e. of ΔV photogeneration) with that of the dye absorption. The value of ΔV grew linearly with increasing light intensity J_{exc} and reached the value of 10 mV at $J_{exc} = 5\ \mathrm{Wt} \times \mathrm{cm}^{-2}$.

The scheme of the processes occurring in such a multilayer structure can be represented as

$$D + h\nu \longrightarrow {}^{*}D \tag{45}$$

$$^{*}D + A \longrightarrow D^{+} + A^{-} \tag{46}$$

$$A^{-} \xrightarrow{(Ba)} A + e^{-} \tag{47}$$

$$D^{+} + e^{-} \xrightarrow{(Al)} D \tag{48}$$

Following the excitation of the dye molecule D (reaction 45), there occurs the transfer of an electron to the acceptor particles A (reaction 46) via electron tunneling mechanism or ordinary collision mechanism, and then the electron tunnels from A^{-} to the barium electrode (reaction 47). The dye molecules are regenerated at the expense of electron tunneling from the aluminium electrode to D^{+}. As a result of reactions (45)–(48), there occurs a photosensitized electron transfer from the aluminium electrode to the barium electrode. As the transfer proceeds, the energy is accumulated and the system of electron carriers returns to its initial state. The introduction of molecules containing π-bonds and oriented along the hydrocarbon chain of fatty acid molecules into the isolating layer located between the layer of the dye and that of the acceptor, resulted in an almost 2-fold increase of ΔV, seemingly due to an increase in the permeability of the barrier to electron transfer.

The scheme of the reactions (45)–(48) is similar in certain aspects to that of electron transfer processes occurring in the membranes of photosynthesizing organisms, where the light absorption also induces a trans-membrane potential difference. The above scheme of processes in multilayer systems explaining the appearance of the photoelectromotive force requires further refinement. But even the data available at present allow these systems to be regarded firstly as convenient models for analyzing the processes occurring in complicated biological systems and secondly as potentially promising objects for developing various systems of converting light energy into electrical and chemical energies.

Enhanced photovoltage and photocurrent signals were observed by the authors of Refs. [183, 184] with linked porphyrin-quinone molecules in planar bilayer lipid membranes (BLM) as compared with preparations containing the non-linked components. They interposed BLM between two aqueous compartments containing a secondary electron donor on one side and a secondary acceptor on the other side. The efficiency of PET increased when the P-L-Q molecules were oriented in the membrane.

PET across a phospholipid bilayer was also shown to occur when triad molecules of the type pictured in Fig. 21a, were brought into BLM [167, 185]. A steady-state photocurrent across the membrane was observed in these experiments.

In a number of publications, there have been reports of laminated molecular systems being used for modelling the initial stages of photosynthesis. Electron phototransfer from chlorophyll to quinone has been studied [186, 187] in a molecular assembly consisting of two layers: (1) a layer of chlorophyll molecules and (2) a layer of chloranil, bromanil, or *N,N'*-distearoyl-1,4'-diaminoanthraquinone molecules. The structure was so arranged that the phytol fragments of the chlorophyll molecules were directed towards the layer of quinone molecules. Thus, phytol fragments served, in fact, as an isolating layer between the porphyrin and the quinone layers. Illumination of such a structure in the chlorophyll absorption bands induced electron transfer from chlorophyll to quinone. Because of the short lifetime of the initial singlet-excited and triplet-excited state of chlorophyll and the long distance between the donor fragment of the chlorophyll molecule and the acceptor fragment of the quinone, the charge separation process was found to proceed via a two-quantum mechanism. This fact suggests that electron phototransfer proceeds perhaps via an over-barrier non-tunneling mechanism. But the characteristic time $\tau \simeq 6 \times 10^{-4}$ s of the back reaction, i.e. recombination of P^+ and Q^- particles, which proceeds in these systems after the light is switched off, was practically independent of temperature in the range 100–300 K. This suggests the tunneling mechanism of the recombination.

When ubiquinone and plastoquinone were used as electron acceptors, as well as when compounds with double bonds such as squalene were incorporated into the isolating layer, the electron phototransfer was found to proceed according to the one-quantum mechanism. This suggests the tunneling mechanism of electron phototransfer. Thus, the efficiency of electron tunneling increases upon incorporating molecules with double bonds into the isolating layer, which consists of phytol fragments of chlorophyll molecules. Insertion of compounds with a saturated hydrocarbon chain into the isolating layer did not cause any acceleration of electron phototransfer.

Systems close to the above multilayer systems in composition and properties are those consisting of dye molecules chemically bound to electrode surfaces through an isolating chain of chemical bonds. In Ref. [187], a one-quantum excitation of porphyrin molecules bound to the electrode by chains of the $-O-Si-(CH_2)_n-CO$-kind has been found to result in the ionization of porphyrins. The photoionization is assumed to be due to electron tunneling from electron-excited porphyrin molecules to the electrode. The probability of tunneling decreases by a factor of 2.5 on passing from a chain of the $-O-Si-CO$-type to that of $-O-Si-(CH_2)_3-CO$-type. Such a small decrease in the efficiency of tunneling upon increasing the chain length, l, by ~ 4 Å seems to be due to the appearance, for long chains, of conformations in which the pigment molecules are located close enough to the electrode surface. In this case, the electron tunneling proceeds not along the whole chain, but by a shorter path.

6.2 Electron Tunneling in Micelles and Vesicles

Other interesting examples of the organized molecular structures used to increase the quantum yield of charge photoseparation are micelles and vesicles.

The increase in the quantum yield of charge separation in micellar systems compared with homogeneous solutions, just as in the case of the systems with isolating molecular layers, is obtained by erecting obstacles to the direct collisions of the D^+ and A^- particles formed during phototransfer due to the predominant localization of one of them inside and the other outside the micelle. Electron phototransfer from excited donor molecules located inside micelles to acceptors located outside has been revealed for a number of donors and acceptors [188–190]. Thus, micelles have been found [189, 190] to promote drastically a one-quantum ionization of tetramethylbenzidine and phenothiazine located inside them. This ionization results in electron transfer into traps localized in the water phase. The promotion effect is assumed to be due to a decrease in the rate of the dark recombination of D^+ and A^- particles formed as a result of electron phototransfer owing to their spatial separation and hindrances to their direct collision. Numerous other examples demonstrating the possibility of decreasing the rate of recombination of strong one-electron reductants and oxidants in micellar solutions can be found in Ref. [189a] and references cited therein.

The possibility of electron tunneling through a vesicle membrane has been repeatedly suggested to account for electron phototransfer between molecules localized on the different sides of a membrane [191, 192]. The typical thickness of vesicle membranes is ca. 50 Å. But considering the proper sizes of the particles participating in phototransfer and the possibility of these particles to penetrating into the subsurface layers of the membrane walls, the actual distance of electron transfer may prove to be as short as ca. 25 Å. Electron tunneling to this distance within the lifetime, τ, of the excited triplet state of a photocatalyst embedded into the membrane, such as, say, ZnTPP ($\tau \simeq 10^{-3}$ s), appears to be quite reasonable. Electron tunneling phototransfer over 9–12 Å from amphiphilic porphyrin molecules embedded in vesicle membranes to methyl viologen molecules placed in the outer water phase has been also discussed [192].

The mechanism of PET in membranes including those based on electron tunneling are discussed in more detail in Chapter 159 by K. I. Zamaraev, V. N. Parmon, S. V. Lymar.

7 Photoinduced Electron Tunneling in Biological Systems

Natural photosynthesis is a unique photoconversion system that transforms sunlight into stored chemical energy in the form of carbohydrates and other products. The high efficiency of this system is due to the elegant macromolecular structure of the reaction centre protein. This protein performs the dual function of harvesting photons and storing some of the sun's energy by synthesizing molecules rich in chemical energy. The reactions are highly specific and are optimized for the survival of the plant or microorganism.

The reaction centre (RC) of purple bacteria seems to be a suitable object for elucidating the mechanisms providing extremely high efficiency of electron transfer in photosynthetic systems.

The reactions centres have been purified from a variety of bacteria, and it is well established that the in vitro rates of the electron transfer reactions and the yields of the various charge-separated intermediates are essentially the same as in vivo [193]. In addition, structures of the RC from *Rhodopseudomonas viridis* [194, 195] and *Rhodobacter sphaeroides* [196, 197] are becoming available at atomic resolution, permitting sophisticated analysis of the relationship between the RC's structure and its function [198, 199]. Figure 25 summarizes the pathways, rates and temperature dependence of the early stages of photosynthesis in RC isolated from *Rhodobacter sphaeroides* strain R-26. According to present-day notions [200–202],. the reaction centre of bacterial photosynthesis, within which the initial conversion of light energy into the chemical energy of separated charges occurs, consists of the following fragments providing for the electron transfer: (1) the dimer of the bacteriochlorophyll $(BChl)_2$, which serves as the primary electron donor and is commonly denoted as P or P890 (the figure 890 indicates the position, in nanometers, of the main peak in the differential optical spectrum of the particle P^+); (2) the intermediate primary electron acceptor J (a molecule of bacteriopheophytin); (3) the primary acceptor Q_A (presumably the iron-ubiquinone complex); a few molecules of cytochrome *c* (secondary donors) and one or several molecules of ubiquinone (secondary acceptor) (not shown in Fig. 25). Electron transfer occurs between these tightly bound-to-protein redox sites, which are held

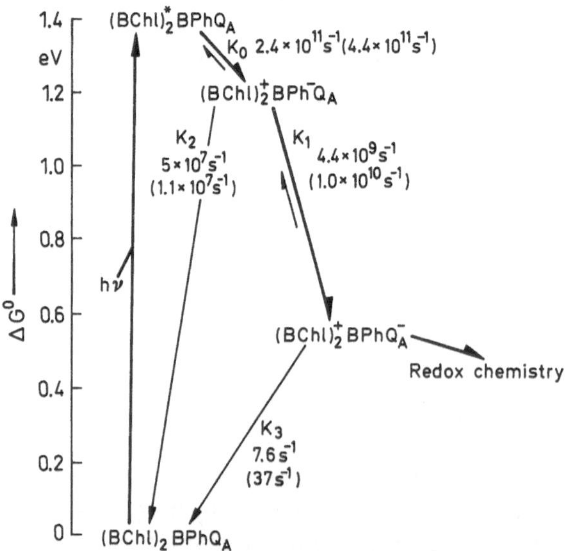

Fig. 25. Electron transfer pathways in the RC isolated from *Rhodobacter Sphaeroides* strain R-26. $(BChl)_2$ is dimer of the bacteriochlorophyll, BPh is the bacteriopheophytin, Q_A-ubiquinone. The electron transfer rates are for the native RC with ubiquinone −10 as Q_A, at room temperature. The rates given in parenthesis were determined below 100 K [194–197]

by surrounding protein in a well-defined orientation and separation [194–197]. The reaction sequence is triggered by light absorption by light collecting pigments, producing excited states of the pigments. The energy of excitation is then transmitted to pigments of another type which serve as an antenna collecting the energy and transmitting it further to the dimer of bacteriochlorophyll, yielding the excited singlet state ($(BChl)_2^*$). Forward charge-separating electron transfer consists of $(BChl)_2^*$ reducing bacteriopheophytin (BPh), which in turn reduces ubiquinone (Q_A). Figure 25 also describes backward charge-recombining electron transfer that competes with the forward reactions and so returns protein to the ground state from $(BChl)_2^*$, $(BChl)_2^+Bph^-$, or $(BChl)^+Q_A^-$.

The electron localized on the particle Q_A is subsequently used, through a complicated chain of chemical reactions, to reduce CO_2 to the carbohydrates $(CH_2O)_6$, while the "hole" is finally used to oxidize some certain substrate, say hydrogen sulphide to sulphur. This results in the regeneration of the active centre.

As distinct from the purple bacteria, plants, when photosynthesizing carboxydrates from CO_2, use water rather than H_2S as an oxidized substrate. The oxidation of water requires far more energy than that of hydrogen sulphide. This circumstance plus the necessity for the sites of formation of oxidized and reduced particles to be spatially separated (in order to avoid their rapid recombination) seems to be the reason for the formation in plants of two consecutive photosystems. A two-stage electron transport scheme, also referred to as the Z-scheme (Fig. 26), is now generally recognized for photosynthesis in plants [203]. According to this scheme, the excitation of the P700 pigment of photosystem 1 (PS1) results in electron transfer from P700* through the chain of carriers to the oxidized form of nicotineamide-adenine dinucleotide phosphate (NADP). $P700^+$ is in turn, reduced by an electron coming to PS1 through the chain of electron carriers from PS2.

Early experimental studies on the role of electron tunneling in biological systems were typically carried out on subchromatophore and subchloroplast fragments at low temperatures. The operation of photosynthetic organisms at low temperatures

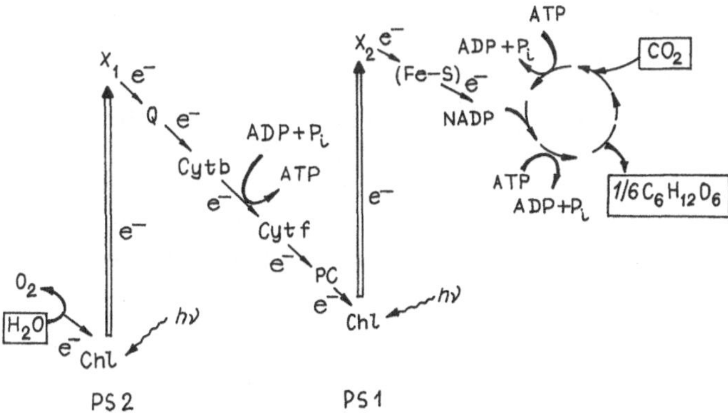

Fig. 26. Z-Scheme of photosynthesis in plants. Chl is chlorophyll, cyt *b*, *f* is cytochrome *b*, *f*; PC is plastocyanine, (Fe–S) is iron sulfer protein. ATP is adenosine triphosphate; ADP is adenosine diphosphate; P_i is the phosphate ion and NADP is the nicotinamide adenine dinucleotide phosphate ion [203]

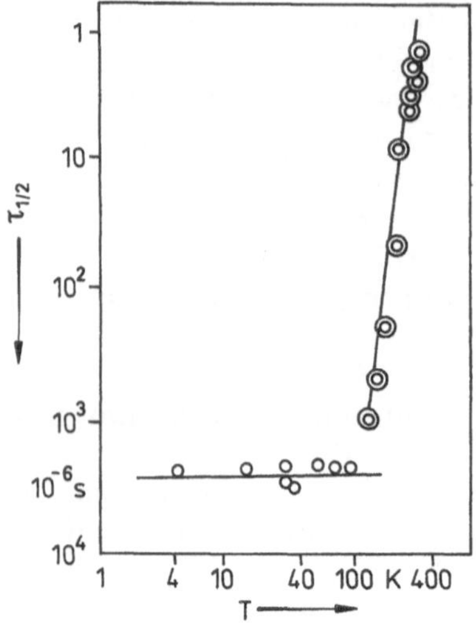

Fig. 27. Temperature dependence [9] of the characteristic time, $\tau_{1/2}$, of electron transfer from cytochrome *c* to the photooxidized form of chlorophyll for *Chromatium* bacteria

was first discovered by Chance and Nishimura [204] who reported the oxidation of cytochrome *c* under the action of light in photosynthetic bacteria *Chromatium* to occur at 77 K. Immediately following this, some data were obtained indicating electron transfer to occur in photosynthesizing bacteria even at 1 K [205]. Subsequently, electron transfer in the process of photo-oxidation of cytochrome b_{559} was also found to occur at 77 K in green leaves [206, 207]. The most interesting of the early studies on electron tunneling in biological systems is that, in which the time of cytochrome oxidation with the oxidized primary donor $P870^+$ in *Chromatium* within the temperature range 4.2–100 K (Fig. 27) has been found to be independent of temperature [9]. In contrast to this, at $T > 100$ K the characteristic time of the process does depend on temperature. The activation energy found from this dependence amounts to about 3.3 kcal mol^{-1}. The authors assume the temperature-independent reaction observed at low temperatures to proceed via electron tunneling.

After [9], numerous papers were published related to the low-temperature photooxidation of P700 in the PS1 reaction centres of plants and the reverse process of recombination of $P700^+$ and a reduced electron acceptor (see e.g. Refs. [208–211]). In these works controversial data on the kinetics of the dark decay of $P700^+$ were obtained. Therefore in Ref. [212] the kinetics of charge recombination in the PS1 reaction centres was investigated in detail over a broad range of times and temperatures.

7.1 Recombination of Charges in the Reaction Centres of the Photosystem 1 of Plants

The kinetics of the decay of $P700^+$ centres due to their recombination with acceptor centres, which most likely are the primary acceptors A^-, i.e. iron-sulphur proteins, has been studied [212, 214] within about a three-order time interval at temperatures of 5–300 K. It was found that in the range 294–240 K the decay of $P700^+$ is characterized by a monotonous decrease in the fraction of particles decaying within a fixed time interval with decreasing temperature and is described by the usual exponential law. The temperature dependence of the rate constant for the decay process k satisfies the Arrhenius equation $k = k_0 \cdot \exp(-E_\alpha/RT)$, where E_α and k_0 have values of 16 kcal mol^{-1} and $10^{10}\ s^{-1}$, respectively.

An essential difference between the data of Ref. [212] and those of the earlier works was that the kinetics of $P700^+$ decay at $T < 240$ K was found not to follow a first-order law, but to show linear or near-linear dependences in the coordinates n(t) vs log t (see Fig. 28). The curves for the ranges 220–160, 160–80 and 80–5 K are seen to have different characters. Actually, as seen from Fig. 28, below 80 K the curves show linear dependencies in the coordinates n(t) vs log t over the whole time range investigated. At $160\ K > T > 80\ K$ the kinetic curves demonstrate linear dependencies only in the initial section while a decrease in the slope of the curves is observed at longer times. In the temperature range 220–160 K notable deviations from linearity are observed throughout the whole time interval. The slope of the initial sections of the kinetic curves in the coordinates n(t) vs log t remains practically constant at $80\ K < T < 160\ K$ and becomes temperature-dependent at $T < 80$ K and at $T > 160$ K.

The temperature range (240–220 K) of the unusual behaviour of the process rate (increase of the rate with decrease in T) appears to correspond to the temperature range of vitrification of the solution. The increase in the decay rate of $P700^+$ upon transition from a liquid solution to a vitreous sample has been explained on the assumption that the decay of $P700^+$ proceeds by reactions with two types of reduced primary acceptors A^- and A'^-,

$$P700^+ + A^- \longrightarrow P700 + A \tag{49}$$

which predominates at $T > 240$ K and has an activation energy $E_\alpha = 16$ kcal mol^{-1}, and

$$P700^+ + A'^- \longrightarrow P700 + A' \tag{50}$$

which predominates at $T < 220$ K and has a lower activation energy, E'_α. In the temperature range 220–240 K both types of reaction are possible. The difference between A^- and A'^- was assumed in Ref. 212 to be due to different chemical compositions, structures or conformational states of the surrounding medium of the "P700-acceptor" pairs. The lower activation energy of reaction (50) as compared with reaction (49) accounts formally for the increase in the process rate upon switching from reaction (49) to reaction (50) as the main channel of $P700^+$ decay with decreasing temperature from 240 to 220 K.

Due to the exponential character of the kinetics of decay of $P700^+$ at $T > 240$ K, reaction (49) was concluded [212] to be the only process responsible for the decrease in the concentration of $P700^+$ at high temperatures.

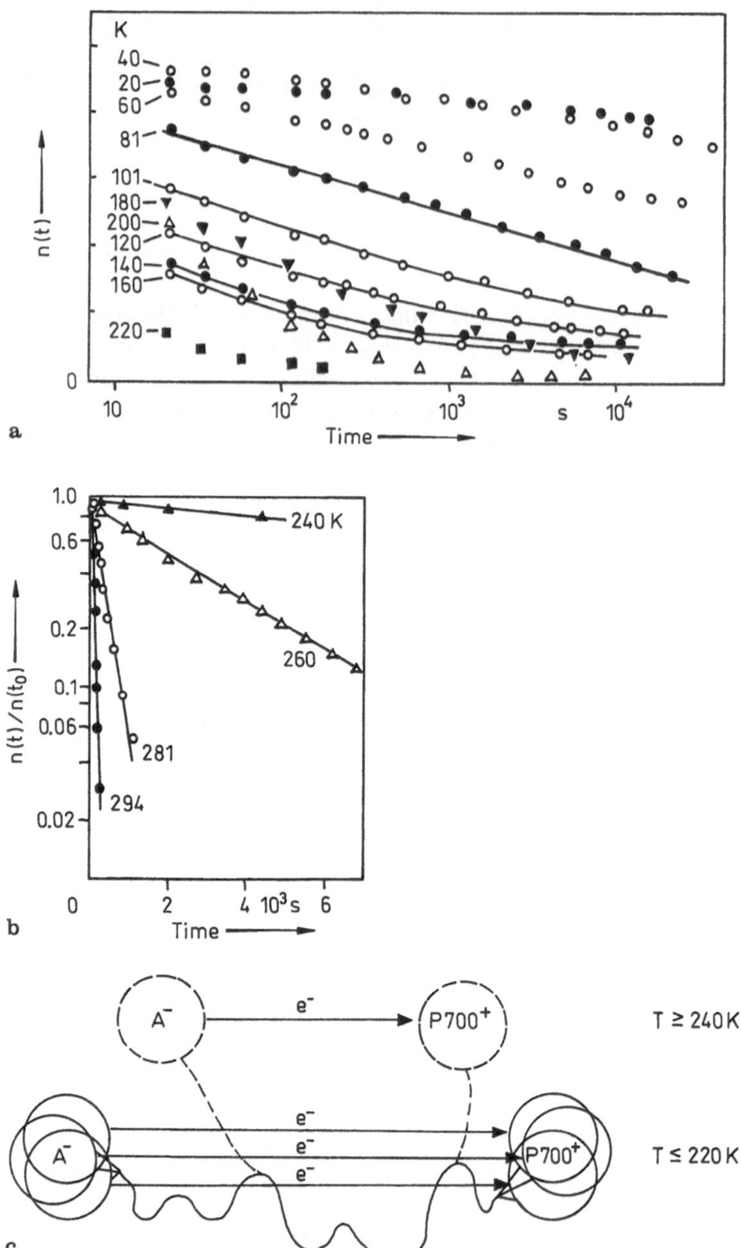

Fig. 28. Dark decay [212] of $P700^+$ in the reaction centre of photosystem 1 of subchloroplasts at T < 240 K (**a**) and at T ≧ 240 K (**b**). Proposed schematic structures of the reaction centre at low and high temperatures are shown at the bottom of the Fig. (**c**)

As has been mentioned above (see also [10, 26, 31]) the logarithmic or close to logarithmic kinetics of the recombination of charges in pairs is characteristic of electron tunneling processes with an exponential dependence of the probability of tunneling per unit time on the distance, $k = W(R) = \nu_e \exp(-2R/a_e)$, and with a rectangular or close to rectangular distribution of pairs over the distances. In the general case, however, the pairs can differ not only in distance but also, for example, in activation energy, reagents orientation, etc. The data obtained in Ref. [212] definitely indicate a tunneling mechanism of electron transfer in reaction (50) since the diffusion of $P700^+$ and A'^- in chloroplasts at very low temperatures can be disregarded.

Analysis of the experimental data on the decay of $P700^+$ was, therefore carried out assuming the $P700^+$-A'^- pairs to be distributed over the value of some parameter ξ in the equation for the rate constant of recombination $k = k^0 \exp(-\xi)$, where k^0 is assumed to be the same for all reagent pairs. If the pairs differ in their values of the activation energy of recombination E'_a, then $\xi = E'_a/RT$, and if they differ in their values of tunneling distance, then $\xi = 2R/a_e$ (in this case $k^0 = \nu_e$).

It is interesting to note that, in the temperature range 80–160 K the initial slope of the kinetic curves in coordinates n(t) vs log t is independent of temperature. Hence, in this temperature range, the observed logarithmic or close to logarithmic character of the $P700^+$ decay kinetics cannot be due to a distribution with respect to the value of the activation energy, because in the case of such a distribution, the slope of kinetic curves would be temperature dependent [31, 212, 213].

From the logarithmic character of the $P700^+$ decay curves it follows that the observed kinetics of the tunneling decay of P^+-A^- pairs in the range T = 80–160 K may be due to the rectangular distribution with respect to the values of R, a_e^{-1} or to the exponential distribution with respect to the values of ν_e. The distribution with respect to ν_e or a_e may be due, for example, to a difference in mutual orientation or to a different environment of the reacting particles in membranes.

The activation energy of the decay of $P700^+$ in the range 80–160 K was found in Ref. [212] from the data of Fig. 28. Assuming that the distribution of pairs over the values of the parameter ξ at the initial time t = 0 is described by the rectangular function, it is easy to derive an equation for the ratio of the current concentration of pairs, n(t), to their concentration at time t_0 [212]:

$$\frac{n(t)}{n(t_0)} = 1 - (\xi_{max} - \ln k^0 t_0)^{-1} \ln t/t_0 \tag{51}$$

where $\xi_{max} = \ln(k^0/k_{min})$, k_{min} is the minimum value of rate constant. Since, as previously noted, the decay of pairs in the interval 80–160 K is characterized by a single value of activation energy, E'_a, the rate constant of the process can be represented as $k = k_0 \exp(-E'_a/RT)$, where $k_0 = k_0^0 \exp(-\xi)$ and k_0^0 is a multiplier which is independent or only slightly dependent (in a non-exponential manner) on temperature. The factor k^0, which is present in Eq. (51), can be related

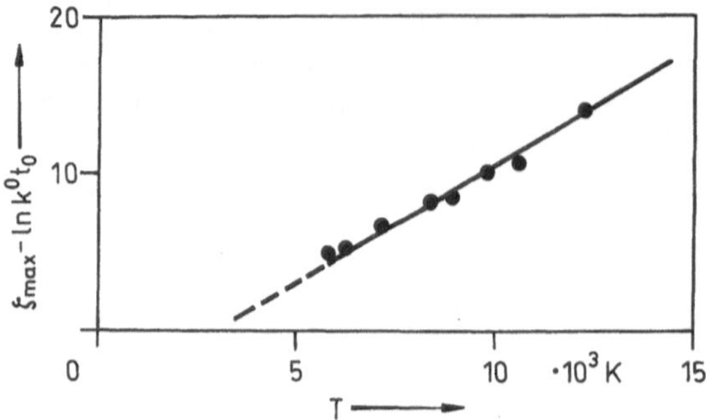

Fig. 29. Temperature dependence [212] of the value of the tangent of the slope of the kinetics curves of Fig. 28 in the coordinates of Eq. (52)

to the values k_0^0 and E_a' as $k^0 = k_0^0 \exp(-E_a'/RT)$. Thus for the inverse of the slope of the kinetic curves in coordinates $n(t)/n(t_0)$ vs $\ln(t/t_0)$ we have

$$\xi_{max} - \ln k^0 t_0 = \xi_{max} - \ln k_0^0 t_0 + E_a'/RT \tag{52}$$

As seen from Fig. 29, $\xi_{max} - \ln k^0 t_0$ is actually linearly dependent on 1/T. The activation energy for the recombination, of P^+-A^- pairs via electron tunneling, which was found from the slope of the line of Fig. 29 by means of Eq. (52) has a value of $E_a' = 2.9$ kcal mol^{-1}.

It should be noted that the above procedure for calculating the activation energy of the pair-wise recombination of P^+ and A'^- species is quite different from the generally used method for calculating E_a from the dependence of the characteristic reaction time on the temperature in Arrhenius coordinates. The latter method is applicable only to reactions characterized by a single value of the rate constant. The use of this method for calculating the activation energy of reactions characterized by a set of rate constants is incorrect and may lead to erroneous conclusions. As an example, Fig. 30 shows the dependence $\tau(T)$ in Arrhenius coordinates for different extents of conversion of $P700^+$ (e.g. $\tau_{0.3}$ corresponds to a decay of 30% of the initial amount of $P700^+$). It is seen that this dependence has a complicated character and cannot be described by a linear function. One can single out three ranges of temperature dependence of τ: (1) in the range 220–180 K, τ decreases with increasing temperature; (2) in the range 180–140 K, τ increases with increasing temperature; and (3) at $T < 140$ K, τ again decreases with increasing temperature. As seen from the Fig., the effective activation energy of the process formally calculated from the Arrhenius dependence of $\ln \tau$ on T^{-1} for $T < 140$ K and $T > 180$ K depends on the fraction of the decaying pairs, and in the range 140–180 K, it has a negative sign. The activation energy found from the Arrhenius dependences for τ differs significantly from the true activation energy. A particularly large difference is observed at small extents of conversion (cf. $E_a = 1.25$ kcal mol^{-1} for $\tau_{0.3}$ (see Fig. 29) with the activation energy $E_a' =$ 2.9 kcal mol^{-1} found above). Also, for all extents of conversion at temperatures below 80 K and above 160 K (up to 220 K) the slope of the initial fraction of kinetic

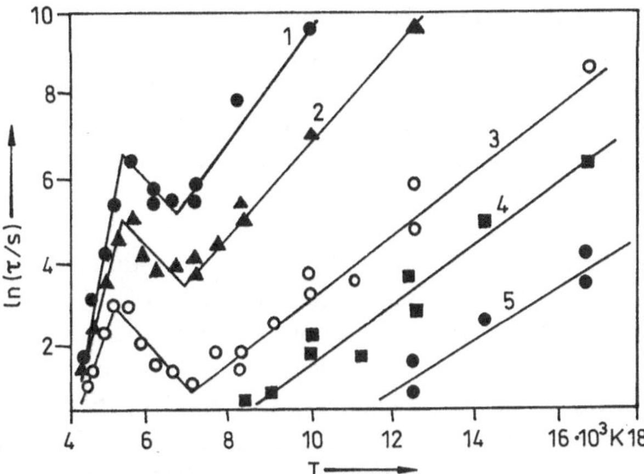

Fig. 30. Temperature dependence [212] of the value of ln τ calculated from the $P700^+$ decay curves. Curves *1–5* correspond to 80, 70, 50, 40 and 30% decay of $P700^+$ from the initial number of $P700^+$ particles

curves depends on the temperature (decreases with decreasing temperature). On this basis it has been concluded that, for the $P700^+$ decay reaction, there are at least two more temperature-dependent channels of tunneling, one of which has an activation energy $E_a'' < E_a'$ and manifests itself at lower temperatures, while the other has an activation energy $E_a''' > E_a'$ and displays itself at higher temperatures.

The appearance, at T > 160 K, of a channel of decay with a relatively high activation energy, E_a''', can be accounted for, e.g. by a defreezing of some additional type of motion (rotation, vibration or conformational transition) creating still more favourable conditions for electron tunneling than the channel with activation energy E_a'.

At still higher temperatures, when the rotations, vibrations and conformational transitions in chloroplasts become sufficiently rapid, the differences in the values of R, a_e or v_e for different $P700^+$-A'^- pairs are averaged out by the motion and do not manifest themselves in the kinetics of the decay of $P700^+$. As a result of the motion, all the reagent pairs can get into one and the same state, the most favourable for electron tunneling and characterized by a definite value of the probability of tunneling, W. Under these conditions the scatter in the values of recombination rate constants for different $P700^+$-A'^- pairs disappears and the kinetics of recombination is described by the usual first-order kinetic law. Under these conditions $P700^+$-A'^- pairs, according to our above made notations, should be designated as $P700^+$-A^- pairs. However, since the motion of the functional groups in chloroplasts may be restricted by their being part of the membrane composition, this most favourable state for electron tunneling may correspond to a still considerable distance between reacting $P700^+$ and A^- particles.

Thus, the study [212, 214] of the kinetics of the charge recombination in the reaction centres of photosystem 1 of subchloroplasts over wide time and temperature intervals has shown an essential difference in the kinetics of the tunneling decay of $P700^+$ at high and low temperatures. The quantitative description of the electron transfer kinetics has proved possible in terms of the assumption of a difference in charge recombination rate constants for different reaction centres. Such a difference may be due, for example, to a non-coincidence, for different reaction centres, of electron tunneling distances or to different conformational states of these centres.

Interesting data about the effect of deuterated water molecules on electron transfer processes in reaction centres of PS1 of plants have been obtained [215, 216]. Earlier [217], it was found out that, in lyophilically dried leaves of plants which were kept in the dark, there occurs a recombination of charges whose efficiency in the temperature range 77–300 K depends on the water content in the leaves. A detailed study of the effect of replacing H_2O by D_2O in the leaves containing 10% of the normal water content on the kinetics of recombination reaction (1) in the temperature range 77–300 K has been made [215, 216].

In this temperature range one can single out two regions with different characters of $P700^+$ decay kinetics. At $273 \leq T \leq 300$ K the exponential dependence of $P700^+$ concentration on time is observed. For this temperature range, the activation energies calculated from the temperature dependence of the recombination rate constant in the Arrhenius coordinates amounts to 2.4 ± 0.5 and 3.1 ± 0.5 kcal mol^{-1} for samples with H_2O and D_2O, respectively, i.e. they coincide within experimental error.

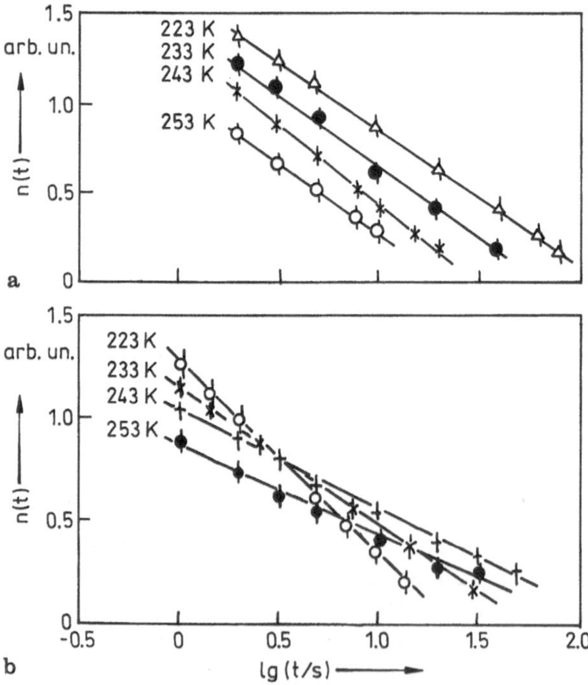

Fig. 31. Decay of $P700^+$ in lyophilized leaves at various temperatures [215, 216]. **(a)** D_2O-containing samples; **(b)** H_2O-containing samples

At low temperatures, $T < 260$ K, the charge recombination kinetics both for H_2O- and D_2O-containing samples corresponds to a linear dependence of the number of $P700^+$-A^- pairs on the logarithm of observation time (see Fig. 31). However, the character of the temperature dependence of the kinetic curves for H_2O- and D_2O-containing samples is essentially different. As seen from Fig. 31, in the samples with D_2O the slope of kinetic curves is virtually independent of temperature. This indicates the equality of the recombination activation energy values for different $P700^+$-A^- pairs [212]. The value of E_a found from Eq. (52) amounts to 3.3 ± 0.4 kcal mol^{-1}.

For the samples with H_2O the slope of kinetic curves depends on temperature. This fact evidences the fact that the activation energy is not the same for the decay of various $P700^+$-A^- pairs.

The efficiency of charge recombination for the samples containing normal water and those containing deuterated water were also different. Thus the relative changes in the amount of $P700^+$-A^- pairs within the first 15 s after the light has been switched off, for the D_2O-containing samples exceeds that for the samples containing H_2O.

The data obtained have been explained [215, 216] by the participation of intramolecular vibrations of water molecules in the process of electron tunneling. Upon deuteration of water molecules the frequencies of intramolecular vibrations decrease. According to the theory [20–24], deuteration can decrease the efficiency of electron tunneling due to the decrease of the Franck-Condon factor (the normal isotopic effect). In the case at issue a decrease in the probability of tunneling for the charge separation reaction upon passing from H_2O- to D_2O-containing samples may lead to a decrease in the efficiency of the reduction of those acceptor particles, A, which are more distant from the *P700 particles. If this is the case, then the observed differences in the kinetics of $P700^+$ decay for the H_2O- and D_2O-containing samples can be explained in terms of the following simplest scheme, taking into account the presence of only two types of acceptor particles, A_1 and A_2, of which A_1 is the closer to *P:

D_2O-containing samples

$$\text{P700-A}_1\text{-A}_2 \underset{}{\overset{h\nu}{\rightleftharpoons}} \text{P700}^+\text{-A}_1^-\text{-A}_2 \tag{53}$$

H_2O-containing samples

$$\begin{array}{ccc} \text{P700-A}_1\text{-A}_2 & \overset{h\nu}{\rightleftharpoons} & \text{P700}^+\text{-A}_1^-\text{-A}_2 \\ & {\scriptstyle h\nu}\searrow\!\!\nwarrow \quad \swarrow & \\ & \text{P700}^+\text{-A}_1\text{-A}_2^- & \end{array} \tag{54}$$

The participation of only one type of the nearest acceptor particles A_1 in the electron transfer processes in D_2O-containing samples ensures both a constant slope of the straight lines in Fig. 31a at different temperatures and a higher efficiency of charge recombination than that in the case of H_2O.

The non-coincidence of the slopes of straight lines in Fig. 31b for various temperatures for the samples with H_2O results, according to Refs. [215, 216], from the participation in the charge recombination process of two (or more) types of $P700^+$-A^- pairs with different values of recombination activation energies.

At $T > 273$ K, when the molecular mobility is unfrozen, the differences between various $P700^+$-A^- pairs in the samples containing H_2O or D_2O can be averaged out by thermal motions of the molecular fragments. In this situation the kinetics of $P700^+$ decay at $T > 273$ K obeys the usual exponential equation.

7.2 Kinetics of Electron Transfer in the Reaction Centre Proteins from Photosynthetic Bacteria

Photochemical reaction centre proteins of photosynthetic bacteria are enzymes that extend across the intracytoplasmic membrane of bacteria. The reaction centre protein from bacteria can be isolated from membrane and purified. Recently the reaction centre from *Rhodopseudomonas sphaeroides* [218, 219] and *Rhodopseudomonas viridis* have been crystallized [220]; for the latter species the three-dimensional structure has been established with about 3 Å spatial resolution [194, 220–222]. These data confirmed results of a variety of earlier structural studies of RC such as measurements of the dichroism [223, 224], magnetic interactions between unpaired electrons in various combinations of redox states of the protein [225, 226], transmembrane voltage generated by each electron transfer [227–230] and resonance X-ray diffraction [231]. These structural investigations permitted to establish that electron transfer in RC occurs indeed over relatively long distances: the centre-to-centre distance between Q_A and $(BChl)_2$ in *Rhodopseudomonas sphaeroides* has been estimated to exceed 23 Å and to be equal to 27 Å in *Rhodopseudomonas viridis* while the distance between BPh and Q_A has been estimated to be within the range 8–12 Å in *Rhodopseudomonas sphaeroides* and to be equal to 14.3 Å in *Rhodopseudomonas viridis*. For such long distances between the participants of PET in RC, electron tunneling seems to be the only possible mechanism, at least at low temperatures.

Early works on electron transfer in RC from *Chromatium vinosum* [204] and *Rhodopseudomonas sphaeroides* [205] demonstrated that these could occur down to 1 K. Then authors of [9] recognized that the oxidation of cytochrome c in *Chromatium vinosum* at low temperatures occurs by a quantum mechanical tunneling mechanism, providing one of the first demonstrations of this phenomenon (see above and Refs. [232–237] for reviews). Figure 25 which represents the rates of the reactions at ambient temperatures and below 100 K, shows the remarkable capability of reaction centres to support nearly temperature-independent electron transfer at low temperatures. Indeed, for electron transfer from BPh^- to Q_A and from Q_A^- to $(BChl)_2^+$ the rates have been shown to become independent of temperature below 100 K [238–243]. But it is necessary to note that electron transfer can occur in a RC also by an alternative thermally activated route which becomes dominant at high enough temperatures [244–248].

Numerous theoretical studies have been made with the aim of explaining the unusual temperature dependence of the electron transfer rates in the RCs of the photosynthetic bacteria [19, 22, 232–237, 249–252]. It is clear that these PCs provide a unique experimental tool for the study of electron transfer processes within a protein environment. However, for the natural RCs, studies are restricted to a detailed analysis of the temperature dependence of the reaction kinetics. One way to overcome this restriction is provided by the technique of replacing the native ubiquinone with other quinone molecules providing different energetics of electron transfer. This yields new RC in which the position of the $(BChl)_2^+ Q_A^-$ energy level relative to the other energy levels of the reaction centre have been changed [225, 246, 253–257]. Detailed investigation was made in Ref. 258 of the kinetics of electron transfer to and from Q_A in RC protein of *Rhododobacter sphaeroides* not only at different temperatures but at $-\Delta G^0$ values varying by about 0.8 eV.

7.3 Kinetics of Electron Transfer in RC Protein from *Rhodobacter Sphaeroides*

Figure 32 shows typical measurements of the kinetics of light induced $(BChl)_2$ oxidation and dark $(BChl)_2^+$ reduction monitored by EPR at 14 K [258]. Data is presented for the native RC (Q_A is ubiquinone-10, UQ_{10}) and for the RC reconstructed with ubiquinone-7, UQ_7. Practically no $(BChl)_2$ oxidation was observed upon illumination of the RC from which UQ_{10} has been extracted and no UQ_7 has been added. The rate constants for $(BChl)_2^+$ and Q_A^- recombination for the native RC and RC reconstructed with UQ_7 were practically the same: $37.2\ s^{-1}$ and $38.0\ s^{-1}$, respectively. Thus, UQ_{10} extraction and its reconstruction with UQ_7 can be carried out without any effect on the reactivity of RC.

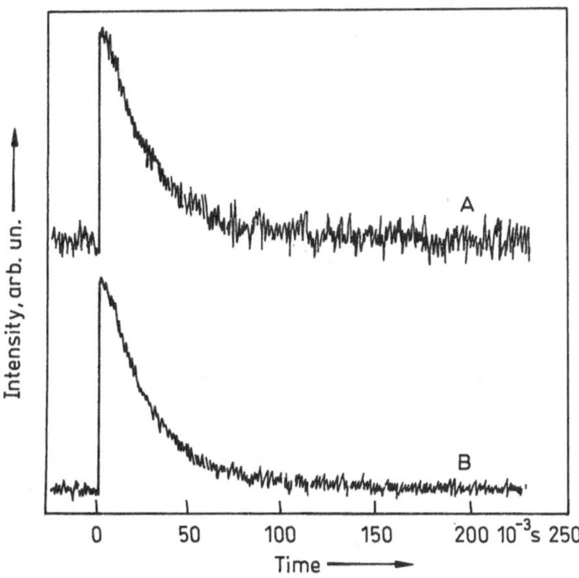

Fig. 32. Flash-induced $(BChl)_2^+$ formation and decay in *Rhodopseudomonas sphaeroides* reaction centre protein: (*A*) – for native RC, (*B*) – for RC, reconstituted with ubiquinone-7. T = 14 K. From Ref. [258]

Eleven 9,10-anthraquinones with various substituents, seven 1,4-naphthoquinones, 1,2-naphthaquinone and five 1,4-benzoquinones were used as Q_A. These quinones provide a series of RCs with a variation of the reaction exothermicity, $-\Delta G^\circ$, from 0.11 to 0.9 eV. The rates of intraprotein electron transfer from various Q_A^- to $(BChl)_2^+$ were found to be virtually temperature independent from 5 to 100 K and to decrease severalfold from 100 to 300 K. Only a small change of the rate upon the $-\Delta G^\circ$ variation was found when reaction was made more exothermic than in the native RC. As the reaction was made less exothermic, the rate decreased notably without becoming temperature dependent.

A study of long-range electron transfer from BPh^- to Q_A in both the native and reconstructed RCs was reported in Ref. [259]. The rate of electron transfer from BPh^- to Q_A was determined at 14 K, 35 K, 113 K and 298 K. The majority of quinones used for reconstruction has in situ polarographic midpoints lower than that for UQ_{10}. The electron transfer rate was determined from EPR measurements of the quantum yield of $(BChl)_2^+Q_A^-$ radical-ion pair.

As for the electron transfer from Q_A^- to $(BChl)_2^+$, the rate of electron transfer from BPh^- to Q_A slows down with decreasing exothermicity. The details of the structure of the quinones reconstructing Q_A function, were found to be relatively unimportant in determining the rate constant, unless they perturbed the reaction exothermicity. Over the range of $-\Delta G^\circ$ studied (230–760 meV), only a small change of the rate was observed as the system was cooled from 295 to 14 K. In the framework of the simple theory of electron transfer this result can be explained by various reasons. One of them is that the total reorganization energy E_r is so dependent on Q_A in the modified RCs that $-\Delta G^\circ$ always equals E_r. But it is rather difficult to understand why this should be so. A more attractive explanation was found when analyzing nuclear vibrations that can be coupled to electron transfer [259].

The character of the rate constant vs $-\Delta G^\circ$ dependence observed in Ref. [259] for electron transfer from BPh^- to Q_A was quantitatively explained in terms of Eq. (23) assuming that: (1) vibrations with $\hbar\omega \leqq 100$ meV are coupled to electron transfer, the total reorganization energy, E_r for these vibrations being 600 ± 100 eV; (2) vibrations with $\hbar\omega \simeq 15$ meV are coupled to electron transfer; (3) E_r for the vibrations with $\hbar\omega < 1$ meV being less than 300 meV.

Thus the studies of the Q_A-involved electron tunneling reactions have shown that analysis of the reaction rate vs $-\Delta G^\circ$ dependences provides information about the character of nuclear motions coupled to electron transfer in RC of photosynthetic bacteria. In particular, as kT becomes smaller than $\hbar\omega$ the dependence of electron transfer rate on $-\Delta G^\circ$ becomes sensitive to the magnitude of $\hbar\omega$. In other words, it is possible to find out from these experiments what aspects of protein and cofactor dynamics are important for the reaction.

7.4 Electron Tunneling Between Molecules Attached to Proteins

Protein matrices are believed to play an important role in the accomplishment of photochemical electron transfer reactions in biological systems. First, the fragments of proteins hinder direct collisions of the reactions centres. Due to this, electron tunneling at large distances often seems to be the only possible way of carrying out electron transfer reactions in biological systems. Special structural organization of protein molecules is often also assumed to provide certain specific favourable channels for electron tunneling within the system donor-mediator (a protein)-acceptor [260, 261]. The necessity to understand better the factors which control the rates of PET in biological systems has stimulated numerous works on the study of electron tunneling between redox centres attached to protein molecules.

There are several requirements for an "ideal" protein system for such studies [262–268]. First, in order to facilitate structure vs function comparisons, the proteins studied should be of known structure. This structure may be obtained, for example, by high-resolution X-ray spectroscopy. Second, physiological redox protein couples are preferred, since such systems are more likely to provide information on biological design than studies of non-physiological redox couples.

In order to utilize photochemical electron transfer, investigators have replaced the Fe-containing porphyrin active sites in several redox proteins with equivalent photoactive porphyrins.

Mixed metal (Zn, Fe) hybrid hemoglobin was used [265, 266] to study long-range electron tunneling between chromophores that are rigidly held at a fixed and crystallographically known distance and orientation. In these experiments iron porphyrin was substituted by zinc proto-porphyrin, ZnP, in hemoglobin chains of one type (α or β) and chains of the opposite type were oxidized to the aquo-ferriheme state, Fe(III)P. Electron phototransfer was found to occur between ZnP and Fe(III)P located within α_1 and β_2 protein chains of hemoglobin. In this case, ZnP and Fe(III)P are separated by two heme pocket walls, the metal-metal distance being equal to 25 Å. Flash photoexcitation of ZnP to its triplet state initiates the primary process

$$^{T}ZnP + Fe(III)P \xrightarrow{k_t} (ZnP)^+ + Fe(II)P \tag{55}$$

The system then returns to its initial state by back electron transfer from Fe(II)P to $(ZnP)^+$. The decay of the ^{T}ZnP triplet state is enhanced only when the β subunit contains Fe(III)P. When the β subunit contains Fe(II)P, the triplet state is unaffected. The rate constant of electron transfer, k_t, for [α(Zn) β(Fe)] hybrid hemoglobin was found to be $10^2\ s^{-1}$ at room temperature [265]. The same value of k_t was obtained for [α(Fe) β(Zn)] hybrid hemoglobin [264].

The temperature dependence of k_t was investigated and reported in Ref. [266]. To obtain low-temperature data, samples were prepared in a 50% glycerol-water mixture. Figure 33 presents the temperature variation of k_t. One can see from this Fig. that the electron transfer rate falls smoothly from the room temperature value to a non-zero value, $k_t = 9 \pm 4\ s^{-1}$, which does not vary further from 170 down to 77 K. Data in the temperature-dependent region (T > 253 K) give the value $E_a \simeq 2$ kcal mol^{-1} for the Arrhenius activation energy.

An analogous approach has been applied to study the electron transfer reaction between yeast cytochrome *c* peroxidase (CCP) and cytochrome *c* (cyt *c*) by

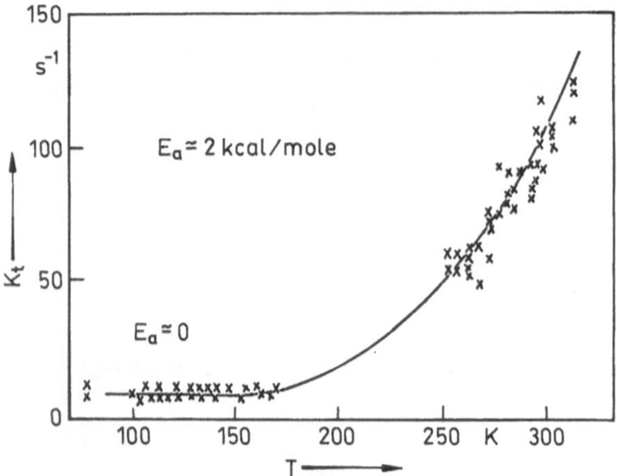

Fig. 33. Temperature dependence of the electron transfer rate, k_t, in the [αFe(III) βZn] hybrid hemoglobin. Data taken from Ref. [266]

employing the complex between zincsubstituted CCP (ZnCCP) and native cyt *c* [264]. The two heme planes in this complex are nearly parallel at a metal-metal distance of 25 Å and an edge-to-edge distance of 17–18 Å.

The decay curves of the triplet excited state of zinc porphyrin for ^{T}ZnCCP and the ^{T}ZnCCP/cyt *c* complex with reduced horse cytochrome *c* heme are exponential with the same decay rate. Upon addition of horse cyt *c* with the oxidized heme to the solution of ZnCCP, decay remains exponential but the decay rate increases until a 1 : 1 ratio is reached and then remains constant. The form of the dependence between the rate and the concentration of cyt *c* indicates that ZnCCP and the cyt *c* form a strong 1 : 1 complex. These results indicate that electron tunneling at the distance of 25 Å is the reason for the enhancement of ^{T}ZnCCP decay in the presence of cyt *c*. The rate of electron transfer from ^{T}ZnCCP to the low-spin ferriheme within the [ZnCCP/horse cyt *c*] complex was found to be $17 \pm 3\ s^{-1}$ at 293 K [264].

The rate of electron tunneling from ^{T}ZnCCP to the ferriheme of the yeast cytochrome *c* was found to be roughly 10 times higher than that observed in the case of the homologous horse cytochrome *c*. This difference demonstrates the fine degree of species specificity involved in biological electron transfer and must reflect subtle structural differences between horse and yeast cytochromes [264].

Electron phototransfer in zinc-substituted cytochrome *c*/cytochrome b_5 complex [Zn(II) cyt *c*/Fe(III) cyt b_5] has been studied [267]. Porphyrin rings lie parallel in this complex with the distances of ca. 8 Å edge-to-edge, 18 Å centre-to-centre. The Zn cyt *c* triplet excited state is quenched by Fe(III) cyt b_5 with the rate constant $k_t = 5 \times 10^5\ s^{-1}$. Neither Fe(II) cyt b_5 nor Zn(II) cyt b_5 quench ^{T}Zn cyt *c*.

The rate of electron tunneling in the topologically similar [Zn hemoglobin/Fe(III) cyt b_5] complex was found to be $8 \times 10^3\ s^{-1}$ [268]. Note that this is an example of electron transfer within the protein complex formed between two physiological redox partners: Zn-substituted hemoglobin and cytochrome b_5 [269].

[Zn cyt *c*/Fe(III) cyt b_5] and [Zn hemoglobin/Fe(III) cyt b_5] complexes have a similar structure. However, the rates of electron tunneling in these complexes are 10^2-fold different. This difference in rate constants may reflect direct differences in the "conductivity" towards electron tunneling of Hb vs cyt *c*. Clearly, subtle structural differences between these complexes are sufficient to cause large rate differences in the case of the tunneling mechanism.

Many other papers on long-range electron transfer between two reactive sites of modified proteins were published [270–288] after the above mentioned pioneering works. Most of them dealt with photoinduced electron tunneling from triplet states of closed shell Mg(II) and Zn(II) porphyrins to Fe(III) or Ru(III). In agreement with the prediction of Marcus theory the rate constants for the majority of these intraprotein electron transfer reactions were found to increase as the free energy of reaction decreased. However for one of the reactions disagreement with this theory was observed [285].

In Ref. [279] the technique of protein modification was used to study the dependence of the rate of photoinduced electron tunneling on the distance between ^{T}ZnP and Ru(III) sites in modified myoglobins. The modified proteins were prepared by substitution of zinc mesoporphyrin IX diacid for the heme in four various pentaammineruthenium (III) derivatives of sperm whale myoglobin: $(NH_3)_5$Ru(His-48)Mb, $(NH_3)_5$Ru(His-12)Mb, $(NH_3)_5$Ru(His-116)Mb and $(NH_3)_5$Ru(His-81)Mb. Metal-to-metal distance between ZnP and $(NH_3)_5$Ru(His) ranges in this seria from 16.1–18.8 Å for His-48 to 27.8–30.5 for His-12. The rate constant of electron tunneling decreases in this series in accordance with Eq. (1) with $\nu_e = 7.8 \times 10^8\ s^{-1}$ and $a_e = 2.2$ Å at T = 298 K.

Temperature dependences of the rate for direct photoinduced electron transfer process and reverse charge recombination reaction were studied in some works. As a rule both processes were found to be temperature dependent. However for [β(MP), α(Fe(III)P hemoglobin hybrid (M = Zn(II), Mg(II)) the rate constants of both processes were found to be temperature independent in the temperature interval 273–293 K [285].

An interesting system for study the mechanism of electron tunneling in proteins was synthesized in Ref. [286]. In this work donor, *p*-dimethylanilino group, and photosensitizer, pyrenyl group were attached to a specially synthesized protein molecule. Electron transfer at the edge-to-edge distance 9.6 Å was observed upon electronic excitation of the sensitizer fragment.

7.5 Electron Transfer from Redox Sites of Proteins to Excited Simple Molecules

The other way to study the "conductivity" of protein molecules towards electron tunneling is to investigate the quenching of luminescence of electron-excited simple molecules by redox sites of proteins [289, 290]. Experiments of this sort on reduced blue copper proteins have involved electron-excited Ru(III) $(bpy)_3$, Cr(III) $(phen)_3$, and Co(III) $(phen)_3$ as oxidants. The kinetics of these reactions exhibit saturation at protein concentrations of 10^{-3} M, suggesting that, at high protein concentrations, the excited reagent is bound to reduced protein in an electron transfer precursor complex. Extensive data have been obtained for the reaction of reduced bean plastocyanin Pl(Cu(I)) with *Cr(III) $(phen)_3$. To analyze quenching experimental data, a mechanistic model that includes both 1:1 and 2:1 Pl(Cu(I))/ *Cr(III) $(phen)_3$ complexes was considered [290].

$$
\begin{array}{ccc}
\mathrm{Pl(Cu(I)) + {}^*Cr(III)(phen)_3} & & \\
\uparrow\downarrow & & \\
\mathrm{[Pl(Cu(I))/{}^*Cr(III)(phen)_3]} & \underset{-\mathrm{Pl(Cu(I))}}{\overset{+\mathrm{Pl(Cu(I))}}{\rightleftharpoons}} & \mathrm{\{[Pl(Cu(I))]_2/{}^*Cr(III)(phen)_3\}} \\
\searrow & & \swarrow \\
& \mathrm{Pl(Cu(II)) + Cr(II)(phen)_3} &
\end{array}
\qquad (56)
$$

In this scheme, [Pl(Cu(I))/*Cr(III) $(phen)_3$] is a complex in which the excited Cr(III) $(phen)_3$ is bound at a site that is relatively distant from the copper atom. NMR experiments have indicated that Cr(III) $(phen)_3$ binds to Pl(Cu(I)) near tyrosine-83 [291–293]. A computer-generated model of this complex showed that the closes donor-acceptor contact is 10.3 Å, which is the distance from the coordinated sulphur atom of cysteine-84 to the nearest phenanthroline carbon of the bound Cr(III) $(phen)_3$. This is the edge-to-edge distance for the long-range electron tunneling pathway. The Cu–Cr distance in this complex is 18.4 Å. The long-range electron tunneling rate constant, k_t, from Pl(Cu(I)) to *Cr(III) $(phen)_3$ was found to be $3 \times 10^6\ s^{-1}$.

The same model has been used to fit the data obtained for the reactions of azurin, Az(Cu(I)) and *Rhus vernicifere* stellacyanin, St(Cu(I)), with *Cr(III) $(phen)_3$ and *Ru(II) $(bpy)_3$ oxidants. Values of the rate constants k_t and stability constants K_1 and K_2 of the 1:1 and 2:1 complexes that have been extracted from the analysis are given in Table 5.

It is interesting to compare the data listed in Ref. [290] for the long-range electron tunneling from Pl(Cu(I)) to *Cr(III) $(phen)_3$ with those of Refs. [294] and [295] for electron tunneling from Pl(Cu(I)) to Co(III) $(phen)_3$. In Refs. [294] and [295] it was found that the kinetics of oxidation of Pl(Cu(I)) by Co(III) $(phen)_3$ exhibited signs

Table 5. Electron transfer rate constants, k_t, free energy changes, $-\Delta G°$, and stability constants, K_1 and K_2, for the reactions of *Cr(III) $(phen)_3$ and *Ru(II) $(bpy)_3$ with reduced blue copper proteins at 295 K [290]

Protein	Oxidant	$10^{-6}k_t$, s^{-1}	$-\Delta G°$, eV	K_1, M^{-1}	$K_2 \times 10^{-3}$, M^{-1}
Pl(Cu(I))	*Cr(III) $(phen)_3$	2.5	1.06	250	4.2
Pl(Cu(I))	*Ru(II) $(bpy)_3$	3.0	0.48	100	0.2
Az(Cu(I))	*Cr(III) $(phen)_3$	1.2	1.11	60	
Az(Cu(I))	*Ru(II) $(bpy)_3$	1.2	0.53	40	2.3
St(Cu(I))	*Cr(III) $(phen)_3$	0.20	1.23	60	1.3
St(Cu(I))	*Ru(II) $(bpy)_3$	0.20	0.65	40	1.4

of saturation behaviour, thereby suggesting that electron transfer might occur within a relatively stable Pl(Cu(I)): Co(III) $(phen)_3$ precursor complex. The rate constant $k_t \simeq 20\, s^{-1}$ of this process is five orders of magnitude slower than for the Pl(Cu(I)) → *Cr(III) $(phen)_3$ electron transfer. One of the reasons of such a difference in the rate constants of electron tunneling may be the difference in the free energy barrier, namely $\Delta G^\circ = -1.06$ eV for Pl(Cu(I)) → *Cr(III) $(phen)_3$ and $\Delta G^\circ = -0.02$ eV for Pl(Cu(I)) → Co(III) $(phen)_3$.

A simple approach to understanding the factors which control the "conductivity" of proteins towards photoinduced electron tunneling is to develop "small molecule" model systems to mimic intramolecular electron transfer in the protein systems. Appropriate models obviously require that the donor and acceptor be held at fixed distances and orientations which correspond to those in the protein-protein complexes. Models of this type have recently been obtained and investigated [296, 297]. In these models the protein matrix is replaced by a simple synthetic spacer which separates two porphyrin molecules. By changing the chemical structure of the spacer, a series of molecules with different reaction distances and geometries has been synthesized. Typical examples of such molecules are presented in Fig. 34.

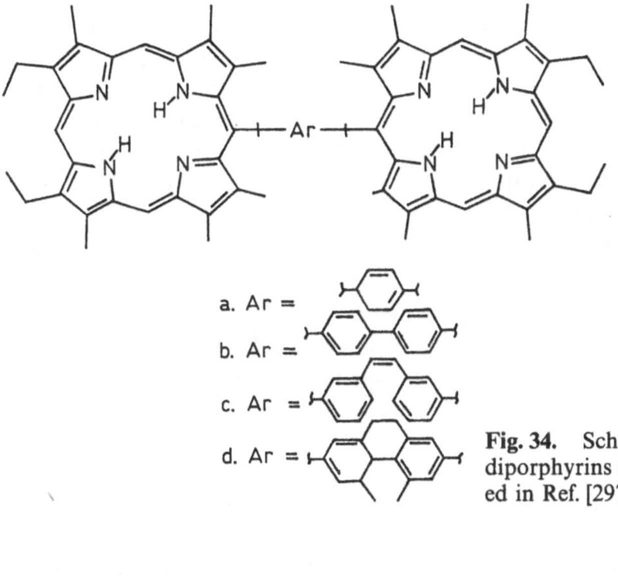

Fig. 34. Schematic picture of coplanar diporphyrins with some spacers synthetized in Ref. [297]

The key finding of the preliminary investigations of such molecules is that the non-protein diporphyrin models can react many times faster (ca. 10^4) than similar protein-protein systems at similar distances between the reaction sites and similar ΔG°. This result suggests that the protein matrix does not accelerate the electron transfer rate [297], although a final conclusion cannot be drawn on the basis of these very limited data. Further studies in this area are necessary.

7.6 Tunneling Charge Transfer Bands of Donor-Acceptor Pairs Attached to Proteins

It has been noted that electron tunneling in the donor-acceptor pair D-A may lead to the appearance of a charge transfer band in the absorption spectrum of this pair [298]. The following formula was obtained describing the dependence of the extinction coefficient, ε, of this band on the energy, E, of the absorbed light quantum

$$\varepsilon(E) = 1.22 \times 10^{19} \frac{V^2R^2}{E\sigma} \exp\left[\frac{-(E - E_{max})^2}{2\sigma^2}\right] \tag{57}$$

In this formula, V is the electron coupling energy, R is the distance between the centres of the D and A particles, σ is the width of the charge transfer band, and $E_{max} = E_D - E_A + \Delta_v$, where $(E_D - E_A)$ is the difference of the redox potentials of the donor and the acceptor and Δ_v is the energy spent on the excitation of the vibrational degrees of freedom.

A band of this type has been observed for an enzyme-substrate complex ES where the enzyme was represented by the oxidized form of peroxidase cytochrome *c*, cyt(Fe(III)) and the substrate was the reduced form of cytochrome *c*, cyt_1 (Fe(II)) [298]. Indeed, on mixing the solution of cyt(Fe(III)) and cyt_1(Fe(II)) there appeared a new absorption band with the absorption maximum at $E_{max} = 1.4$ eV, the extinction coefficient $\varepsilon = 0.35\ M^{-1}\ cm^{-1}$, and the width $\sigma = 0.2$ eV. This band was referred [298] to charge transfer via electron tunneling, [cyt(Fe(III))/cyt_1(Fe(II))] → [cyt(Fe(II))/cyt_1(Fe(III))]. From a comparison of the data on the intensity of this band with the results of fluorescence measurements, the distance between the iron atoms Fe(III) and Fe(II) in the [cyt(Fe(III))/cyt_1(Fe(II))] complex has been estimated to be $R \simeq 15-20$ Å and the edge-to-edge tunneling distance $R_t = 7$ Å.

The appearance of similar absorption bands has also been observed upon the formation of a complex between the reduced form of cytochrome c and the simple inorganic acceptor Fe(III) $(CN)_6$ [299]. The tunneling distance evaluated from the intensity of this band amounts to 7–10 Å. However, more recent experiments have failed to detect such a band [300]. The situation is more favourable in the system [cytochrome *c*/P870] of the *Chromatium* reaction centre, where the intensity of the charge transfer band centred at 200 nm could be correlated with the data obtained in kinetic experiments [301].

The results reported in Refs. [298] and [301] show that, in principle, it is possible to use the data on the absorption spectra of a donor-acceptor pair for estimating the distances of electron tunneling stimulated by light. It should be emphasized that, in this case, the illumination is performed in the band of the tunneling charge transfer from the donor to the acceptor without exciting the electron transitions within the donor and the acceptor molecules themselves.

8 Conclusions

The above data show, that photoinduced long-range electron tunneling reactions are rather widespread in photochemistry and photobiology. The ability to participate in such reactions is inherent in excited states of versatile chemical compounds including organic and inorganic molecules, transition metal complexes and clusters, porphyrins, redox sites of proteins. This makes electron tunneling important for primary processes of PET. In a similar way, electron tunneling is also inherent in ground states of versatile organic and inorganic radicals, radical-ions and molecules, and transition metal complexes. This makes electron tunneling important for secondary processes of PET as well.

In this section we shall:

1) summarize the regularities of electron tunneling in primary and secondary processes of PET as well as indicate unresolved problems in the area, and

2) discuss the new opportunities in designing photochemical redox transformations opened up by electron tunneling.

8.1 Regularities of Photoinduced Tunnel Electron Transfer Processes

The following regularities of electron tunneling in primary and secondary processes of PET can be mentioned.

1) Photoinduced long-range electron tunneling is a special type of non-radiative electron transition in condensed media and demonstrates features characteristic of these processes. In terms of the existing theory of the elementary act of electron tunneling, one can qualitatively explain virtually all the available experimental data on the kinetics of photoinduced tunnel electron transfer reactions. Unfortunately, however, this theory does not yet possess sufficient predicting power and does not allow one to unambiguously answer a priori the question whether this or that electron tunneling reaction will or will not proceed at a sufficient rate at this or that distance between the reagents. Currently this question can only be answered unambiguously by experiment. However the theory gives quite definite ideas about the scale of influence of the various factors on the probability of tunneling, W. Sometimes this allows us to make useful qualitative predictions concerning the expected trends in the changes of W with regard to certain variations in the reaction system.

2) The main feature of electron tunneling is that it can provide the occurrence of both primary and secondary reactions of PET between remote electron donor and electron acceptor sites, at distances sometimes as great as several tens of Angströms.

3) PET via electron tunneling mechanism can take place both from the excited donor to the non-excited acceptor and from the non-excited donor to the excited acceptor. An example of the process of the latter type was given in Sect. 2.2.

4) The dependence of the probability of electron tunneling in PET reactions on the distance R between the reagents can be well discribed by a simple exponential law: $W = \nu_e \exp(-2R/a_e)$. This law is suggested by the theory and allows us to describe quantitatively all the experimental data available so far.

5) The frequency factor ν_e in the above expression for W depends on the characteristics of the nuclear vibrational modes that are coupled to electron transfer. It is not the frequency of electron motion in the reacting particles. For various reactions ν_e can vary over a very broad range and can reach $10^{20}\ s^{-1}$ at maximum. Depending on a particular reaction and at temperature at which kinetic studies are made, ν_e can be both temperature dependent or temperature independent. At elevated temperatures, when at least some of the vibrational modes can be considered as classical (that is at $k_BT \gg \hbar\omega$), the dependence between ν_e and T has the usual Arrhenius form, unless the reaction exothermicity $-\Delta G^\circ$ is equal to the reorganization energy E_r for these degrees of freedom. At low enough temperatures, when all the nuclear vibrational modes are quantum (that is at $k_BT \ll \hbar\omega$), ν_e becomes practically independent of T. ν_e can be also practically independent of T even when some of the nuclear vibrational modes are classical, but E_r for them equals $-\Delta G^\circ$. It explains the unusual temperature dependence of the reaction rate, which is often observed for PET reactions: the Arrhenius dependence at higher temperatures and temperature independence – at lower temperatures. It explains also why many PET reactions can proceed at extremely low temperatures: there is no activation energy for electron tunneling at those temperatures.

6) Under the high temperature approximation, when at least some of the nuclear vibrations are classical ($kT \gg \hbar\omega$), for both primary and secondary reactions of PET W increases as $-\Delta G^\circ$ increases at low values of exothermicity ($-\Delta G^\circ < E_r$), reaches the maximum at $-\Delta G^\circ = E_r$ and then, at $-\Delta G^\circ > E_r$, either decreases or remains almost unchanged with a further increase of $-\Delta G^\circ$. The former situation is observed when reaction products are formed in the ground electronic state, so that the actual exothermicity of the electron tunneling process coincides with $-\Delta G^\circ$. Examples of photoinduced tunnel electron transfer reactions obeying this regularity can be found in Sects. 5.3 and 7.2. The latter situation can sometimes be observed when at large $-\Delta G^\circ$ values reaction products can be formed initially in the excited electronic states, so that the actual exothermicity of the elementary act of electron tunneling is smaller than $-\Delta G^\circ$, or when the excitation of quantum (with $\hbar\omega \gg k_BT$) vibrational modes is involved in the dissipation of the reaction exothermicity. An example of photoinduced tunnel electron transfer reactions obeying this regularity was given in Sect. 2.2.

7) Parameter a_e in the above expression for W characterizes the overlap of those wave functions of the donor and acceptor, which participate in electron transfer. This parameter depends on the nature of the both of the reagents and the medium. The values of a_e for secondary reactions of PET typically fall within the range $a_e = 1-2$ Å, i.e. they are close to those values for usual non-photochemical electron tunneling processes. The values of a_e for primary PET reactions typically exceed those for secondary ones and can sometimes be as high as 4 Å or even more. A reason for this can be a lower binding energy for the electron that tunnels from the excited state of the donor molecule.

8) The rate of photoinduced tunnel electron transfer may depend on the mutual orientation of the reagents. This dependence may result from the angular dependence of the electronic wave functions involved in electron transfer. Examples

of the angular dependences of electron tunneling rates during PET were given in Sect. 3.3.

9) The rate of PET reactions may strongly increase when mediator centres are placed in the space between the donor and the acceptor. The effect becomes more pronounced as the gaps between the energy of the tunneling electron and the energies of the levels of the mediator taking part in tunneling become smaller. Since the density of electronic levels increases on approaching the ionization continuum, mediators are expected to increase the rate of electron tunneling for reactions of excited molecules more than for reactions of the particles in the ground electronic states. Alongside the lower binding energy of the tunneling electron, this effect may also contribute to bigger a_e values for electron tunneling reactions with excited molecules.

10) Electron tunneling can provide a new type of photoinduced electron transfer reaction, i.e. PET reactions stimulated by illumination into the tunnel electron transfer band. In this process the electron tunnels, upon absorption of a light quantum, from the energy level of the donor to the higher level of the acceptor, the energy of which exceeds the energy of the donor by the energy of the light quantum. The ability of tunneling to provide electron transfer via such light absorption results in the appearance of a new rather weak band in the absorption spectrum of the system. The intensity of this band exponentially decreases with increasing distance between the reagents.

11) The kinetics of both primary and secondary reactions of PET may be rather unusual and obey the equations of the type presented in Eqs. (4), (6), (7), (8) and (9). Characteristic kinetic features of tunneling PET reactions may be: (i) the presence of a non-exponential (close to logarithmic) segment with a short observation time followed by an exponential segment, for primary electron transfer processes involving excited states; (ii) a non-exponential (close to logarithmic) kinetics over the whole range of the observation times for secondary dark processes between immobile reagents in crystals, vitreous solutions and biological systems at low temperatures; (iii) the presence of a non-exponential (close to logarithmic) segment with a short observation time followed by a segment corresponding to conventional diffusion controlled kinetics with longer observation time, for both primary and secondary reactions of PET between mobile reagents.

Note that equipment with high enough time resolution is needed in order to observe the initial non-exponential segment of the kinetic curves in cases (i) and (iii).

Thus, by the moment several quite important regularities have already been revealed for both primary and secondary electron tunneling reactions of PET. However many important features of these reactions are still insufficiently studied. In particular, much more detailed studies are needed concerning the influence of the effects of mediators, energetics of the reagents, their mutual orientation, coupling of electron tunneling to nuclear vibrations in the reagents and the media, etc. on the rate of tunneling. Convenient objects for such studies seem to be specially designed bridge molecules of the D-L-A type. With these molecules it is possible to study systematically homologous series within which various factors that are important for tunneling, can often be studied separately.

8.2 Electron Tunneling and New Opportunities in Designing Photochemical Redox Transformations

Long-range electron tunneling allows PET processes to be accomplished under conditions in which reagents are spatially far apart from each other. This opens up new opportunities in organizing photochemical conversions on the molecular level and in controlling these conversions. As was shown in Sect. 7, these opportunities are widely used in photosynthesis. The substances taking part in these very complicated biological processes are frequently fixed in membranes and so cannot collide with one another. This allows them to avoid undesirable side reactions. Nevertheless, these substances can exchange electrons via the electron tunneling mechanism. The products formed upon such electron tunneling reactions may initiate chains of further chemical conversions, each product at its own site in the membrane. Owing to such an arrangement of chemical conversions, in natural photosynthesizing systems, one can prevent a rapid recombination of very reactive primary species formed by light and thus minimize energy losses while converting the energy of light quanta into chemical energy. This provides a high efficiency of energy conversion in the process of photosynthesis. The rapid recombination of reactive species formed by light is known to be one of the most serious obstacles in the way of creating artificial photochemical solar energy converters. By analogy with natural photosynthesis it would be interesting to try to use electron tunneling processes so as to overcome this difficulty. The possible approaches to solving this problem are discussed in Sect. 6 of this Chapter and Chapter 5.9 by K. I. Zamaraev, V. N. Parmon and S. V. Lymar.

With the aid of electron tunneling it appears possible to regulate the selectivity of redox conversions. For practically important reactions this has not been realized so far, but that this approach may prove to be useful is demonstrated, e.g. by the data presented in Table 6. In this table, a comparison is made between the rate constants for reactions of three different acceptors with hydrated electrons in liquid water at 298 K and the characteristic times, τ, for reactions of the same acceptors with trapped electrons in solid water-alkaline glasses at 77 K. The values of τ have been calculated using the values of ν_e and a_e from Ref. [21]. It can be seen in the liquid, when due to diffusion the reagents can approach to within short distances of each other (direct collisions), that the rate constants for all three

Table 6. Rate constants k_e of the reactions of e_{aq}^- with various acceptors in liquid water at 298 K and the characteristic times, τ, for the reactions of e_{tr}^- with the same acceptors in vitreous water-alkaline solutions at 77 K $\tau = \nu_e^{-1} \exp(2R/a_e)$, R is the distance between the reagents, ν_e and a_e are the coefficients in Eq. (1)

Acceptor	k_e, $M^{-1}s^{-1}$	τ, s		
		R = 10 Å	R = 15 Å	R = 20 Å
$ClAc^-$	0.12×10^{10}	$10^{4.6}$	10^{17}	
NO_3^-	0.85×10^{10}	$10^{-9.9}$	$10^{-5.6}$	$10^{-1.3}$
$Cu(en)_2^{2+}$	1.85×10^{10}	$10^{-10.5}$	$10^{-8.1}$	$10^{-5.7}$

processes are close to one another and are very high. In contrast to this, in the solid, where there is virtually no diffusion and the reactions proceed via electron tunneling, the characteristic times of the processes vary with the distance between the reagents substantially differently due to differences in the values of v_e and a_e. For the reaction of e_{tr}^- with $Cu(en)_2^{2+}$, even at R = 20 Å, the characteristic time is very short, $\tau \simeq 10^{-6}$ s, while at R = 10 Å, $\tau \simeq 10^{-10.5}$ s. At the same time for reaction with $ClAc^-$ at R = 10 Å, τ is longer by ca. 15 orders of magnitude. Such a large difference between the values of τ suggests that one can use electron tunneling for selective performance of redox conversions by creating, in one way or another, obstacles to direct collisions of reagents. In so doing, the gain in selectivity will be achieved at the expense of a reduction in the rate, but for rapid enough electron transfer processes this loss appears to be not very significant.

In solids, where direct collisions of reagents are often virtually impossible because of their very slow diffusion, the distance of electron tunneling, and hence the time τ, can be regulated by varying the concentration of reagents. In liquids one can try to hinder the approach of reagents to short distances by surrounding them with a shell of chemically inert molecules of the type of bulky ligands, micelles, or vesicles. The experimental data corroborating the correctness of the conclusion about the possibility of photoinduced electron tunneling reactions with participation of molecules surrounded with bulky ligands and between particles located outside and inside micelles and vesicles are presented in Sects. 2.3 and 6.2.

To regulate the rate of electron tunneling one can also try to use its dependence on the electrical field. Such a regulation is probably effected in living nature for electron transport through phospholipid membranes under conditions when there is a difference of electric potential between the sides of the membrane. Interesting prospects in regulating the rates of photoinduced electron tunneling are also opened up by using organized molecular structures of the type of molecular layers and bridge molecules discussed in Sects. 5.1–5.5 and 6.1.

Owing to electron tunneling some photochemical transformations of organic substances in solids can be performed by a one-quantum, rather than two-quantum, mechanism, i.e. with lower energy losses (see Sect. 2.2). Electron tunneling from electron-excited molecules to acceptors in solids results in a non-exponential decay of luminescence intensity with time and may cause a decrease in the quantum yield of luminescence of excited molecules. Long-range intramolecular electron transfer by a tunneling mechanism may be one of the mechanisms of photochromism of bridge donor-acceptor molecules. It can be hoped that in future the rate of photochromic transformations of such compounds will be varied in a controllable fashion by varying the bridge length and thus changing the rate of electron tunneling.

Long-range electron tunneling may also play an important role in the processes of the protection of materials from photodestruction.

The above data indeed suggest that photoinduced long-range electron tunneling reactions are important for various fields of photochemistry and photobiology and open up new opportunities in designing photochemical redox transformations.

9 Notes Added in Proof

Since this paper had been presented to the editor some new papers concerning photoinduced electron tunneling reactions in chemistry and biology have been published. These papers can be roughly divided into three groups. The papers of the first and second groups present experiments on electron tunneling over large distances in bridge molecules and proteins like those discussed, respectively, in Sects. 4 and 5 and Sect. 7 of this Chapter. The papers of the third group deal with theoretical calculations of the rate constants for tunnel electron transfer via a superexchange mechanism. We will consider the main results obtained in the papers of all three groups.

9.1 Photoinduced Electron Tunneling in Bridge Molecules

In Ref. [302], a difference by a factor of 5 was found for the rate constants of electron tunneling in rigid bridge molecules P-Q that contained porphyrin P and quinone Q fragments at almost the same distances, but had different mutual orientations of P and Q fragments.

In Ref. [303], rates of intramolecular photoinduced electron transfer were given for bichromophoric molecules of the type shown in Fig. 16. The molecules had different mutual orientations of donor and acceptor groups. Different orientations of these groups were achieved by modifying the structure of the hydrocarbon bridge between them. It was found that the intramolecular electron transfer is faster for a less flat cyclopentyl ring attached to the dicyanoethylene group than for a flatter cyclopentyl ring.

Theoretical calculations of the dependence of the probability of electron tunneling on mutual orientation of donor and acceptor fragments in bridge molecules were presented in Ref. [304]. With the use of a cyclic polyene model to construct wave functions of porphyrin fragments, a number of analytical expressions were obtained for orientation dependence of probabilities of tunneling between these fragments.

Photoinduced electron tunneling over large distances between the porphyrin and viologen fragments of P-$(CH_2)_n$-MV^{2+} (n = 3–6) bridge molecules was observed in Refs. [305–307]. The fluorescence decay profiles for the viologen-linked porphyrins were deconvoluted into the sum of two first-order decays with short and long lifetimes. The short lifetime was attributed to quenching of the photoexcited singlet state of the porphyrin by the linked viologen via electron tunneling through the – $(CH_2)_n$-bridge while the long lifetime – to spontaneous deactivation. Strong dependence of the short lifetime on the number n of methylene groups in the bridge was observed.

Photoinduced intramolecular electron tunneling was observed also in some other porphyrin containing bridge molecules, such as porphyrin covalently linked to phenolphthalein [308], dimethylaniline – mesoporphyrin II – quinone triad [309], Zn porphyrin-viologen-quinone triad [310], carotenoid – porphyrin – diquinone tetrad [311]. The influence of conformational state of porphyrin-viologen bridge molecules on the rate of PET reactions was studied in Ref. [312].

In Ref. [313] photoinduced charge separation as well as subsequent charge recombination were studied in bichromophoric molecules containing porphyrin and quinone units interconnected by a rigid saturated hydrocarbon bridge that unequivocally determined both separation and relative orientation of the chromophores. The rate of charge recombination was found to decrease dramatically at lower solvent polarity, thus indicating the importance of a relatively apolar environment for achieving long-lived charge separation.

The lifetimes of charge-separated states formed on photoexcitation of rigid donor-spacer-acceptor molecules D-L-A (where D and A are organic fragments and L is linkage) like those pictured in Fig. 16 with variable edge-to-edge separation of D and A units from 5 to 15 Å, were measured in Ref. [314] by using the time-resolved microwave conductivity technique. The lifetime was found to be governed by two recombination pathways: direct recombination to the ground state and indirect recombination via locally excited donor. The latter was found to become increasingly important as the separation distance increased. The reason for this effect was proposed to be the decrease in the energy difference between the local excited donor state and the charge separated state as the latter gradually lost the energy of Coulomb attraction between the charged centers with inreasing distance between them.

In Refs. [315–316] a synthetic methodology was developed that allows to use transition metal chromophores such as $(bipy)Re^{I}(CO)_3$ in studies of PET across rigid organic spacers. Intramolecular PET from 1,3-benzodithiafulvene to Re^{I} over 16 Å via trans-1,4-cyclohexane spacer at 298 K followed by dark charge recombination was found to occur in the nanosecond time scale.

In Ref. [317] the temperature independence of the intramolecular electron transfer reaction in a cofacial Zinc porphyrin-quinone cage molecule was observed in the range 80–300 K and interpreted in terms of non-adiabatic electron tunneling.

9.2 Photoinduced Electron Tunneling in Protein Molecules

The temperature dependence of the rate constant of electron transfer over large distance from the first triplet state of Zn porphyrin to $Ru^{III}(NH_3)_5$ covalently attached to histidine-33 in Zn-substituted cyt *c* was studied in Ref. [318]. A temperature independent triplet quenching process with the rate constant 3.6 s^{-1}, was observed at 10–100 K and tentatively attributed to electron transfer facilitated by nuclear tunneling.

Temperature independent electron tunneling was observed also in Ref. [319], where the rate of electron transfer over large distance in mixed-metal hemoglobin hybrids [MP, $Fe^{III}(CN^{-})$P], where M = Zn or Mg, was measured in the temperature range from ambient to 100 K. The electron transfer from the triplet state of MP to Fe^{III} was not effected by the freezing of the cryosolvent, which may indicate that coupling of electron transfer to low-frequency solvent modes may be minimal. For both M, but especially for M = Mg, the rate constant of the back reaction is nearly temperature independent.

The temperature dependence of the intramolecular electron transfer from Ru^{II} to Cu^{II} in Ru-modified stellacyanin from *Rhus vernisifera* was studied in Ref. [320] in the temperature interval 298–112.7 K. The activation enthalpy $\Delta H^{\#}$ was found to be 19.1 ± 3.1 kJ/mol. The entropy of activation found for the intramolecular electron transfer process $\Delta S^{\#} = -201 \pm 40$ J/mol K was in a good agreement with the calculated value $\Delta S^{\#} = -193$ J/mol K for intramolecular electron transfer in Ru-stellacyanin-Cu over the distance R = 16 Å estimated from the tentative three-dimensional model of stellacyanin molecule [320].

In recent publications [321–324] parameters a_e and v_e that characterize the decay of the probability of electron tunneling with the distance (see Eq. (1)) were evaluated experimentally. In Ref. [321] light induced intramolecular charge separation followed by dark recombination was studied in $ZnP-Fe^{III}P_1$ hybrid diporphyrins with various bridges. It was found that the rate constant of charge separation k_{cs} depends exponentially on the center-to-center distance R between two porphyrins: $k_{cs} = v_e \exp(-2R/a_e)$, where $a_e = 5$ Å, $v_e = 1.4\ 10^{13}\ s^{-1}$. Note a big value obtained for a_e. No apparent orientation dependence of k_{cs} was observed. Charge recombination was found to be practically independent of the distance and orientation. One of the possible explanations of this result may be that the rate determining step in the charge recombination process is not an inter-site electron transfer reaction, but some reaction within the $Fe^{II}P_1$ complex site, such as the change of ligation.

Big values of parameter a_e for electron tunneling in peptide molecules were obtained also in Refs. [322–324]. In Refs. [322–323] intramolecular electron transfer from tirozine to the tryptophane radical generated by the pulse radiolysis of peptide molecules, was studied for such molecules with various number of proline fragments between the donor and acceptor groups. The rate constants were found to fall exponentially with the distance with the value of $a_e \simeq 6$ Å [322]. For PET over large distances in osmium-ruthenium binuclear complexes bridged with oligoproline peptides, parameter a_e was found to be 3 Å [324].

9.3 Theoretical Calculations

Experimental studies of electron transfer over large distances in peptides stimulated theoretical studies of the influence of polypeptide chains on the rate of electron tunneling. In recent publications [325–337], calculations were made of the rate of superexchange electron tunneling through the system of chemical bonds in order to explain the large values of a_e (i.e. the large values of rate constants for large values of R) for long-range electron transfer processes in protein systems discussed above. Consider some of the results obtained in these works.

In Refs. [325, 332] electron transfer and energy transfer were discussed through bridged systems where the bridge was resonant with the donor and acceptor. It was shown that in contrast to non-resonant situation, in resonant case exponential decay of the rate constant with increasing bridge length breaks down and the reaction rate constant can even oscillate as bridge length increases.

Direct and superexchange electron tunneling between adjacent and remote sites of plastocyanins were considered in Ref. [326]. Exchange matrix elements for direct and superexchange tunneling were calculated using extended Huckel theory and superexchange theory, respectively. The results suggest that electron transfer through the protein is provided by selective through-bond pathways rather than simple direct transfer over the geometric distance between the reagents.

In Refs. [327–329] a numerical algorithm was used to survey proteins for electron tunneling. The dominance of pathways through covalent bonds and hydrogen bonds was shown. The method predicted the relative electronic couplings in ruthenated myoglobin and cytochrome *c* that occured to be consistent with measured electron transfer rates. Presumably, it can be also used to design and optimize molecular systems for electron transfer over large distances.

An artificial intelligence search method was used in Ref. [330] to select a subset of 15–20 aminoacid residues among ca. 150 such residues as being most relevant for electron transfer.

Rates of non-adiabatic intramolecular electron transfer were calculated in Ref. [331] using a self-consistent perturbation method for the calculation of electron-transfer matrix elements based on Lippman-Schwinger equation for the effective scattering matrix. Iteration of this perturbation equation provides the data that show the competition between the through-bond and through-space coupling in bridge structures.

10 List of Abbreviations

Parameters Characterizing Long-Range Electron Tunneling

W – tunneling probability per unit time

ν_e, a_e – parameters in the dependence of the tunneling probability on the distance, $W = \nu_e \exp(-2R/a_e)$

R – distance between the donor and the acceptor

R_t – distance of tunneling during the time t

R_D – distance at which the characteristic time of tunneling becomes equal to the time of a diffusion jump

R_q – Perrin quenching radius

$f(R)$ – distribution function over the distance R

Kinetic Parameters

k, k_1, k_2 – rate constants of monomolecular decay of the donor particles

k_q – rate constant of donor luminescence quenching

E_a, E'_a, E''_a – activation energies

$\tau_{1/2}$ – characteristic time of reaction

τ_s – characteristic time of secondary dark processes

τ_D – characteristic time of a diffusion jump

D – diffusion coefficient

ξ – mathematical parameter characterizing distribution over the values of the rate constant

β – tangent of the logarithmic decay kinetic curve

Energetic Parameters

J – reaction exothermicity
ΔG° – free energy change in the reaction of electron transfer
E_r – reorganization energy
V(R) – electronic coupling energy
Δ – energy gap
$E(S_1)$ – excitation energy of the donor singlet state S_1
$E(D/D^+)$ – oxidation potential of the donor
$E(B/B^-)$ – reduction potential of the acceptor
E(R) – energy of the formation of the oxidized donor and reduced acceptor from the initial donor and acceptor located at distance R
$\hbar\omega$ – energy of nuclear vibration

Concentrations

n(t) – concentration of the donor particles at time t
N – concentration of the acceptor particles
α – fraction of the donor molecules that have formed a complex with acceptor molecules

Spectroscopy and Luminescence

λ_{max} – maximum of the absorption band
λ_{exc} – excitation wavelength
σ – width of the charge transfer band
I – intensity of luminescence
J_{exc} – intensity of the exciting light
$\vec{E}$ – electric field vector of the light
EPR – Electron Paramagnetic Resonance
$\vec{H}$ – direction of the magnetic field of the EPR spectrometer

Chemical Compounds

A, B – electron acceptors
ADP – adenosine diphosphate
AH – ascorbic acid
AQ – anthraquinone
Az – azurin
ATP – adenosine triphosphate
BChl – bacteriochlorophyll
BPh – bacteriopheophytin
BQ – benzoquinone
CCP – cytochrome c peroxidase
Chl – chlorophyll
D – electron donor
DEA – *N,N′*-diethylaniline
D-L-A – bifunctional bridge molecule
DPA – diphenylamine
ES – enzyme-substrate complex

(Fe–S) – iron sulphur protein
Hb – hemoglobin
HMTI – hexametyltriindan
His – histidine
Mb – myoglobin
MgEtio-1 – magnesium Etio-1-porphyrin
MP – metalloporphyrin
MV^{2+} – *N*,*N*′-dimethyl-4,4′-bipyridinium
MTHF – 2-methyltetrahydrofuran
NADP – nicotinamide adenine dinucleotide phosphate ion
Nh – naphthalene
Nh-d_8 – deuterated naphthalene
NQ – naphthoquinone
P – porphyrin
P_i – phosphate ion
P700,
P890 – primary electron donors
PA – phthalic anhydride
Pl – plastocyanin
P-L-Q – porphyrin-quinone bridge molecule
PMA – pyromellitic dianhydride
PS1 – photosystem 1
PS2 – photosystem 2
Q – quinone
RC – reaction centre
St – stellacyanin
TMPD – *N*,*N*,*N*′,*N*′-tetramethyl-*p*-phenylenediamine
THF – tetrahydrofuran
TPP – tetraphenylporphin
bpy – 2,2′-bipyridine
cyt *c*, cyt *b* – cytochromes
e_{tr}^- – trapped electron
en – ethylendiamine
phen – phen-phenanthroline

11 References

1. Gamow G (1928) Z Phys 51: 204
2. Gurney RW, Condon EV (1928) Nature (London) 122: 439
3. Fowler RH, Nordheim L (1928) Proc R Soc London A 119: 173
4. Oppenheimer JR (1928) Phys Rev 1: 66
5. Bourgin DG (1929) Proc Natl Acad Sci USA 15: 357
6. Roginsky S, Rozenkewitsch L (1930) Z. Phys Chem Abf B 10: 47
7. Bell RP (1937) Proc Roy Soc London A 158: 128
8. Wigner E (1932) Zschr Phys Chem Abf B 19: 1903
9. De Vault D, Chance B (1966) Biophys J 6: 825

10. Zamaraev KI, Khairutdinov RF, Mikhailov AI, Goldanskii VI (1971) Dokl Akad Nauk SSSR 199: 640
11. Zamaraev KI, Khairutdinov RF, Miller JR (1980) Kinet Katal 21: 616
12. Khairutdinov RF, Zamaraev KI (1975) Izv Akad Nauk SSSR, Ser Khim 2782
13. Miller JR (1975) Science 189: 221
14. Khairutdinov RF, Zamaraev KI (1975) Dokl Akad Nauk SSSR 222: 654
15. Khairutdinov RF, Zhutkovskii RB, Zamaraev KI (1976) Khim Vys Energ 10: 38
16 a. Brikenstein EK (1984) Electron tunneling in reactions of porphyrins. Thesis, Institute of Chemical Physics, Moscow
16 b. Kong JLY, Loach PA (1980) J Heterocycl Chem 17: 734
16 c. Yocom KM, Shelton JB, Shelton JR, Schroeder WA, Worosila G, Isied SS, Bordignon E, Gray HB (1982) Proc Natl Acad Sci USA 79: 7052
17. Aristov YI, Volkov AI, Parmon VN, Zamaraev KI (1984) React Kinet Catal Lett 25: 627
18. Miller JR (1987) Nouv J Chim 11: 83
19 a. Grigorov LN, Chernavskii DS (1972) Biofizika 12: 202
19 b. Blumenfeld LA, Chernavskii DS (1973) J Theor Biol 39: 1
20 a. Kestner NR, Logan J, Jortner J (1974) J Phys Chem 78: 2148
20 b. Ulstrup J, Jortner J (1975) J Chem Phys 63: 4358
20 c. Kestner NR (1980) J Am Chem Soc 84: 1270
20 d. Dogonadze RR, Kuznetsov AM, Levich VG (1968) Electrochim Acta 13: 1025
20 e. Dogonadze RR, Ulstrup J, Kharkats YuI (1972) J Electroanal Chem 39: 47
20 f. Dogonadze RR (1971) Ber Bunsenges Phys Chem 75: 628
21. Alexandrov IV, Khairutdinov RF, Zamaraev KI (1978) Chem Phys 32: 123
22. Hopfield JJ (1974) Proc Natl Acad Sci USA 71: 3640
23. Jortner J (1976) J Chem Phys 64: 4860
24 a. Ivanov GK, Kozhushner MA (1978) Fiz Tverd Tela 20: 9
24 b. Ivanov GK, Kozhushner MA (1982) Khim Fiz 2: 1039
25. Kakitani T, Mataga N (1988) J Phys Chem 92: 5059
26 a. Parmon VN, Khairutdinov RF, Zamaraev KI (1974) Fiz Tverd Tela 16: 2572
26 b. Tachiya M, Mozumder A (1974) Chem Phys Letters 28: 87
26 c. Gailitis AA (1975) Uchen Zap Latv Univ 234: 42
26 d. Dainton FS, Pilling MJ, Rice SA (1975) J Chem Soc Faraday Trans 2 71: 1311
26 e. Tachiya M, Mozumder A (1975) Chem Phys Letters 34: 77
26 f. Doktorov AB, Khairutdinov RF, Zamaraev KI (1981) Chem Phys 61: 351
26 g. Khairutdinov RF, Berlin YuA, Zamaraev KI (1977) Izv Akad Nauk SSSR, ser khim 1977
26 h. Berlin YA (1975) Dokl Akad Nauk SSSR 223: 1387
26 j. Khairutdinov RF (1976) Khim Vys Energ 10: 556
27 a. Doktorov AB, Kotomin EA (1982) Phys Status Solidi B 114: 9
27 b. Kotomin EA, Doktorov AB (1982) Phys Status Solidi B 114: 287
27 c. Fabrikant I, Kotomin EA (1975) J Limin 9: 502
28. Doheny AJ, Albrecht AC (1977) Canad J Chem 55: 2065
29. Pilling MJ, Rice SA (1975) J Chem Soc Faraday Trans 2 71: 1563
30 a. Zhdanov VP (1985) Fiz Tverd Tela 27: 733
30 b. Zhdanov VP (1987) Khim Fiz 6: 278
31. Zamaraev KI, Khairutdinov RF, Zhdanov VP (1989) Electron tunneling in chemistry. Comprehensive chemical kinetics (Ed Compton RG) vol 30. Elsevier, Amsterdam
32. Inokuti M, Hirayama F (1965) J Chem Phys 43: 1978
33. Miller JR (1975) J Phys Chem 79: 1070
34. Zamaraev KI, Khairutdinov RF (1980) Sov Sci Rev B Chem Rev 2: 357
35. Leonhardt H, Weller A (1963) Ber Bunseges Phys Chem 67: 791
36. Okada T, Oohari H, Mataga N (1970) Bull Chem Soc Japan 43: 2750
37. Leiman VI (1972) Fiz Tverd Tela 14: 3650
38. Zhutkovskii RB, Khairutdinov RF, Zamaraev KI (1973) Khim Vys Energ 7: 558
39. Khairutdinov RF, Sadovskii NA, Parmon VN, Kuzmin MG, Zamaraev KI (1975) Dokl Akad Nauk SSSR 220: 888

40. Sahai R, Hofeldt RH, Lin SH (1971) Trans Faraday Soc 67: 1690
41. Parmon VN, Fiksel AI (1985) private communication
42. Fiksel AI, Parmon VN, Zamaraev KI (1982) Chem Phys 69: 135
43. Miller JR, Peeples JA, Schmidt MJ, Closs GL (1982) J Am Chem Soc 104: 6488
44. Marcus RA (1957) J Chem Phys 26: 867
45. Miller JR, Hartmann KW, Abrash S (1982) J Am Chem Soc 104: 4268
46. Murtag J, Thomas JK (1987) Chem Phys Lett 139: 437
47. Dominque RP, Fayer MD (1985) J Chem Phys 83: 2242
48. Korolev VV, Bazhin NM (1974) Khim Vys Energ 8: 506
49. Korolev VV, Bazhin NM (1978) Khim Vys Energ 12: 421
50. Bazhin NM, Korolev VV (1978) Khim Vys Energ 12: 425
51. Guarr T, McGuire M, Strauch S, McLendon G (1983) J Am Chem Soc 105: 616
52. Strauch S, McLendon G, McGuire M, Guarr T (1983) J Phys Chem 87: 3579
53. Milosavljevic BH, Thomas JK (1985) J Phys Chem 89: 1830
54. Milosavljevic BH, Thomas JK (1986) J Am Chem Soc 108: 2513
55. Guarr T, McGuire ME, McLendon G (1985) J Am Chem Soc 107: 5104
56. Musser RD, Nocera DG (1988) J Am Chem Soc 110: 2764
57. Rau H, Frank R, Greiner G (1986) J Phys Chem 90: 2476
58. Stern O, Volmer M (1919) Phys Z 20: 183
59 a. Ershov BG, Pikaev AK (1967) Zh Fiz Khim 41: 2573
59 b. Zimbrick J, Kevan L (1967) J Am Chem Soc 89: 2483
60. Khairutdinov RF (1986) Dokl Akad Nauk SSSR 228: 143
61. Krasnovskii AA (1948) Dokl Akad Nauk SSSR 60: 421
62. Dolphin D (ed) (1979) The porphyrins. Academic Press, New York
63. Carapellucci PA, Mauzerall D (1975) Ann NY Acad Sci 244: 214
64. Debye P (1942) Trans Electrochem Soc 82: 265
65. Kholmogorov VE, Bobrovskii AP (1973) in: Rubin AB, Samuilov VD (eds) Problems of Biophotochemistry, Nauka, Moscow p 92
66. Strekova LN, Brickenstein EKh, Asanov AN, Sadovskii NA, Khairutdinov RF (1981) Izv Akad Nauk SSSR, ser khim 1127
67 a. Shelimov BN, Vinogradova VT, Maltsev VI, Fok NV (1967) Dokl Akad Nauk SSSR 172: 655
67 b. Suwalski JP (1981) Radiat Phys Chem 17: 397
68 a. Yusupov RG, Khairutdinov RF (1987) Dokl Akad Nauk SSSR 295: 665
68 b. Yusupov RG, Khairutdinov RF (1988) Khim Fiz 7: 1161
69. Brickenstein EKh, Strekova LN, Asanov AN, Khairutdinov RF (1982) Khim Vys Energ 16: 54
70. Brickenstein EKh, Ivanov GK, Kojushner MA, Khairutdinov RF (1984) Chem Phys 91: 133
71 a. Khairutdinov RF, Lazarev GG, Brickenstein EKh, Lebedev YaS (1987) Khim Fiz 6: 183
71 b. Brickenstein EKh, Yusupov RG, Khairutdinov RF (1987) in: Proc. 13th International Conference on Photochemistry. Budapest, p 367
72. Khairutdinov RF, Asanov AN, Brickenstein EKh, Strekova LN (1985) Khim Fiz 4: 1210
73. Marcus RA (1987) Chem Phys Letters 133: 471
74. Marcus RA (1988) Chem Phys Letters 146: 13
75. Bixon M, Jortner J, Michel-Beyerle ME, Ogrodnik A, Lersch N (1987) Chem Phys Letters 140: 626
76. Plato M, Möbius K, Michel-Beyerle ME, Bixon M, Jortner J (1988) J Am Chem Soc 110: 7279
77. Lockhart DJ, Goldstein RF, Boxer SG (1988) J Chem Phys 89: 1408
78. Michel-Beyerle ME, Plato M, Deisenhofer J, Michel H, Bixon M, Jortner J (1988) Biochim Biophys Acta 932: 52
79. Creighton S, Hwang J-K, Warshel A, Parson WW, Norris J (1988) Biochemistry 27: 774
80. Ogrodnik A, Remy-Richter N, Michel-Beyerle ME, Flick R (1987) Chem Phys Letters 135: 576

81. Warshel A, Creighton S, Parson WW (1988) J Phys Chem 92: 2696
82. Petrov EG (1979) Intern J Quant Chem 16: 133
83. Kharkianen VA (1983) Khim Fiz 2: 192
84. Ivanov GK, Kozhushner MA (1989) Khim Fiz 8: 500
85. Brickenstein EKh, Kozhushner MA, Penkov DN, Strekova LN, Khairutdinov RF (1989) Chem Phys 135: 209
86. Passman P, Verhoeven JW, De Boer Th (1978) Chem Phys Letters 59: 381
87. Creutz C, Kroger P, Matsubara T, Netzel TL, Sutin NJ (1979) J Am Chem Soc 101: 5442
88. Mazur S, Dixit VM, Gerson F (1980) J Am Chem Soc 102: 5343
89. Khairutdinov RF (1980) Electron tunneling in chemistry. Thesis, Institute of Chemical Physics of the USSR Academy of Sciences, Moscow
90. Fujita I, Fajer J, Chang C-K, Wang CB, Bergkamp MA, Netzel TL (1982) J Phys Chem 86: 3754
91. Calcaterra LT, Closs GL, Miller JR (1983) J Am Chem Soc 105: 670
92. Passman P, Koper NW, Verhoeven JW (1983) Recl.: J R Neth Chem Soc 102: 55
93. Huddleston RK, Miller JR (1983) J Phys Chem 79: 5337
94. Roach KJ, Weedon AC, Bolton JR, Connolly JS (1983) J Am Chem Soc 105: 7224
95. Isied SS (1984) Progr Inorg Chem 32: 443
96. Beratan DN (1986) J Am Chem Soc 108: 4321
97. Heitele H, Michel-Beyerle ME, Finch P (1987) Chem Phys Letters 134: 273
98. Isied SS, Vassilian A (1984) J Am Chem Soc 106: 1732
99. Gonzales MC, McIntosh AR, Bolton JR, Weedon AC (1984) J Chem Soc, Chem Commun 1183
100. Mataga N, Karen A, Okada T, Nishitani S, Kurata N, Sakata Y, Misumi S (1984) J Phys Chem 88: 5138
101. Joran AD, Leland BA, Geller GG, Hopfield JJ, Dervan PB (1984) J Am Chem Soc 106: 6090
102. Hush NS, Paddon-Rowe MN, Cotsans E, Overing H, Verhoeven JW (1985) Chem Phys Lett 117: 8
103. Bolton JR, Ho T-F, Liauw S, Siemiarczuk A, Wan CSK, Weedon AC (1985) J Chem Soc Chem Commun 559
104. Heitile H, Michel-Beyerle ME (1985) in: Michel-Beyerle ME (ed) Antennas and reaction centers of photosynthetic bacteria, Springer-Verlag, West Berlin, p 250
105. Wasielewski MR, Niemczyk MP, Svec WA, Pewitt EB (1985) J Am Chem Soc 107: 5562
106. Schmidt JA, Siemiarczuk A, Weedon AC, Bolton JR (1985) J Am Chem Soc 107: 6112
107. Miller JR, Calcaterra LT, Closs GL (1984) J Am Chem Soc 106: 3047
108. Warman JM, De Haas MP, Paddon-Row MN, Cotsaris E, Hush NS, Oevering H, Verhoeven JW (1986) Nature (London) 320: 615
109. Warman JM, De Haas MP, Oevering H, Verhoeven JW, Paddon-Row MN, Oliver AM, Hush NS (1986) Chem Phys Letters 128: 95
110. De Haas MP, Warman JM (1982) Chem Phys 73: 53
111. Verhoeven JW, Paddon-Row MN, Hush NS, Oevering H, Heppener M (1986) Pure Appl Chem 58: 1285
112. Oevering H, Paddon-Row MN, Heppener M, Oliver AM, Cotsaris E, Verhoeven JW, Hush NS (1987) J Am Chem Soc 109: 3258
113. Paddon-Row MN, Oliver AM, Warman JM, Smit KJ, De Haas MP, Oevering H, Verhoeven JW (1988) J Phys Chem 92: 6598
114. Penfield KW, Miller JR, Paddon-Row MN, Cotsaris E, Oliver AM, Hush NS (1987) J Am Chem Soc 109: 5061
115. Schroff LG, Zsom RLJ, van der Weerdt AJA, Schrier PI, Nibberin NMM, Verhoeven JW, de Boer ThJ (1976) J Roy Neth Chem Soc 95: 89
116. Schroff LG, van der Weerdt AJA, Staalman DJH, Verhoeven JM, de Boer ThJ (1973) Tetrahed Lett 1649
117. Verhoeven JW, Dirkx IP, De Boer ThJ (1969) Tetrahed 25: 4037
118. Okada T, Fujita T, Kubota M, Masaki S, Mataga N, Ide R, Sakata Y, Misumi S (1972) Chem Phys Letters 14: 563

119. Pasman P, Mes GF, Koper NW, Verhoeven JW (1985) J Am Chem Soc 107: 5839
120. Kong JLY, Loach PA (1978) in: Dutton PL, Leigh JS, Scarpa A (eds) Frontiers of biological energetics (vol 1) Academic Press, New York, p 77
121. Kong JLY, Loach PA (1980) J Heterocycl Chem 17: 734
122. Tabushi I, Koga N, Yanagita M (1979) Tetrahedron Letters 20: 257
123. Dalton J, Milgrom LR (1979) J Chem Soc Chem Commun 609
124. Kong JLY, Spears KG, Loach PA (1982) Photochem Photobiol 35: 545
125. Nishitani S, Kurata N, Sakata Y, Misumi SM, Migita M, Okada T, Mataga N (1981) Tetrahedron Letters 22: 2099
126. Migita M, Okada T, Mataga N, Nishitani S, Kurata N, Sokota S, Misumi S (1981) Chem Phys Letters 84: 263
127. Quinlan KP (1968) J Phys Chem 72: 1797
128. Shakhverdov PA (1968) Dokl Akad Nauk SSSR 174: 1141
129. Ho T-F, McIntosh AR, Bolton JR (1980) Nature (London) 286: 254
130. Kong JLY, Loach PA (1981) in: Connolly JS (ed) Photochemical conversion and storage of solar energy, Academic Press, New York, p 350
131. McIntosh AR, Siemiarczuk A, Bolton JR, Stillman MJ, Ho T-F, Weedon AC, Connolly JS (1983) J Am Chem Soc 105: 7215
132. Siemiarczuk A, McIntosh AR, Ho T-F, Stillman MJ, Roach KJ, Weedon AC, Bolton JR, Connolly JS (1983) J Am Chem Soc 105: 7224
133. Chandross EA, Thomas HT (1971) Chem Phys Letters 6: 393
134. Migita M, Ada T, Mataga N, Nakashina N, Yoshihara K, Sakata Y, Misumi S (1980) Chem Phys Letters 72: 229
135. Mataga N, Okada T, Masuhara H, Nakashima N, Sakata Y, Misumi S (1976) J Lumin 12/13: 159
136. Mataga N, Migita M, Nishimura T (1978) J Mol Struct 471: 109
137. Mishitani S, Kurata N, Sakata O, Misumi S, Karen A, Okada T, Mataga N (1983) J Am Chem Soc 105: 7771
138. Vorotinzev MA, Itskovich EM (1980) J Theor Biol 86: 223
139. Ganesh KN, Sanders JKM (1982) J Chem Soc Perkin Trans 1 1611
140. Ganesh KN, Sanders JKM, Waterton JC (1982) J Chem Soc Perkin Trans 1 1617
141. Ganesh KN, Sanders JKM (1980) J Chem Soc Chem Commun 1129
142. Irvine MP, Harrison RJ, Beddard GS, Leighton P, Sanders JKM (1986) Chem Phys 104: 315
143. Lindsey JS, Mauzerall DC (1982) J Am Chem Soc 104: 4498
144. Lindsey JS, Mauzerall DC (1983) J Am Chem Soc 105: 6528
145. Lindsey JS, Delaney JK, Mauzerall DC, Linschitz H (1988) J Am Chem Soc 110: 3610
146. Weiser J, Staab HA (1984) Angew Chem Int Ed Engl 23: 623
147. Weiser J, Staab HA (1985) Tetrahed Letters 26: 6059
148. Wasielewski MR, Niemczyk MP (1984) J Am Chem Soc 106: 5043
149. Wasielewski MR, Niemczyk MP, Svec WA, Pewitt EB (1985) J Am Chem Soc 107: 1080
150. Leland BA, Joran AD, Felker PM, Hopfield JJ, Zewail AH, Dervan PB (1985) J Phys Chem 89: 5571
151. Wasielewski MR, Niemczyk MP (1986) in: Gouterman M, Rentzepis PM, Straub KD (eds) Porphyrins-excited states and dynamics: ACS symposium series, No 321, American Chemical Society Washington, DC, p 154
152. Gonzaler MC, McIntosh AR, Bolton JR, Weedon AC (1984) J Chem Soc Chem Commun 1138
153. Harriman A, Porter G, Wilowska A (1984) J Chem Soc Faraday Trans 2 80: 193
154. Milgrom LR (1983) J Chem Soc Perkin Trans 1 2325
155. Leighton P, Sanders JKM (1984) J Chem Soc Chem Commun 856
156. Harriman A (1984) Inorg Chim Acta 88: 213
157. Blondeel G, De Keukeleire D, Harriman A, Milgrom LR (1983) Chem Phys Letters 118: 77
158. Kanda Y, Sato H, Okada T, Mataga N (1986) Chem Phys Letters 129: 306

159. Harrison RJ, Beddard GS, Cowan JA, Sanders JKM (1986) In: Fleming GR, Siegman AE (eds) Springer series in chemical physics, vol 46. Springer, Berlin Heidelberg New York, p 322
160. Harrison RJ, Pearce B, Beddard GS, Cowan JA, Sanders JKM (1987) Chem Phys 116: 429
161. Cowan JA, Sanders JKM (1985) J Chem Soc Perkin Trans 1 2435
162. Cowan JA, Sanders JKM, Beddard GS, Harrison RJ (1987) J Chem Soc Chem Commun 55
163. Moore TA, Gust D, Mathis P, Mialocq J-C, Chachaty C, Bensasson RV, Land EJ, Doiri D, Liddell PA, Lehman WR, Nemeth GA, Moore AL (1984) Nature (London) 307: 630
164. Gust D, Moore TA (1985) J Photochem 29: 173
165. Gust D, Moore TA (1987) In: Balzani V (ed) Supramolecular photochemistry. Reidel, Dordrecht, The Netherlands, NATO ASI Series C: Mathematical and Physical Sciences, vol 214, p 267
166. Gust D, Moore TA, Liddell PA, Nemeth GA, Makings LR, Moore AL, Barrett D, Pessiki PJ, Bensasson RV, Rougee M, Chachaty C, De Schryver FC, Van der Auweraer M, Holzwarth AR, Connolly JS (1987) J Am Chem Soc 109: 846
167. Seta P, Bienvenue E, Moore A, Mathis P, Bensasson RV, Liddel P, Pessiki PJ, Joy A, Moore TA, Gust D (1985) Nature (London) 316: 653
168. Jortner J (1980) J Am Chem Soc 102: 6676
169. Liddell PA, Barrett D, Makings LK, Pessiki PJ, Gust D, Moore TA (1986) J Am Chem Soc 108: 5350
170. Clayton RK (1980) Photosynthesis: physical mechanisms and chemical patterns. Cambridge University Press, Cambridge
171. Khairutdinov RF, Strekova LN, Borovkov VV, Filippovich EI, Evstigneeva RP (1987) Proceedings of the 13th international conference on photochemistry, 1987, Budapest, p 136
172. Khairutdinov RF, Strekova LN, Borovkov VV, Filippovich EI, Evstigneeva RP (1988) Dokl Akad Nauk SSSR 298: 487
173. Zamaraev KI, Khairutdinov RF (1978) Usp Khim 47: 992
174. Seefeld KP, Möbius D, Kuhn H (1977) Helv Chim Acta 60: 2608
175. Zamaraev KI, Parmon VN (1980) Catal Rev Sci Eng 22: 261
176. Grätzel M (1981) Acc Chem Res 14: 376
177. Bagdasaryan KhS (1982) Khim Fiz 1: 391
178. Möbius D (1978) Ber Buns Phys Chem 82: 848
179. Cunningham J, Polymeropoulos EE, Möbius D, Baer F (1980) In: Faissard JP, Resing HA (eds) Magnetic resonance in colloid and interface science. Reidel, Dordrecht, p 603
180. Kuhn H (1979) Pure Appl Chem 51: 341
181. Möbius D (1981) Acc Chem Res 14: 63
182. Polymeropoulos EE, Möbius D, Kuhn H (1980) Thin solid films 68: 173
183. Joshi NB, Lopez JR, Tien HT, Wang C-B, Lin Q-Y (1982) J Photochem 20: 139
184. Wang C-B, Tien HT, Lopez JR, Lin Q-Y, Joshi NB, Hu Q-Y (1982) Photobiochem Photobiophys 4: 177
185. Moore TA, Gust D, Moore AL, Bensasson RV, Seta P, Bienvenue E (1987) in Balzani V (ed) Supramolecular photochemistry. Reidel, Dordrecht, The Netherlands, NATO ASI Series C: Mathematical and physical sciences, vol 214, p 283
186 a. Janzen AF, Bolton JR, Stillman MJ (1979) J Am Chem Soc 101: 6337
186 b. Jansen AF, Bolton JR (1979) J Am Chem Soc 101: 6342
187. Fujihira M, Kuboto T, Osa T (1981) J Electroanal Chem 119: 379
188. Fromherz P (1977) Chem Phys Letters 77: 460
189 a. Turro NJ, Grätzel M, Braun AM (1980) Angew Chem Int Ed Eng 19: 675
189 b. Alkaitis SA, Grätzel M (1976) J Am Chem Soc 98: 3549
190. Maestri M, Infelta PP, Grätzel M (1978) J Chem Phys 69: 1522
191 a. Zamaraev KI, Lymar SV, Khramov MI, Parmon VN (1988) Pure Appl Chem 60: 1039
191 b. Tunuli MS, Fendler JH (1981) J Am Chem Soc 103: 2507

192. Tsuchida E, Kaneko M, Nishide H, Hoshino M (1986) J Chem Phys 90: 2283
193. Clayton RK, Sistrom WR (1978) The photosynthetic bacteria. Plenum, New York
194. Deisenhofer J, Epp O, Miki R, Michel H (1985) Nature 318: 618
195. Michel H, Epp O, Deisenhofer J (1986) The EMBO Journal 5: 2445
196. Chang CH, Tiede D, Tang J, Smith U, Norris J, Schiffer M (1986) FEBS Letters 205: 82
197. Allen JP, Feher G, Yeates TO, Komiya H, Rees DC (1987) Proc Natl Acad Sci USA 84: 5730
198. Michel-Beyerle ME (ed) (1985) Antennas and reaction centers of photosynthetic bacteria. Springer, Berlin Heidelberg New York
199. Austin R, Buhks E, Chance B, DeVault D, Dutton PL, Frauenfelder H, Goldanskii VI (eds) (1987) Protein structure: molecular and electronic reactivity. Springer, Berlin Heidelberg New York
200. Parson WW, Ke B (1982) In: Govindjee (ed) Photosynthesis: energy conversion in plants and bacteria. Academic Press, New York, p 331
201. Okamura MY, Feher G, Nelson N (1982) In: Govindjee (ed) Photosynthesis: energy conversion in plants and bacteria. Academic Press, New York, p 195
202. Budil DE, Gast P, Chang C-H, Schiffer M, Norris JR (1987) Ann Rev Phys Chem 38: 561
203. Govindjee (ed) (1982) Photosynthesis. Academic Press, New York
204. Change B, Nishimura M (1960) Proc Natl Acad Sci USA 46: 19
205. Arnold W, Clayton RK (1960) Proc Natl Acad Sci USA 46: 763
206. Knaff DB, Arnon DI (1969) Proc Natl Acad Sci USA 64: 115
207. Floyd RA, Chance B, DeVault D (1971) Biochim Biophys Acta 226: 103
208. Cost K, Bolton JR, Frenkel AW (1969) Photochem Photobiol 10: 251
209. Warden JT, Mohanty P, Bolton JR (1974) Biochim Biophys Res Commun 59: 872
210. Visser JWM, Rijgersberg KP, Amesz J (1974) Biochim Biophys Acta 368: 236
211. Ke B, Sugahara K, Shaw ER, Hansen ER, Hamilton WD, Beinert H (1974) Biochim Biophys Acta 368: 401
212. Ke B, Demeter S, Zamaraev KI, Khairutdinov RF (1979) Biochim Biophys Acta 545: 265
213. Khairutdinov RF, Zamaraev KI (1976) Khim Vys Energ 10: 195
214. Ke B, Demeter S, Zamaraev KI, Khairutdinov RF (1979) In: Chance B, DeVault D, Frauenfelder H, Marcus RA, Schrieffer JR, Sutin N (eds) Tunneling in biological systems. Academic Press, New York, p 371
215. Strekova LN, Khairutdinov RF, Goldfeld MG, Mikoyan AD, Timofeev VP, Chetverikov AG (1988) Biofizika 33: 495
216. Goldfeld MG, Strekova LN, Khairutdinov RF (1987) Proc 13th International Conference on Photochemistry, Budapest
217. Chetverikov AG, Goldfeld MG (1985) Biofizika 30: 1022
218. Gast P, Norris JR (1984) FEBS Letters 177: 277
219. Allen JP, Theiler R, Feher G (1985) Biophys J 47: 4a
220. Deisenhofer J, Epp O, Miki K, Huber R, Michel H (1984) J Mol Biol 180: 385
221. Deisenhofer J, Michel H, Huber R (1985) Trends Biochem Sci 10: 243
222. Michel-Beyerle ME (ed) (1985) Antenna and reaction centers of photosynthetic bacteria. Springer, Berlin Heidelberg New York (Springer Ser Chem Phys, vol 42)
223. Kirmaier C, Holton D, Parson EE (1983) Biochim Biophys Acta 725: 190
224. Vermeglio A, Clayton RK (1976) Biochim Biophys Acta 449: 500
225. Okamura MY, Isaacson RA, Feher G (1979) Biochim Biophys Acta 546: 394
226. Butler WF, Calvo R, Fredkin DR, Isaacson RA, Okamura MY, Feher G (1984) Biophys J 45: 947
227. Packham NK, Dutton PL, Mueller P (1982) Biophys J 37: 465
228. Trissel HW (1983) Proc Natl Acad Sci USA 80: 7173
229. Gopher A, Blatz Y, Schonfeld M, Okamura MY, Feher G, Montal M (1985) Biophys J 48: 311
230. Tiede DM (1985) Biochim Biophys Acta 811: 357
231. Blasie JK, Machence JM, Tavormina A, Dutton PL, Stamatoff J, Eisenberger P, Brown G (1983) Biochim Biophys Acta 723: 350

232. Blankenship RE, Parson WW (1979) In: Barber J (ed) Photosynthesis in relation to model systems. Elsevier/North-Holland, New York, p 71
233. Chance B, DeVault, Frauenfelder H, Marcus RA, Schreffer JR, Sutin N (eds) Tunneling in biological systems. Academic Press, New York
234. Jortner J (1980) Biochim Biophys Acta 594: 193
235. De Vault D (1980) Q Rev Biophys 13: 387
236. De Vault D (1984) Quantum mechanical tunneling in biological systems. Cambridge University Press
237. Marcus RA, Sutin N (1985) Biochim Biophys Acta 811: 265
238. Clayton RK, Yau HF (1972) Biophys J 12: 867
239. McElroy JD, Mauzerall DC, Feher G (1974) Biochim Biophys Acta 333: 261
240. Hsi ESP, Bolton JR (1974) Biochim Biophys Acta 347: 126
241. Schenck CC, Parson WW, Holton D, Windsor MW, Sarai A (1981) Biophys J 36: 479
242. Marr T, Vadeboncoeur C, Gingras G (1983) Biochim Biophys Acta 724: 317
243. Kirmaier C, Holton D, Parson WW (1985) Biochim Biophys Acta 810: 33
244. Parson WW, Clayton RK, Cogdell RJ (1975) Biochim Biophys Acta 387: 265
245. Schenck CC, Blankenship RE, Parson WW (1982) Biochim Biophys Acta 680: 44
246. Gunner MR, Tiede DM, Prince RC, Dutton PL (1982) In: Trumpower BL (ed) Function of quinones in energy conserving systems. Academic Press, New York, p 271
247. Chidsey CED, Takiff L, Goldstein R, Boxer SG (1985) Proc Natl Acad Sci USA 82: 6850
248. Boxer SG (1985) In: Michel-Beyerle ME (ed) Antennas and reaction centers of photosynthetic bacteria. Springer, Berlin Heidelberg New York, p 306
249. Kuznetsov AM, Sondergart NC, Ulstrup J (1978) Chem Phys 29: 383
250. Kuznetsov AM, Ulstrup J (1981) Biochim Biophys Acta 636: 498
251. Kakatani T, Kakatani H (1981) Biochim Biophys Acta 635: 498
252. Warshel A (1980) Proc Natl Acad Sci USA 77: 3105
253. Gunner MR, Liang Y, Nagus DK, Hochstrasser RM, Dutton PL (1982) Biophys J 37: 226a
254. Pocinki AG, Blankenship RE (1982) FEBS Letters 147: 115
255. Gopher A, Blatt Y, Schonfeld M, Okamura MY, Feher G, Montal M (1985) Biophys J 48: 311
256. Gunner MR, Braun BS, Bruce JM, Dutton PL (1985) In: Michel-Beyerle ME (ed) Antennas and reaction centers of photosynthetic bacteria. Springer, Berlin Heidelberg New York, p 298
257. Kleinfeld D, Okamura MY, Feher G (1985) Biophys J 48: 849
258. Gunner MR, Robertson DE, Dutton PL (1986) J Phys Chem 90: 3783
259. Gunner MR, Dutton PL (1989) J Am Chem Soc 111: 3400
260. Petrov EG (1984) Physics of charge transfer in biological systems. Naukova Dumka, Kiev
261. Petrov EG (1986) Khim Fiz 5: 1193
262. Gray HB (1986) Chem Soc Rev 15: 17
263. McLendon G, Guarr T, McGuire M, Simolo K, Straugh S, Taylor K (1985) Coord Chem Rev 64: 113
264. Peterson-Kennedy SE, McGourthy JL, Ho PS, Sutoris CJ, Liang N, Zemel H, Blough NV, Margoliash E, Hoffman BH (1985) Coord Chem Rev 64: 125
265. McGorthy JL, Blough NV, Hoffman BM (1983) J Am Chem Soc 105: 4470
266. Peterson-Kennedy SE, McGorthy JL, Hoffman BM (1984) J Am Chem Soc 106: 5010
267. McLendon GL, Winkler JR, Nocera DG, Nauk MR, Nauk AG, Gray HB (1985) J Am Chem Soc 107: 739
268. Simolo KP, McLendon GL, Nauk MR, Nauk AG (1984) J Am Chem Soc 106: 5012
269. Hultquist DE, Passon PG (1971) Nature (London) 229: 252
270. Peterson-Kennedy SE, McGourty JL, Kalweit JA, Hoffman BM (1986) J Am Chem Soc 108: 1739
271. Liang N, Kang CH, Ho PS, Margoliash E, Hoffman BM (1986) J Am Chem Soc 108: 4665
272. Mayo SSL, Ekis WR, Crutchley RJ, Gray HB (1986) Science 233: 948

273. Crutchley RJ, Ellis WR, Gray HB (1986) J Am Chem Soc 108: 5002
274. Lieber CM, Karas JK, Gray HB (1987) J Am Chem Soc 109: 3778
275. McLendon G, Pardue K, Bak P (1987) J Am Chem Soc 109: 7540
276. Liang N, Pielak GJ, Nauk AG, Smith M, Hoffman BM (1987) Proc Natl Acad Sci USA 84: 1249
277. McGourty JM, Peterson-Kennedy SE, Ruo WY, Hoffman BM (1987) Biochemistry 26: 8302
278. Elias H, Chou MH, Winkler JR (1988) J Am Chem Soc 110: 429
279. Axup AW, Albin M, Mayo SL, Crutchley RJ, Gray HB (1988) J Am Chem Soc 110: 435
280. Karas JL, Lieber CM, Gray HB (1988) J Am Chem Soc 110: 599
281. McLendon G (1988) Acc Chem Res 21: 160
282. Conklin KT, McLendon G (1988) J Am Chem Soc 110: 3345
283. Jackman MP, McGinnis J, Powls R, Salmon GA, Sykes AG (1988) J Am Chem Soc 110: 5880
284. Osvath P, Salmon GA, Sykes AG (1988) J Am Chem Soc 110: 7114
285. Natan MJ, Hoffman BM (1989) J Am Chem Soc 111: 6468
286. Sisido M, Tanaka R, Inai Y, Imanishi Y (1989) J Am Chem Soc 111: 6790
287. Liang N, Mauk AG, Pielak GJ, Johnson JA, Smith M, Hoffman BM (1988) Science, 240: 311
288. Shosheva ACh, Christova PK, Atanasov BP (1988) Biochim Biophys Acta 957: 202–206
289. English AM, Lum VR, De Laive PJ, Gray HB (1982) J Am Chem Soc 104: 870
290. Brunschwig BS, De Laive PJ, Goldberg M, Gray HB, Mayo SL, Sutin N (1985) Inorg Chem 24: 3743
291. Handford PM, Hill HAO, Lee RWK, Henderson RA, Sykes AG (1980) J Inorg Biochem 13: 83
292. Cookson DJ, Hayes MT, Wright PE (1980) Nature (London) 283: 683
293. Cookson DJ, Hayes MT, Wright PE (1980) Biochim Biophys Acta 591: 162
294. Segal MG, Sykes AG (1977) J Chem Soc Chem Commun 764
295. Segal MG, Sykes AG (1978) J Am Chem Soc 100: 4585
296. Dixon DW, Barbush M, Shirazi A (1985) Inorg Chem 24: 1081
297. Heiler D, McLendon G, Rogalskyi P (1987) J Am Chem Soc 109: 604
298. Potasek MJ (1978) Science 201: 53
299. Potasek MJ, Hopfield JJ (1977) Proc Natl Acad Sci USA 74: 3817
300. Chang AM, Austin RA (1982) J Chem Phys 77: 5272
301. Goldstein RF, Bearien A (1984) Proc Natl Acad Sci USA 81: 135
302. Sakata Y, Nakashima S, Goto Y, Tatemitsu H, Misumi S, Asahi T, Hagihara M, Nishikawa S, Okada T, Mataga N (1989) J Am Chem Soc 111: 8979
303. Lawson JM, Craig DC, Paddon-Row MN, Kroon J, Verhoeven JW (1989) Chem Phys Letters 164: 120
304. Dewidenko AA (1989) Fiz Mnogochastichnykh Sist 16: 98
305. Noda S, Hosono H, Okura I, Yamamoto Y, Inoue Y (1990) J Mol Catal 59: L21
306. Noda S, Hosono H, Okura I, Yamamoto Y, Inoue Y (1990) J Photochem Photobiol, A: Chemistry 53: 423
307. Noda S, Hosono H, Okura I, Yamamoto Y, Inoue Y (1990) J Chem Soc Faraday Trans 186: 811
308. D'Souza F, Krishnan V (1989) Photochem Photobiol 51: 285
309. Borovkov VV, Evstigneeva RP, Kamalov VF, Struganova IA, Toleutaev BN (1990) Opt Spektrosk 68: 81
310. Batova EE, Levin PP, Shafirovich VYa (1990) Izv Akad Nauk SSSR, Ser Khim 1287
311. Hasharoni K, Levanon H, Tang J, Bowman MK, Norris JR, Gust D, Moore TA, Moore AL (1989) J Am Chem Soc 112: 6477
312. Harriman A, Novak AK (1990) Pure Appl Chem 62: 1107
313. Autolovich M, Keyte P, Oliver AM, Paddon-Row MN, Kroon J, Verhoeven JW, Jonker SA, Warman JM (1991) J Phys Chem 95: 1933
314. Warman JM, Smit KJ, de Haas MP, Jonker SA, Paddon-Row MN, Oliver AM, Kroon J, Oevering H, Verhoeven JW (1991) J Phys Chem 95: 1979

315. Cabana LA, Schanze KS (1989) Adv Chem Ser 226: 101
316. Perkins TA, Hauser BT, Eyler JR, Schanze KS (1990) J Phys Chem 94: 8745
317. Delaney JK, Mauzerall DC, Lindsey JS (1990) J Am Chem Soc 112: 957
318. Zang LH, Maki AH (1990) J Am Chem Soc 112: 4346
319. Kuila D, Baxter WW, Natan MJ, Hoffman BM (1991) J Phys Chem 95: 1
320. Farver O, Pecht I (1990) Inorg Chem 29: 4855.
321. Osuka A, Maruyama K, Mataga N, Asahi T, Yamazaki I, Tamai N (1990) J Am Chem Soc 112: 4958
322. De Felippis MR, Faraggi M, Klapper M (1990) J Am Chem Soc 112: 5640
323. Bobrovski K, Wierzchowski KL, Holeman J, Ciurak M (1990) Int J Radiat Biol 57: 919
324. Vassilian A, Wishart JF, van Hemelryck B, Schwarz H, Isied SS (1990) J Am Chem Soc 112: 7278
325. Reimers JR, Hush NS (1990) Chem Phys 146: 89
326. Christensen HEM, Conrad LS, Mikkelsen KV, Nielsen MK, Ulstrup J (1990) Inorg Chem 29: 2808
327. Beratan DN, Onuchic JN, Betts JN, Bowler BE, Gray HB (1990) J Am Chem Soc 112: 7915
328. Onuchic JN, Beratan DN (1990) J Chem Phys 92: 722
329. Beratan DN, Onuchic JN (1989) Photosynth Res 22: 173
330. Siddarth P, Marcus RA (1990) J Phys Chem 94: 8430
331. Ratner MA (1990) J Phys Chem 94: 4877
332. Ivanov GK, Kozhushner MA (1989) Khim Fiz 8: 1587
333. Reimers JR, Hush NS (1990) Inorg Chem 29: 4510
334. Da Gama, Arnobio A (1990) J Theor Biol 142: 251
335. Sutin N, Brunschwig BS (1989) Adv Chem Ser 226: 65
336. Kosloff R, Ratner MA (1990) Isr J Chem 30: 45
337. Hu Y, Mukamel S (1989) J Chem Phys 91: 6973

The Marcus Inverted Region

Paul Suppan

Institute of Physical Chemistry, Fribourg University, CH-1700 Fribourg, Switzerland

Table of Contents

Topics in Current Chemistry, Vol. 163

The "Marcus Inverted Region" (MIR) is that part of the function of rate constant versus free energy where a chemical reaction becomes slower as it becomes more exothermic. It has been observed in many thermal electron transfer processes such as neutralization of ion pairs, but not for photoinduced charge separation between neutral molecules. The reasons for this discrepancy have been the object of much controversy in recent years, and the present article gives a critical summary of the theoretical basis of the MIR as well as of the explanations proposed for its absence in photoinduced electron transfer. The role of the solvent receives special attention, notably in view of the possible effects of dielectric saturation in the field of ions. The relationship between the MIR and the theories of radiationless transitions is a topic of current development, although in the Marcus-Hush Model electron transfer is treated as a thermally activated process.

1 Introduction and Overview

1.1 Energetics and Kinetics of Electron Transfer Reactions

There are two major concepts involved in the physico-chemical description of a chemical reaction: the energetics, which determines the "feasibility" of the reaction, and the kinetics which determines its rate. In general these two concepts are independent and the rate of a chemical reaction can be varied according to the mechanism (e.g. catalysis); but within certain assumptions there is a mathematical relationship between the rate constant and the reaction free energy difference. These relationships are either linear (linear free energy relationship, LFE) or quadratic (QFE), the latter being often referred to as the "Marcus model" – a description which should not hide the important contributions of other workers in this field [1].

This opening remark concerning the independence "in principle" of the energetics and kinetics of chemical reactions will be seen to be important in the present context, and the conditions implied in the applicability of LFE or QFE are all too often overlooked. Nor should it be believed that such relationships break down only in the special cases of catalysed reactions; the Woodward-Hoffmann rules which play a considerable role in mechanistic organic chemistry provide a good illustration of the limitations of predictions of rate constants from energetics alone. In a reaction which can lead to two stereoisomers (for instance an electrocyclic ring closure which can proceed through conrotatory or disrotatory motions, Fig. 1) the two pathways have exactly the same free energy difference [2].

Although we shall restrict our attention to electron transfer processes only, the detailed mechanisms of such processes will have to be considered when discussing the applicability of the Marcus QFE relationships. Three types of electron transfer (e.t.) reactions can be distinguished, as illustrated in Scheme 1:

a. The charge separation in which two neutral reactant molecules form an ion pair;
b. the charge shift between an ion and a neutral molecule; and
c. the charge recombination of two ions to form two neutral molecules.

Charge separation	$M + N \rightarrow M^{\cdot\pm} + N^{\cdot\mp}$	thermal
	$M^{*} + N \rightarrow M^{\cdot\pm} + N^{\cdot\mp}$	photoinduced
	$M^{*} + N^{*} \rightarrow M^{\cdot\pm} + N^{\cdot\mp}$	biphotonic
Charge recombination	$M^{\cdot\pm} + N^{\cdot\mp} \rightarrow M + N$ or $M^{*} + N$	thermal
Charge shift	$M^{\pm} + N \rightarrow M + N^{\pm}$	thermal
	$M^{\pm} + N^{*} \rightarrow M + N^{\pm}$	photoinduced
	$M^{\cdot\pm *} + N \rightarrow M + N^{\cdot\pm}$	

Scheme 1

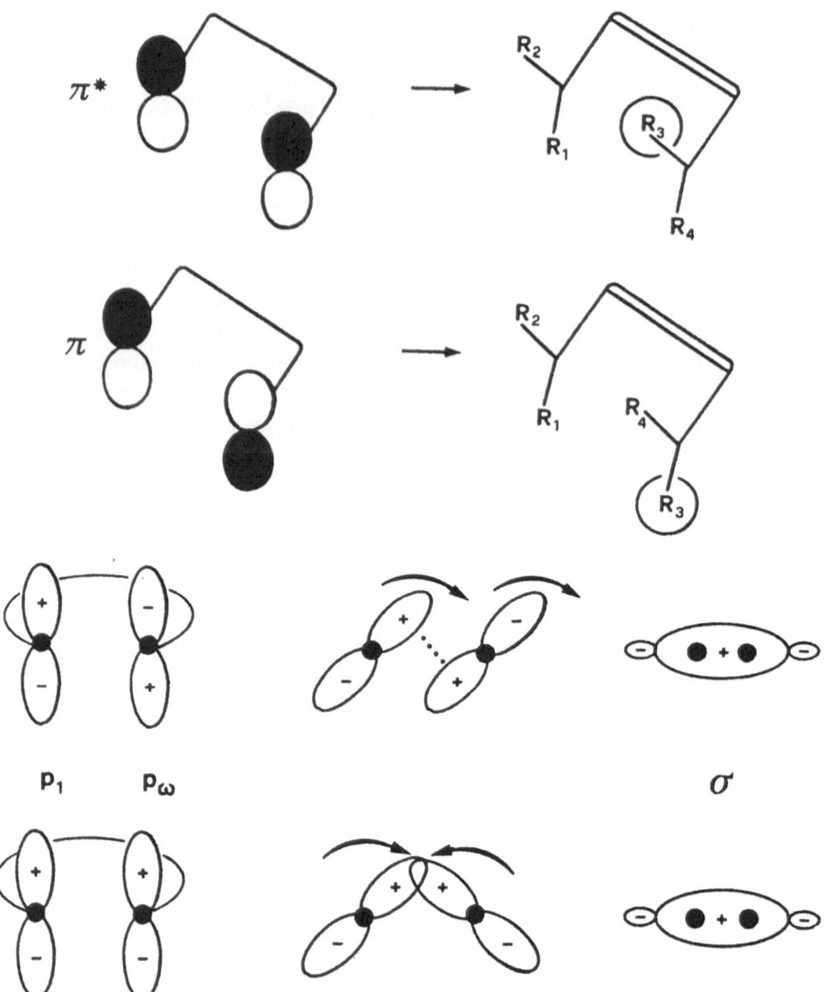

Fig. 1. The conrotatory and disrotatory pathways of an electrocyclic isomerization lead to stereoisomers with different activation barriers

The reactions drawn in this way are intermolecular processes, but they can also be intramolecular when the reactant molecules are linked together by a formally saturated chain. They can be considered in that case as two chromophoric groups within a single molecule, rather than two separate e.t. partners. There are then two distinct mechanisms open in principle for e.t.; the through-space and the through-bond interactions [3].

As shown in Scheme 1, all three types of e.t. can be either thermal or photochemical, when one of the partners reacts in an electronically excited state. We shall be concerned essentially with such light-induced e.t. processes, but reference will also be made to some thermal e.t. which are important in the discussion of the "Marcus inverted region" (M.I.R.). The important characteristics of an excited state are its energy, its lifetime, its multiplicity (spin state) and its electron distribution (e.g. n-π^*, π-π^* states of organic molecules).

A further photoinduced e.t. shown in Scheme 1 represents the interaction of *two* excited molecules, and is given here only as a theoretical possibility. There seems to be no well documented example of such an e.t. reaction, although energy transfer processes involving two excited molecules are known (e.g. triplet-triplet annihilation) [4]. It must be borne in mind that such pathways of very high energy may become important in conditions of high light intensities such as obtained in laser flash photolysis.

1.2 The Marcus Model of Energy Wells

The classical description of a chemical reaction as the crossing from a reactant "well" to a product "well" is the usual starting point to introduce the M.I.R. (Fig. 2) [5]. The abcissa of the diagram is some kind of reaction coordinate x, but it must be realized that the two-dimensional diagram represents only one cross-section of a multidimensional energy well. Depending on the property of interest, the reaction coordinate will be related to the geometry of the molecules, to the state of polarization of the solvent, etc. The ordinate is the energy, the difference ΔG between the bottom of the wells being the reaction free energy. It is now held that the chemical reaction takes place as the system moves from the reactant to the product well, the intersection point X corresponding to the transition state. The activation energy $\Delta G^{\#}$ in the Arrhenius-Eyring equation [6]

$$k = A \exp - (\Delta G^{\#}/RT) \tag{1}$$

is the difference in energy between the point of intersection and the bottom of the reactants well.

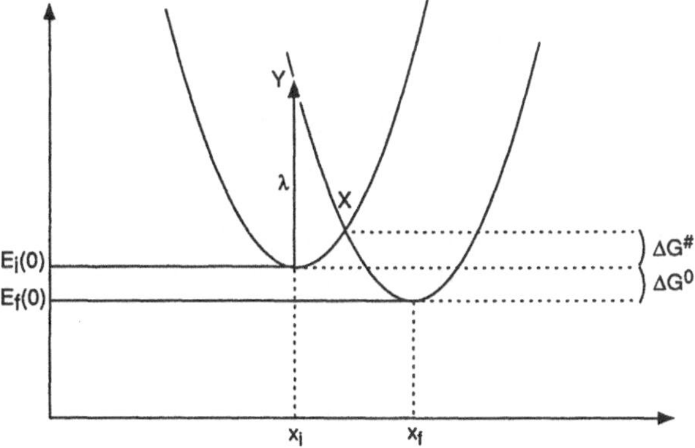

Fig. 2. The model of an electron transfer reaction as the intersection of two parabolic potential energy wells. The abcissa x represents a "reaction coordinate"; such a diagram is only one two-dimensional section through a multidimensional space

The major concepts which will be discussed in this review can be derived from this classical picture. It implies a clear relationship between the activation barrier $\Delta G^{\#}$ and the reaction free energy ΔG, leading to the M.I.R. according to the three different cases of intersections of the energy wells shown in Fig. 3.

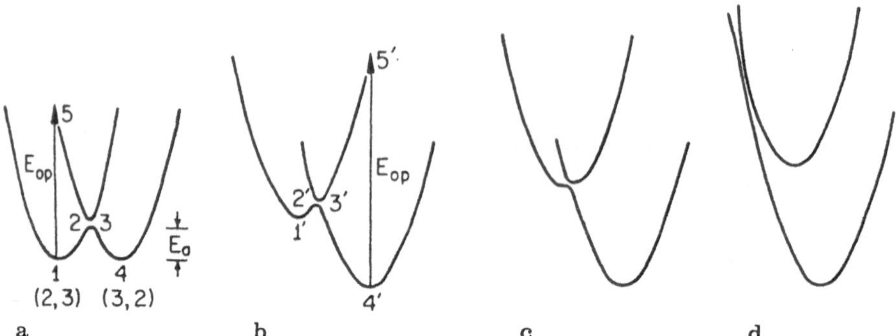

Fig. 3. Reaction coordinate diagrams for electron transfer. (*a*) $\Delta G^0 = 0$; (*b*) ΔG^0 negative; (*c*) diffusion-controlled limit, $-\Delta G^0_{R'} = \lambda$; (*d*) abnormal free-energy region, $-\Delta G^0_{R'} > \lambda$

Keeping the reaction coordinate increment x constant, the activation barrier first decreases as the free energy becomes more favorable, from (a) towards (b) when the reaction is activationless; but as the products well sinks deeper, the point of intersection moves further to the side of the reactants, and the activation barrier rises again. The plot of the rate constant k derived from Eq. 1 against the reaction free energy ΔG^0 is then predicted to be bell-shaped, with a maximum corresponding to the activationless process. The falling part of the plot beyond the maximum is known as the M.I.R.

1.3 The Classical Formulation in Terms of Harmonic Wells [7]

The picture of a chemical reaction as the passage from the reactant to the product energy well implies the existence of the M.I.R., but some assumption is required about the shape of the wells in order to obtain an analytical formulation of the rate constant. The simplest case is that of harmonic energy wells, based on Hooke's law which relates the restoring force F linearly to the displacement Δx,

$$F = -b\,\Delta x \tag{2}$$

b being the force constant. The wells are then parabolas of the form $E = E^0 + b(x - x^0)^2$ and the coordinates of the point of intersection correspond to the common solutions of their respective equations, leading to the simple relationship

$$\Delta G^{\#} = \frac{\lambda}{4}\left(1 + \frac{\Delta G^{0}}{\lambda}\right) \tag{3}$$

The energy difference λ between the bottom of the reactant well and the product well at the same reaction coordinate is known as the reorganization energy; within the assumption of parabolic wells of equal force constants this is related to the activation barrier $\Delta G^{\#}$ simply by $\Delta G^{\#} = \lambda/4$ [8].

1.4 The Concept of Reorganization Along the Reaction Coordinate [9]

The vertical energy gap of the wells at the initial reaction coordinate is described as the reorganization energy, since it implies the rise in energy as the system moves towards the saddle point X along the reaction coordinate (see Fig. 2). Conceptually, the system starts to resemble the situation it will have when the products are finally formed.

So far the Marcus model of the intersection of energy wells has been considered without specific reference to electron transfer processes. Indeed, in these general terms its applicability may appear to extend to all types of chemical reactions, without entering into the semantic argument of whether e.t. is a "chemical" reaction in which no chemical bonds are actually made or broken. Discussions of the Marcus model to processes other than e.t. are outside the scope of this text, reference being made to some recent publications in this field [10], [11].

Electron transfer processes differ from "chemical" reactions in two important respects, and these are highly relevant to the Marcus model. In the first place, the actual e.t. step itself is thought to be usually very fast, since it involves only the jump of an electron from one molecule to another, or from one orbital to another in an intramolecular process. Secondly, e.t. can in principle take place between distant molecules, especially in the intramolecular process which relies on the through-bond interaction of distant chromophores [12]. The "instantaneous" nature of the electron jump leads to the concept of *prior* reorganization which has become a cornerstone of the Marcus model: the motion of the system from the reactant to the product well along the reaction coordinate (Fig. 2) implies that the reactant(s) reorganize *of its own accord* to reach the transition state, before e.t. can actually take place [13]. We shall see later on that this leads to difficulties when the role of the solvent is considered, and these are currently leading to modifications and extensions of the original model.

1.5 Types of Electron Transfer Reactions

In the discussion of the M.I.R., it is useful to distinguish four different types of e.t. processes. The first classification is based on early work with metal complexes, and the terms of "inner-sphere" and "outer-sphere" e.t. are generally used in the broad sense of intramolecular and intermolecular processes [14].

However, some authors consider that an e.t. reaction which takes place within a complex of two molecules qualifies as an inner sphere process [15]. The distinction between inner- and outer-sphere e.t. reactions is then blurred, and only long-range e.t. between distant molecules or chromophores would be truly outersphere processes.

The second classification concerns the adiabaticity of the e.t. reaction. Conceptually, the reaction is "adiabatic" if the probability of reaction for each passage through the intersection region (point X of Fig. 2) between the potential surfaces of the reactants and products is close to unity [16]. If this probability is small, then the system remains on the initial state potential surface and the reaction is "nonadiabatic". In the quantitative formulation of Jortner and Bixon [16], the pre-exponential factor in the rate constant equation is given as follows, for the case of an adiabatic e.t.:

$$A'^{ad}_{bac} = \frac{2\pi}{\hbar} V^2 (4\pi\lambda_0 k_B)^{-1/2} \times \left(\frac{1}{1 + 4\pi V^2 e^{-S}\tau_e/\lambda_0\hbar}\right) e^{-S} \tag{4}$$

Here V is the matrix element which describes the coupling of the electronic states of reactants and products, S is known as the electron-vibration coupling constant which is equal to the inner-shell reorganization energy λ_i, expressed in units of vibrational quanta,

$$S = \lambda_i/h\nu \tag{5}$$

The pre-exponential factor in this case includes the solvent longitudinal relaxation time τ_L, which will be discussed further on when the recent works concerned with the role of the solvent will be considered. This longitudinal relaxation time is related to the usual Debye relaxation time according to

$$\tau_L = \tau_D(D_{op}/D_S) \tag{6}$$

where D_S is the static dielectric constant of the solvent, and D_{op} its "optical" dielectric constant which is approximately equal to the square of its refractive index n, $D_{op} \sim n^2$ [17].

The quantitative criterion for adiabacity is given by a parameter $\varkappa$ defined as

$$\varkappa = \frac{4\pi V^2 e^{-S}\tau_1}{\lambda_0 h} \tag{7}$$

For $\varkappa \ll 1$, the process is nonadiabatic and is expected to follow the Marcus-Hush theory with the rate constant of e.t. given as

$$k_{et} = (2\pi/\hbar) V^2 (4\pi k_B T)^{-1/2} e^{-S} \times \sum_{v=0} \left[\frac{S^v}{v!} \exp\left(-\frac{(\Delta G^0_{et} + \lambda_0 + vh\nu)^2}{4\lambda_0 k_B T}\right)\right] \tag{8}$$

where ν is the mean vibration frequency involved in the process and v is the number of vibrational quanta excited in the final state. The role of the solvent is essentially static in this model, since there is no term which includes the solvent relaxation time. The solvent effect is described by the "outer-sphere reorganization energy" λ_0, which is a time-independent quantity (this is discussed in detail in Section 3.1).

When $\varkappa > 1$, the e.t. reaction is adiabatic and the rate constant can be controlled by the solvent relaxation. There are of course intermediate cases when $\varkappa \sim 1$, and the transition from adiabatic to nonadiabatic regime is not a step function.

The concept of adiabacity in e.t. processes has gained importance in recent years, and the question does arise to what extent it may influence the observation of the M.I.R. In principle, the occurence of the M.I.R. is related only to the quadratic form of the activation energy, not to the form of the pre-exponential factor. The M.I.R. should therefore be observed for both adiabatic and nonadiabatic reactions. However, if the observable rate of an adiabatic process is controlled by the solvent relaxation time, the influence of the exponential factor may be negligible [18].

1.6 The Marcus Inverted Region Before and After 1985

Several review articles state that the existence of the M.I.R. is was at first received with scepticism [19]. This was perhaps inevitable with a generation of chemists schooled in the concept of the *linear* free energy relationships, and the fact that for a long time there was no experimental evidence to confirm the theoretical prediction. This became so disturbing that several papers appeared from the Marcus group itself in order to explain the absence of the M.I.R. particularly in photoinduced e.t. reactions. The latter show what has become to be known as "Rehm-Weller behaviour", as opposed to the predicted "Marcus behaviour" [20]; that is, the plot of the rate constant versus free energy rises rapidly in the "normal" region up to $\Delta G \geqq 0$, then reaches the diffusional limit (in the case of intermolecular reactions) and stays there no matter how exergonic the process becomes. Figure 4 gives an illustration of this Rehm-Weller behaviour which still seems to be the rule for processes of luminescence quenching through e.t. in solution [21].

Of the various explanations suggested for the absence of the M.I.R. in such e.t. processes, the formation of electronically excited products gained the widest acceptance – although it has been questioned in recent years, as we shall see. This is indeed a tempting and reasonable explanation, for it is well-known that molecular ions generally have much lower excited states than the neutral molecules [22]. Thus the radical ions of aromatic molecules such as benzene, naphthalene and anthracene absorb in the visible (VIS) or even the near infrared (NIR) regions of the spectrum, while the neutral molecules absorb only in the UV; the energy gap between the reactant(s) and product(s) wells is then much smaller than that calculated for the e.t. which leads directly to the ground state products, so that the process would not be sufficiently exergonic to reach the M.I.R. (Fig. 5) [23].

The first clear indications of Marcus-type e.t. reactions date back to the mid-'80s, and these proved to be a turning point in our understanding of e.t. processes [24]. These observations did not, however, concern photoinduced e.t. between neutral excited state and ground state partners, but thermal charge shifts and later charge recombinations [25]. In the last five years a split has developed

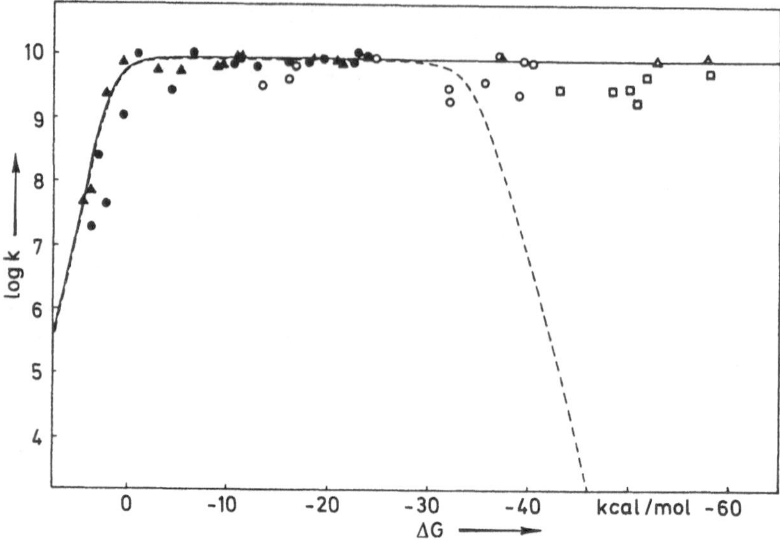

Fig. 4. Plot of log *k* vs. ΔG for the forward (solid symbols) and the back (open symbols) electron-transfer reactions (Eq. 3 and Eq. 4, respectively). *Circles, squares,* and *triangles* refer to $M^{n+} = Cr(bpy)_3^{3+}$, Rh-$(phen)_3^{3+}$, and $Ir(Me_2phen)_2Cl_2^+$, respectively. The curves are calculated by using Eq. (2) *(full line)* and the Marcus free energy relationship (Eq. 1, *broken line*), according to standard kinetic equations [61]

between Marcus-type and Rehm-Weller-type e.t. processes, and much of current research (and controversy) is centred around this particular problem: what is the difference between e.t. reactions which follow the Marcus model and those which do not?

This question is the major concern of this paper, as it has become the most important research area connected with the M.I.R. – in recent years, the role of the solvent or rigid matrix comes only second to it. If the solvent must reorganize before e.t. can take place, the kinetics of solvent relaxation may come to control the overall process; this is the second aspect of current research which has a direct bearing on the assessment of experimental data concerned with the M.I.R., namely the role of solvent relaxation kinetics.

To conclude this introduction, a summary of the following sections is given here as a guide. The M.I.R. is so greatly involved in all e.t. processes that it is

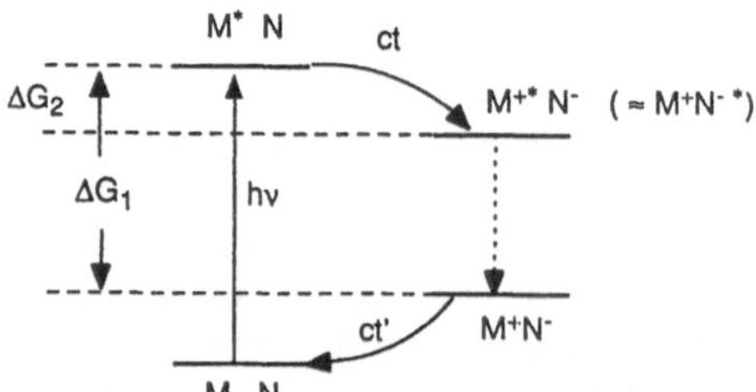

Fig. 5. If the ion pair can be formed with one of the ions in an electronically excited state, the free energy for the reaction (ΔG_2) can be much smaller than that calculated for the ground state products (ΔG_1)

impossible to quote all the papers which are of more or less direct relevance to it – the selection of references cannot claim to be exhaustive and there is always the fear in the reviewer's mind that some important contribution may have been omitted.

2 Calculations and Measurements of Reaction Free Energies and Rate Constants

The M.I.R. is a particular type of relationship between the free energy of an e.t. reaction and its rate constant, as shown graphically by the plots discussed in the introduction. Experimental data used to obtain such plots are the reaction free energies on the one hand, and the reaction rate constants on the other hand, and it is therefore essential to have some understanding of the reliability of these data for any assessment of the validity – and possibly limitations – of the Marcus model of e.t.

The fact is that the reaction free energies are hardly ever determined experimentally, but are simply calculated from the Rehm-Weller equation which will be discussed in detail in the next section [26]. There are still considerable technical problems in direct experimental measurements, because standard methods of calorimetry cannot cope with reactions in time scales of ns or μs; but this is slowly changing with the advent of fast calorimetric techniques such as time-resolved photoacoustic spectroscopy [27] and thermal lensing [28]; these are considered in the following section. Nevertheless, it appears that all the data currently used in the rate constant–energy plots simply use the Rehm-Weller equation (sometimes with various corrections) and it is obviously important to consider the assumptions built into this equation, its limitations, and possible improvements.

The situation with respect to the rate constants is much better, but here again there are some potential pitfalls which must be taken into account. The most obvious of these is the limitation of observable intermolecular reaction rates by the rate of diffusional encounters, but this was recognized many years ago and can now be taken for granted [29]. Other difficulties may arise in the measurements of intermolecular photoinduced e.t. reactions when the rate constant is inferred only from luminescence quenching; the assumption that quenching always involves e.t. may not be valid in some cases. Further problems in the determination of the rate constant of back e.t. within ion pairs formed in a primary photoinduced charge separation should not be overlooked [30]; here the observable quantity is generally the yield of free ions in polar solvents, and the rate constant is calculated on the basis of some assumptions concerning the rate of escape from the solvent cage [31].

2.1 Calculations and Measurements of Free Energies of Electron Transfer

The Rehm-Weller equation which is based on electrochemical and spectroscopic data is used almost universally for the calculation of reaction free energies of e.t. It relates the oxidation potential of the donor D and the reduction potential of the acceptor A, taking into account the energy E* of the excited state involved in a light-induced process:

$$\Delta G = E_{ox}(D) - E_{red}(A) - E^* + C \tag{9}$$

C is the "Coulomb" term which must be introduced to describe the electrostatic interaction of the products. This term vanishes if either of the product molecules is electrically neutral [32]. The Rehm-Weller (R-W) equation is usually considered to be semiempirical, but recently an attempt was made to give it a theoretical basis [33]. This has not questioned its fundamental assumptions, which we shall discuss presently.

1. The redox potentials: These are measured by electrochemical methods such as cyclic voltametry which impose restrictions on the solvents since solvation of ionic species must be obtained. In practice, highly polar solvents such as acetonitrile (MeCN), methanol (MeOH), or water are used in most cases, and corrections must then be made when e.t. takes place in less polar solvents. Here it is assumed that the only difference is due to the solvation energy of the ions — this is calculated from the Born equation which gives the solvation energy of a spherical ion of charge q and radius a in a solvent of static dielectric constant D as [34]

$$E = (1/2)\, q^2 a^{-1} F(D) \tag{10}$$

where the polarity function is $F(D) = (1/D) - 1$.

This equation is based on a very simple classical electrostatic concept, and even its derivation has been questioned [35] but more elaborate calculations still refer to it as the simplest limiting case. It is still used quite generally both for the correction of redox potentials and for the calculation of solvent reorganization energies [36]. However, it is only fair to state that it does not seem to have clear experimental support, probably because of the difficulties of actual measurements of such solvation energies.

2. The excited state energy: The R-W equation assumes that the entire electronic excitation energy of the macroscopic sample is available as chemical *free* energy; this is clearly implied by the form of the equation itself, and has been questioned only recently [37]. At the molecular level, the electronic excitation raises the potential energy of the molecule, but this is entirely available as chemical free energy of the macroscopic sample only if light carries no thermodynamic entropy. There are, however, theories which consider light as a form of heat; the "temperature" and "entropy" of radiation then being derived from the Planck equation [38]. It is a rather remarkable fact that while such theories are considered sound by many eminent workers in this field, they are totally ignored in practice. Yet there are potentially important implications — in particular that the conversion efficiency of light energy into chemical free energy would depend on light intensity — and this introduces another uncertainty in the free energies of photoinduced e.t. reactions calculated from the R-W equation.

3. The Coulomb term: This term represents the electrostatic energy gained (or spent) as the product molecules are brought together from "infinity" to the separation relevant to e.t.; this correction is needed in principle because the redox potentials of molecules are measured independently, so their combination as $E_{ox} - E_{red}$ in the R-W equation would pertain to e.t. between distant (independent) molecules. The simplest formulation of this Coulomb term is found in the point charge approximation neglecting the space occupied by the molecules in the solvent [39]. This may be unrealistic at the microscopic level, especially for e.t. between molecules in contact, for there is in that case no screening of the charges by polar solvents; typical values of the Coulomb term C in various solvents are given in Table 1. This shows that the values calculated according to the molecular model diverge increasingly from those of the point charge model as the polarity of the solvent increases. This is not a trivial effect, because the free energies can differ by several tens of eV [40].

Table 1. Comparison of point-charge C′ and molecular Coulomb terms C

Center to center separation r [Å]	Dielectric constant ε_s	C [eV]	C′ [eV]
6	36.7	−0.632	−0.065
9		−0.295	−0.044
12		−0.174	−0.033
6	7.58	−0.757	−0.316
9		−0.407	−0.211
12		−0.268	−0.158
6	2.38	−1.103	−1.007
9		−0.714	−0.671
12		−0.527	−0.504

2.1.1 Experimental Measurements of Energies of Electron Transfer

The usual calorimetric methods cannot be used for the measurement of reaction enthalpies and entropies of photo-induced e.t. reactions because of their slow response time. This is also a problem in most thermal e.t. processes which follow the light-induced step, such as charge recombination through diffusional encounter of ions; in conditions of laser flash photolysis, the kinetics of such processes occur in the μs time scale.

Two experimental techniques are currently being used for calorimetric measurements of fast photophysical and photochemical processes, and these will be described briefly in order to assess their potentials and limitations; some results which have a direct bearing on the energetics of e.t. reactions will then be examined.

1. Time-resolved photoacoustic spectroscopy [41]: The steady-state technique of photoacoustic spectroscopy (PAS) is widely used for the measurement of absorp-

tion spectra of "opaque" samples; it is based on the detection of the heat evolved when the energy of light absorbed by the sample is degraded through nonradiative transitions and/or chemical reactions [42]. This thermal energy leads to the formation of pressure waves in the surrounding gas, and these are detected as acoustic signals by a microphone.

Figure 6 shows an outline of a PAS instrument designed for fast time-resolved measurements. The excitation light is a laser pulse of some 20 ns duration, at a wavelength which falls within the absorption spectrum of the sample (e.g. 337 nm with a nitrogen laser). Total absorption of this pulse then deposits an energy E in the sample and this will decay in the course of time into heat which will give rise to the pressure sensed by the detector; usual microphones have slow response times, so that piezo-electric devices are used to improve the instrument's time resolution [43].

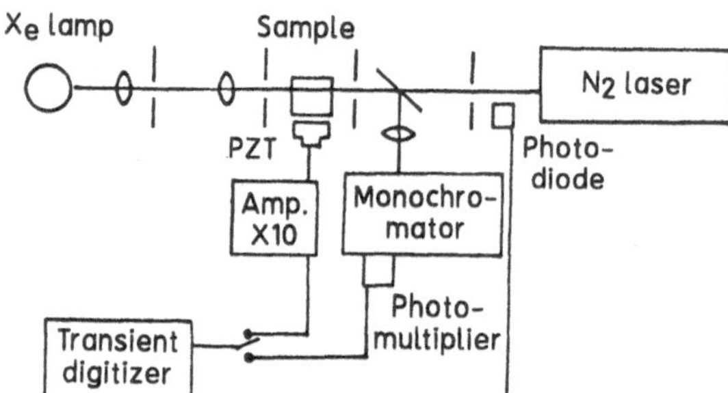

Fig. 6. Experimental setup for the simultaneous measurement of PA and transient absorption.

The time resolution currently accessible is of the order of 1 μs, but recent developments may extend this to around 10 ns [44].

2. Thermal lensing [45]: This method relies on the change of refractive index of a liquid which is heated suddenly. Those regions of the liquid which are at a higher temperature then act as a lens, therefore called the "thermal lens", which can be detected by the refraction of a beam of probe light (Fig. 7).

In this arrangement the pulse and probe beams are colinear; the probe beam is generated by a CW laser, and it falls on the photomultiplier tube D through a pinhole PH. When the thermal lens is formed in C, the divergence of the probe beam increases and the intensity seen by the detector D decreases.

The temporal response of the thermal lensing technique is limited by the "acoustic transit time": the heated liquid must expand in volume for the lens to be formed and this depends in particular on the size of the heated region. In practice the observation time scale is of about 1 μs to a few ms.

There are in general several processes which generate heat in a photophysical or photochemical reaction. Figure 8 shows two examples with an outline of the temporal evolution of the temperature of the sample – this would be a simplified form of the kind of oscillogram observed in a thermal lensing experiment. The

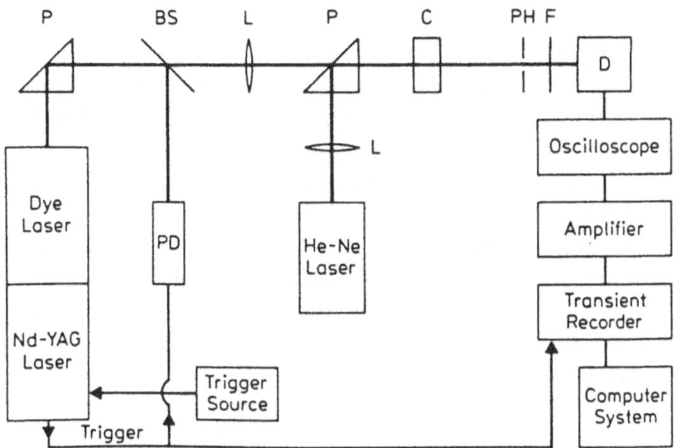

Fig. 7. *TL* spectrometer: *C*, sample cuvette; *PH*, pinhole; *F*, filter; *BS*, beam splitter; *PD*, pyroelectric detector; *D*, diode; *P* 90° prism; *L*, lens

first example is benzophenone in acetonitrile, a solvent in which the photoreduction is very slow. The entire energy of the absorbed laser pulse (e.g. at 355 nm) is then degraded into heat, but this takes place in several stages: Internal conversion will bring the excited molecule from the prepared state S** to the relaxed singlet excited state $^1S^*$, in a time of ps; intersystem crossing then brings the molecule to a higher level of the triplet state, and this also takes place in a few ps for benzophenone; another internal conversion now brings the molecule to the relaxed triplet state ^1T, which will decay back to the ground state in a time of several μs.

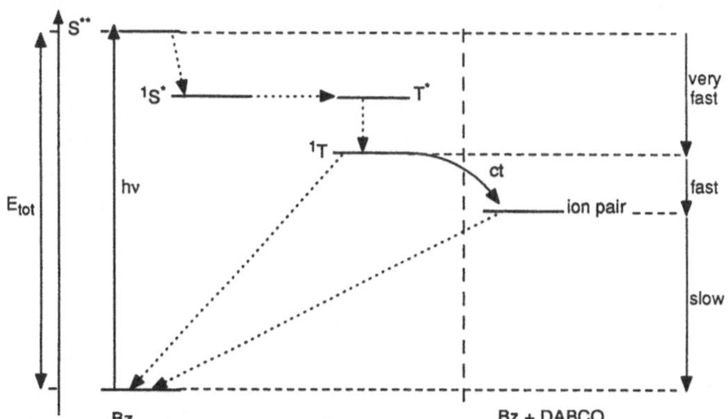

Fig. 8. Outline of the time-scale of the processes observed during an electron transfer reaction observed through thermal lensing. Processes which occur in times below ca. 0.5 μs are "very fast", beyond the temporal resolution of the thermal lensing technique; they would appear as a step function in the kinetics of heat release. The slowest processes which would be observed in this case are the second-order recombinations of free ions, which take place in time scales of μs to several ms.

All the processes which form the relaxed triplet state will appear as a fast temperature rise, followed by the slower return to the ground state.

The second example concerns a photo-induced e.t. with this same chemical system, to which an amine (e.g. DABCO) is added as an electron donor. The fast decay to the relaxed triplet excited state of benzophenone remains unchanged, but this is now followed by two further reactions: the forward e.t. step which forms the radical ions, and the back e.t. of these ions to restore the initial system.

Since the chemical system finally returns to its initial state, and since there is no loss of energy through luminescence, the total temperature rise must represent the energy input of the absorbed laser pulse. If the detector signal is proportional to the temperature rise, the relative heights of the very fast, fast and slow parts of the kinetic trace then correspond to the relative energies.

2.2 Rate Constants of Electron Transfer and Yields of Ion Separation

The measurement of the rate constants of photo-induced e.t. processes relies on both direct and indirect observations. In a direct measurement, the kinetics of formation of a product, or of the disappearance of a reactant, are determined by ns or ps laser flash photolysis techniques [46]. These are the most reliable data, provided e.t. is substantially slower than the diffusional limit – a well recognized complication of intermolecular processes.

Intramolecular photo-induced e.t. can be much faster than the diffusional limit, and in most cases the rate constant is then estimated from the luminescence quantum yield, which is the quantum yield of the excited chromophore in the absence of the quenching process [47]. One potential difficulty of such indirect measurements is that the quenching mechanism may not be established definitely as being e.t.; although vertical energy transfer can usually be ruled out on grounds of the energy gap, other modes of quenching are seldom considered.

Luminescence quenching data apply of course to photo-induced e.t. processes, and while they give the rate constant for the primary process itself they provide no information concerning the separation of the ions. In nonpolar solvents, the presence of separated ions is not expected, but the situation is different in polar (e.g. acetonitrile) and protic (e.g. alcohols) solvents. Measurements of ions yields have been reported using different experimental techniques which fall broadly into three classes:

1. Spectroscopic observation of free ions [48]: The absorption spectra of many aromatic radical ions have been determined in rigid matrices, where they are formed by ionization reactions (X-ray or γ-ray irradiation). On the assumption that these spectra remain practically unchanged in the liquid state, flash photolysis can then be used to measure the ion yields in various solvents. The major problem is that the exact values of the extinction coefficients are seldom known, and this introduces some uncertainty on the absolute ion yields; the relative yields are, however, reliable.

2. Spectroscopic observation of an acceptor ion [49]: The difficulty linked to the values of the extinction coefficients of different ions can be circumvented in some cases with the use of an e.t. step to produce an acceptor ion of well-known extinction coefficient, as in the example of scheme 2. Biphenyl has been used in this case as an electron acceptor, and it must of course be assumed that all donor ions have undergone e.t.

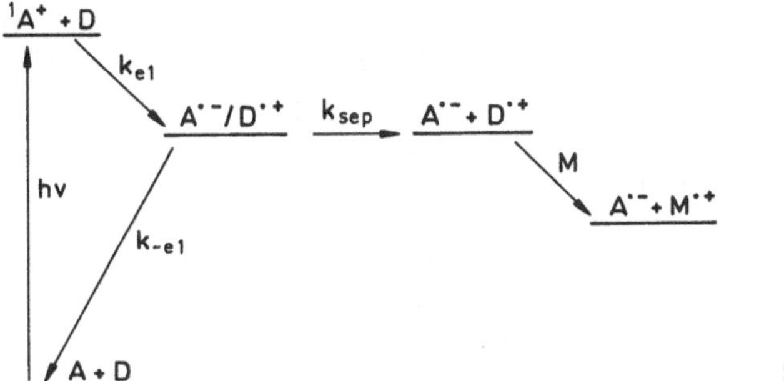

Scheme 2

3. Transient photocurrent measurements [50]: The assignment of spectroscopic transients observed in flash photolysis is seldom straightforward, when absorption spectra are generally rather broad and structureless. There is an additional problem in the case of e.t. reactions, namely that the free ions and the geminate ions (still in contact with the partner ion following e.t.) may not have distinct absorption spectra. It is indeed known that the spectra of most aromatic radical ions are essentially independent of the counterion, although some exceptions exist as well [51]. The observation of electrical conductivity is therefore a welcome additional clue as to the presence of free, solvated ions which alone can act as charge carriers. In some experimental systems, the simultaneous observation of transient photoconductivity and absorption spectra is possible, so that the kinetics can be compared directly [52].

Absolute values of ion yields are difficult to determine through photoconductivity, because the current depends on many factors such as the mobility of the charge carriers, and these are not known accurately. On the other hand, relative ion yields are reliable, and free from the potential ambiguity of direct spectroscopic measurements which do not distinguish between free and geminate ions.

2.2.1 Rate Constants of Geminate Ion Recombination

Many experimental data concerning the M.I.R. pertain to the back e.t. between geminate ions, prior to separation in polar solvents [53]. In some cases direct spectroscopic measurements can be made, although these depend on the correctness of the assignment of the absorption spectra. In principle the geminate ion recombination is expected to follow first order kinetics, whereas the diffusional recombination of free ions follows second order kinetics, so there is here another possibility to distinguish between the two processes [54]. In some cases, the rate constant of geminate back e.t. is estimated from the free ion yield, according to Eq. (11) which relates this yield Φ_{ion} to the rate constants k_{bac} of geminate back e.t. and k_{sep} of separation into free, solvated ions,

$$\Phi_{ion} = \frac{k_{sep}}{k_{bac} + k_{sep}} \tag{11}$$

The rate constant of geminate ion separation is usually calculated according to diffusion equations, which can be considered fairly reliable for partners which are not held together by electrostatic forces; in this case the rate constant is given as [55]

$$k_{sep} = p\,\frac{2D_{AD}}{j^2} = \frac{pk_BT}{3\pi r j^2 \eta} \tag{12}$$

Here j is the distance of each jump of a molecule, taken as the diameter of a solvent molecule, and p is the probability that the product molecules become separated after one jump. The diffusion coefficient D_{AD} is related to the viscosity η and to the temperature T, r being the mean molecular radius of the molecules.

The inclusion of electrostatic attraction depends on the assumed screening effect of the solvent, a problem similar to that of the Coulomb term in the free energy equations. This remains one of the major questions in the theory of e.t. reactions, and it may be safer to avoid any discussion of the kinetics of separation of ion pairs in polar solvents until this is resolved.

3 The Role of the Solvent in Electron Transfer Processes

As mentioned in the introduction, the debate concerning the observation or the absence of the M.I.R. has shifted in recent years to various models of the role of the solvent in e.t. reactions. In this section we shall consider the concept of the "outer sphere reorganization" in the Marcus-Hush theory, its implications and experimental predictions; the more recent concepts of the dynamics of solvent relaxation as a controlling factor in e.t. will then be discussed.

3.1 Solvent Reorganization as an Activation Barrier in the Marcus Theory [56]

According to the model of intersecting energy wells of reactants and products (Figs. 2 and 3), the chemical system must move of its own accord along the reaction coordinate x to reach the saddle point from which it will drop into the products well. If the reaction coordinate is meant to represent the state of the solvent then this implies that the solvent must reorganize so as to resemble its relaxed state around the products – it does not have to reach that state, but it must reach the half-way stage which corresponds to the crossing point X.

This concept of the *prior* reorganization of the solvent is best illustrated by a charge shift e.t., something like the self-exchange reaction of a neutral molecule M with its ion M^+ see scheme 3 [57].

$$M^+ + M \longrightarrow M + M^+$$

Scheme 3

In the relaxed state of the reactants (M, M^+) the solvent is polarized around M^+ and unpolarized around M; these states of polarization being determined by the respective electric fields of M and M^+. In the relaxed state of the products, the situation is simply the mirror image, the polarization of the solvent being again determined by the solute's electric field; but in the transition state, it is held that the solvent is half polarized around M (in spite of the absence of any polarizing field) and half depolarized around M^+ (against the electric field). The energy required to obtain this strange state of solvent polarization is known as the "outer sphere reorganization energy", and it is calculated simply from the Born equation of ion solvation.

The solvent reorganization energy λ_0 is therefore twice the Born solvation energy of an ion of charge q and radius a, for distant partners:

$$\lambda_0 = \frac{q^2}{a}\left(\frac{1}{D_{op}} - \frac{1}{D_S}\right) \tag{13a}$$

The general equation used for partners with a centre to centre distance d is [58]

$$\lambda_0 = q^2\left(\frac{1}{a} - \frac{1}{d}\right)\left(\frac{1}{D_{op}} - \frac{1}{D_S}\right) \tag{13b}$$

This reorganization energy implies complete reversal of solvent polarization around the partners; the fact that the saddle point along the reaction coordinate corresponds to the state of half polarization is taken care of by the factor 4 in the denominator in Eq. 3.

While it is usually accepted that the solvent polarization/depolarization energy is given in this way to a reasonable accuracy – within the limitations of the Born equation – the question of the mechanism whereby this peculiar state of the solvent may be reached has seldom been considered. The concensus has developed in recent years that this must rely on the random fluctuations of the positions of solvent molecules, sometimes called "density fluctuations"; but there is no analytical formulation of the probability of such extraordinarily localized fluctuations [59].

Computer simulations have provided further insight into the model of random fluctuations as a prerequisite for e.t. in polar solvents [60]. It has been shown that spontaneous local polarity fluctuations of the magnitude envisaged by the Marcus model are so improbable as to be statistically insignificant; and it was necessary to assume that the solvent could adjust continuously in order to follow the position of the electron in the course of e.t., as if e.t. would be slow enough to be the rate-determining kinetic step. To what extent such a modification of the model of prior reorganization breaks the spirit and/or the letter of the Marcus model remains to be assessed; but it is becoming clear that random fluctuations of local solvent polarity are unlikely to describe correctly the role of the solvent in e.t. processes, and that here at least some major changes in the original Marcus theory may be at hand. The first indications of such changes are the inclusions of the dynamics of solvent relaxation as the rate controlling step in some e.t. reactions in protic solvents and this deserves special attention.

3.2 Electron Transfer in Polar Glasses [61]

Before coming to the role of solvent relaxation dynamics in liquid solvents, a further complication of the model of prior reorganization must be mentioned, namely photoinduced e.t. processes in rigid, polar matrices. There are many examples of such reactions, based usually on the observation of the quenching of the luminescence of a light-absorbing species by an e.t. partner co-dissolved in a rigid glass [62] or polymer matrix [63]; Table 2 lists some representative example.

Table 2. Electron transfer quenching in rigid matrices

Donor	Acceptor	Matrix	Remarks[a]	Ref
TMPD	Phthalic anhydride	MTHF 77 K	τ_F = 7 ns, R = 17 Å τ_p = 3 s, R = 25 Å	[93]
TMPD	$Ru(Me_4phen)_3$	Polycarbonate film	τ_L = 3.7 µs R = 9 Å	[94]
Trapped e^-	*p*-benzoquinone	MTHF 77 K		[95]
TMPD	$Biphenyl^{\cdot +}$	2-chlorobutane 77 K	$\tau_L \simeq$ 1 µs R = 17 Å	[96]
$Ru(Me_4phen)_3$	MV^{2+}	glycerol 250 K	$\tau_L \simeq$ 1 µs R = 17 Å	[61]
$Ru(Me_4phen)_3$	MV^{2+}	cellophane film 21 °C	$\tau_{et} \simeq$ 2 µs $\tau_{back} \simeq$ 5 ms	[63]
DEA	9-methylanthracene	MTHF 77 K	$\tau_F \simeq$ 4.6 ns R = 8 Å	[97]
Hemato-IX-Zn porphyrin	CCl_4	EtOH 77 K	$\tau_{rec} > 10^3$ s R > 20 Å	[98]

TMPD, *N,N,N'*, *N'*-tetramethyl-*p*-phenylene diamine; MV^{2+}, methyl viologen; DEA, *N,N*-diethylaniline; MTHF, methyl-tetrahydrofuran; [a] The R values in Angstrom units refer to the radius of the Perrin quenching "sphere of action"

This type of static quenching requires relatively high quencher concentrations and it follows the Perrin "action sphere" model [64]. According to this model, each emitter molecule is surrounded by an "active volume" (in the general case it needs not be a sphere), such that if there is one quencher molecule at least within this volume, then quenching takes place instantly; but molecules which have no quencher within the active volume emit just like those in a sample devoid of quencher. The Perrin model leads to two observable results:

1) The logarithmic decrease of luminescence intensity with quencher concentration:

$$\ln (I_o/I_q) = -c[Q] \tag{13}$$

where c is a constant related to the size of the volume of action

2) The invariance of the luminescence decay kinetics in the presence of quenchers

A number of experiments have confirmed these two predictions, although the second one is expected to be only an approximation; it would be strictly true if the boundary of the action volume could be ideally sharp, and this is of course never the case in practice [65]. Luminescence quenching through e.t. follows the Perrin model quite closely because of the exponential decrease of its rate constant with distance – as has been established in a number of instances [66].

The observation of e.t. in rigid, polar glasses and polymers raises the question of the role of the solvent in the Marcus theory, since in these cases any reorganization is ruled out. If e.t. takes place in rigid polar media, why should the reorganization be a prerequisite for e.t. in liquid solvents? One attempt at reconciling the model of prior reorganization with the observations of e.t. in rigid matrices is based on the assumption that the instantaneous local polarity fluctuations are frozen in the matrix [67]. Each emitter molecule would then be trapped in different local environments, some of them corresponding to the reorganized solvent shell which could lead to e.t. This explanation does not seem to be in agreement with at least some of the experimental observations: In many cases quenching takes place in *all* contact pairs, and it is highly unlikely that all would be found in suitably reorganized environments, especially since these correspond to states of high energy – without considering the probality. The problem of reconciling the model of solvent reorganization with the observation of e.t. in rigid matrices therefore appears to have found no solution at the present time.

At this stage we come back to the distinction between adiabatic and nonadiabatic e.t. reactions, alluded to in Sect. 1.4. The most important experimental clues about the role of the solvent are found in the temperature dependence of the rate constants of e.t., which broadly follow four distinct patterns:

1) The rate constants can be fitted to an Arrhenius-Eyring type equation, with positive activation energies in agreement with the formulation of the outer-sphere reorganization energy calculated from the Born equation of ion solvation, or the Onsager equation of dipole (or geminate ion pair) solvation [68]
2) The rate constants can still be fitted to the Arrhenius-Eyring equation, but the magnitude of the activation energy departs significantly from the calculated values [69]

3) The rate constants of e.t. follow an anti-Arrhenius/Eyring behaviour, as they tend to decrease with increasing temperature – this would lead formally to negative activation energies which make no physical sense [70]
4) The rate constants do not follow any type of Arrhenius-Eyring relationship, especially at low temperatures where they level off to some constant value; this is thought to be an indication that electron tunnelling takes over from the thermally activated process, a concept altogether foreign to the classical Marcus model – and there would be no M.I.R. in such cases [71].

3.3 Solvent Relaxation Dynamics in Electron Transfer

In the original Marcus-Hush model, the role of the solvent is described by the static equation of reorganization, Eq. 13b. The time scale of the reorganization does not appear explicitly and this has been included in more recent treatments of e.t. processes. There are indeed several reports of e.t. reactions of which the rates seem to be controlled by solvent relaxation, related to the "longitudinal" relaxation time [72].

This longitudinal relaxation time differs from the usual Debye relaxation time by a factor which depends on the static and optical dielectric constants of the solvent; this is based on the fact that the first solvent shell is subjected to the unscreened electric field of the ionic or dipolar solute molecule, whereas in a macroscopic measurement the external field is reduced by the screening effect of the dielectric [73].

The first example of e.t. controlled by solvent relaxation was probably the TICT state formation in DMABN and similar molecules [74]. The *t*wisted *i*ntramolecular *c*harge *t*ransfer state is, however, formed gradually; the charge separation increases with the twist angle of the dimethylamino substituent

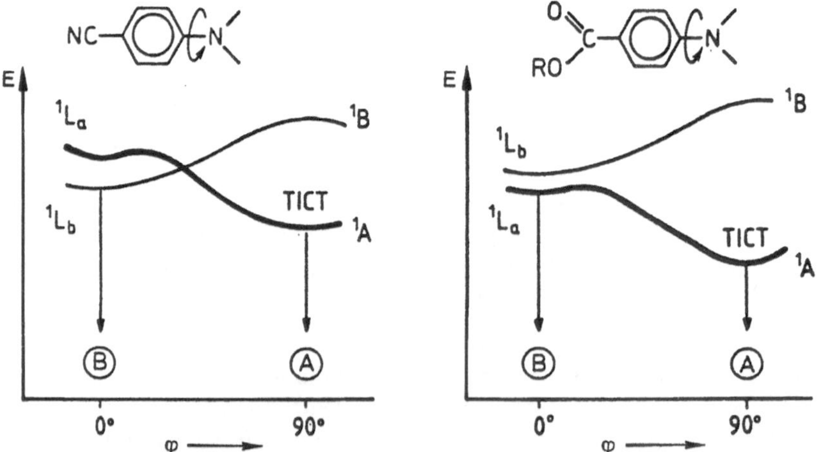

Fig. 9. The energy dependence of the excited states of TICT molecules on the twist angle between the dimethylamino group and the aromatic ring. In the TICT state (at 90°) full charge separation is reached

from the aromatic ring. The solvent can therefore adjust to the e.t. in this case, and there is no need to invoke local polarity fluctuations. Charge transfer and solvent relaxation are cooperative; the latter stabilizing the increasing dipole moment of the molecule (Fig. 9).

In the case of intermolecular e.t. it is generally supposed that the electron jump is quasi instantaneous, so that no such cooperative action of solvent relaxation can take place. However, if a local polarity fluctuation is required for e.t., then the occurrence of such fluctuations can still be controlled by the motion of the solvent. From many experimental data, solvent motion appears to become the rate limiting factor when alcohols are used as polar liquids. Table 3 lists a selection of such data, with those obtained in acetonitrile for comparison [75].

Table 3. Kinetic parameters of electron transfer reactions in protic and aprotic solvents

Molecule(s)	Solvent	Parameter[a]	Ref.
Bichromophoric molecule 1	MeCN (τ_l = 0.3 ps)	$k_{et} = 6.6 \times 10^7$	[81]
	MeOH (τ_l = 3.3 ps)	$k_{et} = 11 \times 10^7$	
Ox$^{\cdot}$/1,4-DMB	MeCN	$E_a = -0.05$	[72]
	MeOH	$E_a = +0.06$	
	EtOH (τ_l = 9.8 ps)	$E_a = +0.24$	
Rh$^{\cdot}$/1,2,4-TMB	MeCN	$E_a = -0.05$	
	MeOH	$E_a = +0.01$	
	EtOH	$E_a = +0.04$	
Bichromophoric molecule A1D	PrCN	$E_a = +0.047$	[56]
	MeOH	$E_a = +0.08$	

[a] Rate constants are given in s^{-1}, activation energies E_a in eV; For details of the molecules concerned, see references

3.4 Direct Involvement of the Solvent in Electron Transfer Reactions [76]

We come here to a rather special concept of condensed-phase e.t. reactions, according to which the solvent itself would act as an intermediary between the actual electron acceptor and electron donor partners. According to this model, each molecule would be surrounded by a solvent cage which would keep them seperated, so that the solvated electron would be the primary species formed by e.t., from the donor, as well as the actual donor to the final electron acceptor. In Fig. 10 this model is illustrated for the case of two spherical molecules embedded in their "specific" solvation shells, although the word "specific" is not clearly defined in this respect.

Although no distinction is made between photo-induced and thermal e.t. reactions, the model predicts that for a given electron donor (or acceptor) which

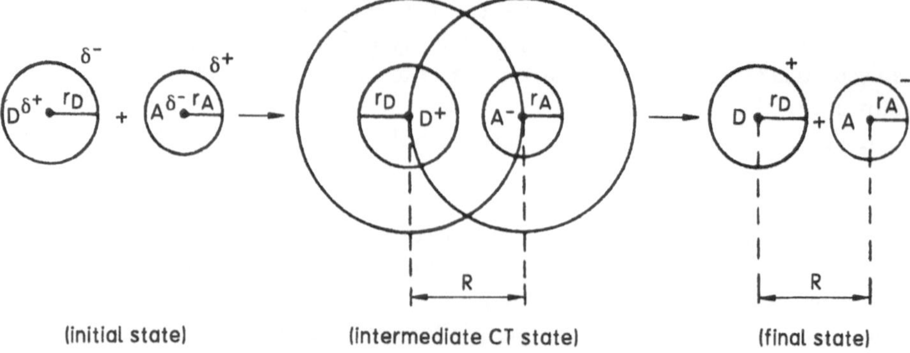

Fig. 10. Solute–solvent cage charge-transfer complex model for an electron-transfer reaction between a redox reactant pair: donor D and acceptor A

reacts with various electron acceptors (or donors) in the same solvent, the solvent reorganization energy should be related linearly to the reaction free energy ΔG; the Marcus quadratic equation which is the basis of the M.I.R. would then reduce to a linear form and no M.I.R. would be observed. This is not in general agreement with experimental data, and the basic assumption of the "charge transfer to solvent state" model is left open to question. It is well established that electron donor/acceptor pairs of molecules often form exciplexes, which are supermolecules resulting from the overlap of the frontier orbitals of the isolated partners. Clearly, this requires the direct contact of these partners, so that there can be no solvent shell between them. In any case, this rather peculiar model cannot explain the absence of the M.I.R. in most photo-induced e.t. processes.

4 "Rehm-Weller" Behaviour and "Marcus" Behaviour

The considerable amount of experimental data concerned with the rates and free energies of e.t. processes can be fitted into two distinct trends, which are referred to commonly as "Rehm-Weller" behaviour and "Marcus" behaviour. The distinction is based on the absence or presence of the M.I.R. in the plots of ln k versus ΔG, and much of the research in the last few years has been concentrated on the theoretical basis of these two opposite behaviours [77].

A rather general observation has been that intermolecular photoinduced e.t. processes did not follow Marcus behaviour; that is, the rate constant rises as expected in the slightly endergonic region, to reach rapidly the diffusional limit, but e.t. then remains diffusion controlled no matter how exergonic the reaction may become. Figure 4 gives a classical example of this Rehm-Weller behaviour [78].

In such cases it is of course impossible to determine the actual rate constant of e.t., since diffusional encounter becomes the rate limiting step. It has therefore been suggested that the failure to observe the M.I.R. results from the truncation of the rate-energy plots in intermolecular e.t. processes, and that a genuine test must rely on intramolecular e.t. [79]; but this is not quite fair, since observations can be extended well beyond the expected truncated region without the rate constant showing any sign of dropping with free energies which should undoubtedly reach into the M.I.R.

Since the Marcus model was initially a classical or later a semi-classical theory, the introduction of quantum effects was considered to account for these observations. In particular, tunnelling pathways of e.t. would increase the rates of some reactions. A closer analysis has, however, led to the conclusion that this could not be a general explanation [80].

4.1 Experimental Data of Rehm-Weller and Marcus Behaviour

Although we are concerned here specifically with photo-induced e.t., it is important to extend the discussion to some thermal e.t. processes for comparison. This is also justified by the fact that in the many instances of fully reversible processes the photo-induced step must be followed by a thermal back e.t. which restores the reactants, so that the entire sequence of events involves both light-induced and thermal e.t. reactions.

Most of the observations of the M.I.R. actually involve thermal e.t. processes; Fig. 11 gives an illustration of the rate constant versus energy plot of what appears to be a complete bell-shaped function, but this consists really of two distinct parts [25b]: the photo-induced forward process is restricted to a "normal" region, because of the restricted range of free energies; and similarly, the thermal back e.t. shows

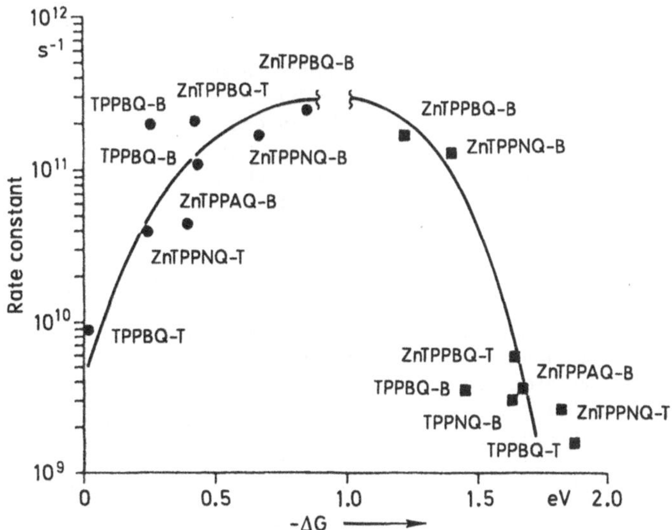

Fig. 11. Plot of rate constant vs. exothermicity for the reaction $^{1*}P+Q \rightarrow P^{+}+Q^{-}$ and for $P^{+}+Q^{-} \rightarrow P+Q$, where P = porphyrin and Q = quinone. Band T denote data obtained in butyronitrile or in toluene, respectively. The maximum uncertainty in any given rate constant is $\pm 20\%$

only the M.I.R., again as a result of the limited values of ΔGs. Nevertheless, this is one of the most convincing examples of the bell-shaped function, even though it does raise the question of the justification of fitting together two plots of e.t. reactions which probably follow quite different mechanisms.

Table 4. Examples of electron transfer reactions which show the Marcus Inverted Region

Acceptor	Donor	Remarks	Ref.
$TMPD^{\cdot +}$	$Perylene^{\cdot -}$	Charge recombin.	[99]
$Pyrene^{\cdot +}$	$TCNE^{\cdot -}$	Charge recombin.	
$DMA^{\cdot +}$	$Perylene^{\cdot -}$	Charge recombin.	
$Biphenyl^{\cdot +}$	$DCA^{\cdot -}$	Charge recombin.	[49]
$Biphenyl^{\cdot +}$	$TCA^{\cdot -}$	Charge recombin.	
2,6-Dimethyl-naphthalene$^{\cdot +}$	$TCA^{\cdot -}$	Charge recombin.	
Bichromophoric mol. Naphth.	$Biphenyl^{\cdot -}$	Intramolecular charge shift	[79]
Bichromophoric mol. Naphth.	$Biphenyl^{\cdot -}$	charge shift	[24]
Phenanth.	$Biphenyl^{\cdot -}$	charge shift	
p-benzoquinone	$Biphenyl^{\cdot -}$	charge shift	
$TMPD^{\cdot +}$	$Ru(bpy)^{\cdot +}$	Charge recombin.	[25a]
p-toluidine$^{\cdot +}$	$Ru(bpy)^{\cdot +}$	Charge recombin.	
N,N,N′,N′-tetra-methylbenzidine	$Ru(bpy)^{\cdot +}$	Charge recombin.	
Bichromophoric mol.			[25b]
Zn-porphyrin	*p*-benzoquinone	Intramol. charge recombination	
2H-porphyrin	*p*-benzoquinone	Intramol. charge recombination	
Zn-porphyrin	anthraquinone	Intramol. charge recombination	
2H-porphyrin	anthraquinone	Intramol. charge recombination	
Triphenyl-amine$^{\cdot +}$	anthraquin.$^{\cdot -}$	Triplet exciplex decay	[29]
p-methoxy-DMA$^{\cdot +}$	anthraquin.$^{\cdot -}$	Triplet exciplex decay	
$TMPD^{\cdot +}$	anthraquin.$^{\cdot -}$	Triplet exciplex decay	
p-benzoquinone Trapped e^-	in electron shift		[95]
Benzonitrile	anthraquin.$^{\cdot -}$	Triplet exciplex decay	
Trimethyl-pyridine	anthraquin.$^{\cdot -}$	Triplet exciplex decay	
Naphthalene	$Biphenyl^{\cdot -}$	Charge shift	[1c]
Nitrobenzene	$Biphenyl^{\cdot -}$	Charge shift	
TCNE	$Biphenyl^{\cdot -}$	Charge shift	
$Biphenyl^{\cdot +}$	$DCA^{\cdot -}$	Charge recombination	[102]
$Naphthalene^{\cdot +}$	$DCA^{\cdot -}$	Charge recombination	
$Phenanthrene^{\cdot +}$	$DCA^{\cdot -}$	Charge recombination	

›nce has been given for the occurrence of dielectric saturation of polar solvents e fields of ions and large dipoles, and here is an area where further research quired.

espective of the occurrence of dielectric saturation, it has been questioned her it would produce the desired result in any case, namely the distortion displacement of the potential wells as suggested by the M–K theory [89]. re 13 shows an outline of the effect of dielectric saturation on the energy wells e.t. process, following the M–K model (a) and the objections of other authors.

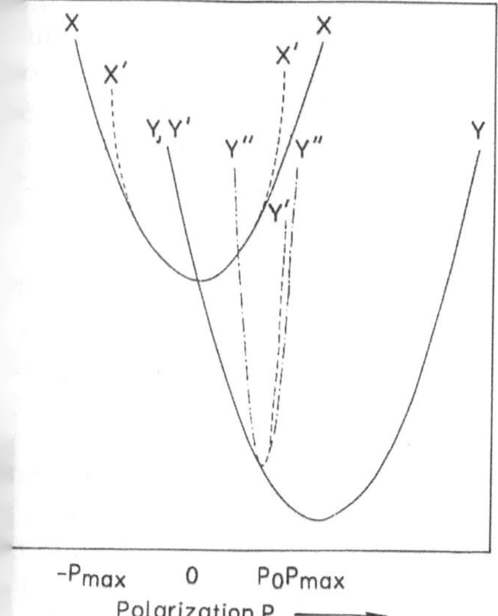

Fig. 13. Configuration coordinate diagram for the donor. The free energies E of the donor before and after electron transfer are shown as functions of the polarization P. The free energies of the neutral donor and the ion are shown by the parabolas denoted by X and Y, respectively. If one takes into account the dielectric saturation effect, these parabolas are modified as the curves denoted by X′ and Y′, respectively. The parabola denoted by Y″ is the free energy curve of the ion used in the Kakitani–Mataga model

From an experimental standpoint, there is a further complication which does t seem to have been discussed within the framework of the M–K theory, namely observations of the M.I.R. in nonpolar solvents. To take one example, bell-shaped rate constant vs energy plot shown in Fig. 11 contains points which resent observations of the same e.t. processes in a nonpolar solvent (toluene) well as a highly polar solvent. Dielectric saturation is usually described as being y of the orientational kind and no suggestion has been made that induction uration in a nonpolar solvent could also take place. It is then difficult to see, lowing the M–K model, why a saturable polar solvent and a nonsaturable vent should lead to similar behaviors with respect to the presence, or the sence, of the M.I.R.

A concensus has developed in recent years that the interesting M–K eory may not provide the right approach towards a better understanding of the fferences between e.t. systems which follow Rehm-Weller or Marcus behaviour. nother suggestion made recently by Rau therefore deserves attention [90]. This still based on the general observation that intermolecular systems limited by

In this case both e.t. processes are intramolecular, the distance between the porphyrin and the quinone being fixed by a short space. The highest rate constants of the photo-induced charge separation reach about $4 \times 10^{11}\ s^{-1}$ and these would not be observable in a diffusion controlled intermolecular reaction. In these measurements, two solvents were used – nonpolar toluene and polar butyronitrile – and the ΔG values are therefore subject to the uncertainties discussed in Sect. 2.1. In particular, the Coulomb term for butyronitrile assumes the point charge model with full solvent screening, which may well underestimate the free energy values, especially in such an intramolecular system where it is clear that the solvent cannot be inserted between the chromophores.

The use of intramolecular systems seems to be particularly indicated when the rate constants of e.t. (either forward, photo-induced or thermal return) are likely to be very large, exceeding the diffusion controlled limit [81]. There are indeed many reports of the observation of the M.I.R. in intramolecular e.t., too many to be analyzed individually. A survey of a selection of such reports will be found in Table 4.

It can be noticed that these processes are essentially charge recombinations, with a few examples of charge shifts [82]. Photoinduced charge separations between neutral chromophores do not appear in this table, either because the range of free energies was not sufficient to reach the M.I.R., or because the M.I.R. was not observed; this has carried on the trend already noted in the early 1980s, that photo-induced charge separations follow Rehm-Weller behaviour.

4.2 The Temperature Dependence of Rate Constants/Free Energies of Electron Transfer

The classical (or semiclassical) equation for the rate constant of e.t. in the Marcus-Hush theory is fundamentally an Arrhenius-Eyring transition state equation, which leads to two quite different temperature effects. The pre-exponential factor implies only the usual square-root dependence related to the activation entropy so that the major temperature effect resides in the exponential term. The quadratic relationship of the activation energy and the reaction free energy then leads to the prediction that the influence of the temperature on the rate constant should go through a minimum when $\Delta G^{\#}$ is zero, and then should increase as ΔG^0 becomes either more negative, or more positive (Fig. 12). In a quantitative formulation, the derivative dk/dT is expected to follow a bell-shaped function [83].

4.3 Theoretical Discussions of Marcus- and Rehm-Weller Behaviour in Electron Transfer Processes

The accumulation of experimental data concerned with the free energy dependence of the rates of e.t. has led to interesting problems of theoretical interpretations and controversies. It can be stated that at the present time the bulk of experimental observations has shown that photo-induced charge separations follow Rehm-

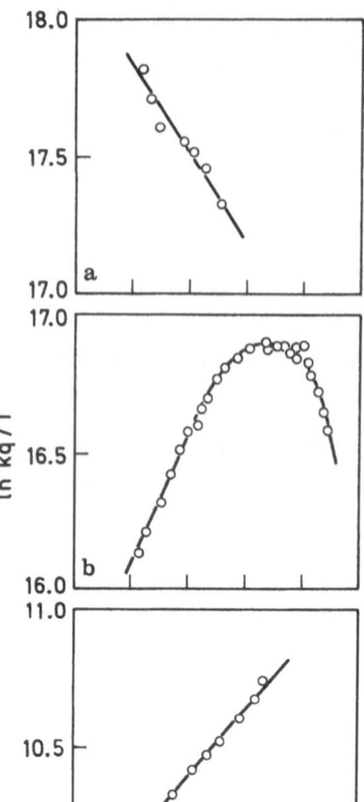

Fig. 12. Eyring plots for the quenching of $^*Ru(bpy)_3^{2+}$ by 1,4-naphthoquinone (**a**), duroquinone (**b**), and methyl m-nitrobenzoate (**c**) in acetonitrile

Weller behaviour, so that in intermolecular reactions the rate constant remains at the diffusion controlled limit in the most exergonic processes; in apparent contradiction with the results of Marcus theory [84]. On the other hand, observations of the M.I.R. have been made essentially in the case of thermal charge recombination reactions, with a few examples of charge shifts. Why should some e.t. systems follow the Marcus model, and others not? This is the fundamental theoretical question which has been addressed to by different authors in recent years and several interesting models have been suggested to account for the different behaviours of the Rehm-Weller and Marcus systems.

In the early 1980s, the classical Marcus theory was reanalyzed to consider the influence of quantum effects, notably electron tunneling [23]. Qualitatively, this gives the right direction, since it increases the observed rate constants when the thermally activated process becomes slow; but it was concluded that it could not account for the quantitative discrepancies of observed Rehm-Weller type plots from Marcus behaviour. For this reason the hypothesis was retained that the

absence of the M.I.R. in photo-induced e.t. between neutral
the formation of products in a electronically excited state – se
this happen, then the actual reaction free energy would be smal
energy of the products and could presumably be insufficient to
This hypothesis also appears quite reasonable on the grounds of
of molecular ions. It is well established that such open-shell
very low-lying excited states – as shown by their electronic
important point will be considered in more detail further on.

The fact remains that the formation of electronically ex
photo-induced e.t. reactions remains a hypothesis to this
evidence is very difficult to obtain, and it does not appear
investigation has been oriented in this direction so far. The presen
excited molecular ions could in principle be ascertained by t
their luminescence, but this is known to be generally so weak
the detection limits of most, if not all, currently available instrur
The conclusion concerning the intermediacy of electronically ex
photo-induced e.t. must therefore be that it is unproven, though no

It is perhaps because of the lack of experimental evidence for
excited products in such e.t. processes that quite different th
proposed in recent years to explain the absence of the M.I.R.
photo-induced e.t. reactions. The most comprehensive, and most
is the Mataga-Kakitani theory which emphasized the role o
controlling the rates of e.t. processes in polar solvents [86].

The Mataga–Kakitani (M–K) theory is based on the rather ger
that e.t. processes which show the M.I.R. are mostly charge rec
charge shifts, whereas the photo-induced charge separations w
neutral reactants follow Rehm-Weller behaviour. It is then su
difference is due to the electric field which acts on the solvent; in
or ion pairs, partial dielectric saturation of polar solvents would
this would restrict solvent motion. No such dielectric saturation e
in the solvent shell of neutral reactants, so that solvent motion remai

The only evidence for this partial dielectric saturation of polar
field of ions is the results of some computer simulations of mole
following the Monte-Carlo method [87]. It is claimed that these
saturation up to a distance of about 1 Å from the solute's mole
Such a thin shell does not accommodate a single solvent molecu
reason the dielectric saturation is called partial.

The M–K theory of the role of dielectric saturation has led to ma
and comments. In the first place, the occurrence of dielectric sat
established beyond doubt by molecular dynamics simulations; eve
are sometimes described as "computer experiments", they cannot
observations of the natural world. The results of such simulations
assumptions made in the model itself; other authors claim that dielec
becomes important only in the neighborhood of very small ions like
but that it is negligible in the solvent shell of the larger molecular
in e.t. processes [88]. At the present time, it appears therefore that no

diffusion usually do not show the M.I.R., whereas intramolecular or "pseudo intramolecular" systems usually do; a pseudo intramolecular system is for instance a contact pair of two ions which have been formed in a solvent cage, so that they cannot change their separation prior to e.t. When the e.t. partners approach from a great distance towards contact distance, the rate constant of e.t. varies according to the separation, because the free energy varies; for example, the Coulomb term becomes increasingly favorable at shorter separations. At very short distances, in particular at Van der Waals contact, the free energy could bring the system within the M.I.R., but e.t. would then take place at some larger distance when the free energy is such that the activation barrier approaches zero. This option is not open to a system of molecules or ions in contact, nor to a system in which the partners are held at a fixed distance – as in the case of many intramolecular e.t. processes between chromophores separated by rigid spacers.

5 General Discussion, Conclusions, and Potential Developments

At the time of writing this review, the major question remains in the understanding of the reasons for absence of the M.I.R. in most photo-induced e.t. processes. Observations of the M.I.R. in thermal charge shift and charge recombination reactions have now become commonplace, especially in intramolecular e.t.; as well as in formally bimolecular geminate ion pair neutralizations – here the molecules which form the ion pair may form a "supermolecule", something like an exciplex.

If it is assumed that the crucial difference between photochemical charge separations and thermal charge shifts or charge recombinations is the availability of low-lying excited state products, in the former, then a fairly obvious test would suggest itself: taking two open-shell reactant partners, e.t. would lead to closed shell products which do not have such low energy excited states. For example, a photo-induced e.t. between two organic radical ions should show the M.I.R., so long as the products are closed-shell species. No doubt such investigations will be carried out in the near future, since the formation of excited state products remains presently the most likely explanation for the absence of the M.I.R. in photo-induced e.t. processes.

Other explanations should not be dismissed lightly. Of these, the M–K theory based on the influence of dielectric saturation of polar solvents is currently the most controversial, and here also a relatively straightforward test can be suggested: If the absence of the M.I.R. in photo-induced e.t. reactions between two neutral reactant partners is due to the absence of dielectric saturation in polar solvents, then the M.I.R. should be observed in a photo-induced e.t. reaction between charged partners, for instance a phenolate donor and an oxonine acceptor. The reverse should be true of a thermal charge recombination between two neutrals when e.t. would lead to the formation of two ions, a very exceptional process – though not an impossible one. Quite apart from the question of the M.I.R. itself, the problem of dielectric saturation of polar solvents in the field of ions and giant dipoles will deserve more attention. In this respect, there already seems to be a weakness in the M–K theory, since the observation of the M.I.R. is not related to the polarity of the solvent (see Fig. 10 and the related discussion). It might be argued that a polar solvent in a state of dielectric saturation comes to resemble a nonpolar solvent, but this would be true only if complete dielectric saturation of at least the first solvation shell would take place; this would require dielectric saturation to extend over some 5 Å at least from the edge of the solute ion, a range well beyond the expectations even of the M–K model.

It is then important to look more closely at the suggestion of Rau et al., based on the distinction between diffusionless and diffusion-controlled e.t. reactions. The apparent weakness of this model is that it would predict the observation of the M.I.R. for photo-induced intramolecular (diffusionless) processes, but so far this has not been confirmed experimentally. Nevertheless, the importance of the distance between the partners in intermolecular e.t. is most interesting. There are many reports of long-range e.t. reactions, though mostly in rigid polar glasses. Extended to diffusion controlled e.t. between freely moving partners, the occurrence of long-range e.t. would affect directly both the reaction free energy (notably through the Coulomb term) and the outer-sphere (solvent) reorganization energy in polar solvents.

5.1 Comparison of the Marcus Model with the Theory of Radiationless Transitions; is the Marcus Inverted Region Related to the Energy Gap Law?

Although the existence of the M.I.R. may have appeared counter-intuitive to many chemists, photophysicists had a different point of view, since an "inverted" relationship of the rate constant of nonradiative transitions and the energy difference between the states is well established [91]. This energy gap law results from the decreasing vibrational overlap of electronic states, the so-called Franck-Condon factor. It predicts an exponential relationship of the rate constant of nonradiative deactivation of excited states with the energy gap, of the form:

$$k_{nr} = C \exp(-k\,|\Delta E|) \tag{14}$$

There are however essential differences between the Marcus model and the theory of radiationless transitions. In the former, the decrease of the rate constant in the inverted region results from an activation barrier which must be overcome by thermal energy, whereas the rates of radiationless transitions are in principle temperature independent. As implied in [14], there is no "normal" region in the case of nonradiative transitions, a no bell-shaped curve is expected from the plot of the rate constant against the energy gap.

The temperature dependence of the rates of e.t. processes is therefore the obvious criterion for a choice between the Marcus model and a model of e.t. as a type of radiationless transition. In this respect, it must, however, be noted that an apparent temperature effect may appear in the theories of radiationless transitions, if the coupling of higher vibrational levels of the initial and final states is more favorable [92]. This is illustrated in Fig. 14, based on a model of hydrogen abstraction reaction treated as radiationless transitions. If the vibrational overlap increases in the higher vibrational levels ($> v^2$), then a temperature effect is expected since such higher levels are increasingly populated at higher temperatures. However, the activation barrier itself would not be constant and the Arrhenius plot of ln k versus T is expected to be curved. For this reason an assessment of the models of e.t. processes as either thermally activated reactions (the Marcus model) or as radiationless transitions must be based on the observation of Arrhenius plots over temperature ranges large enough to yield a clear-cut difference between linear and curved functions.

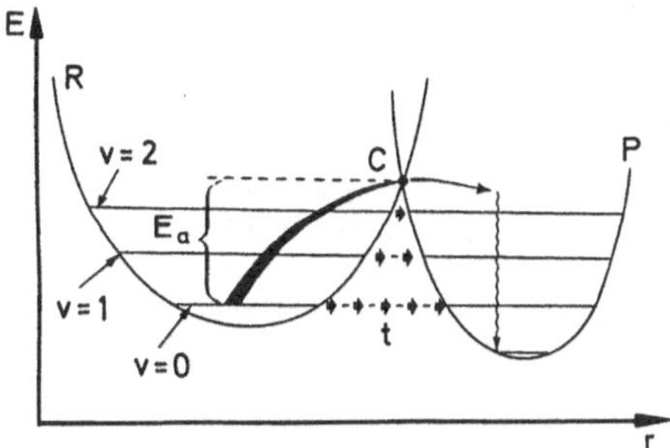

Fig. 14. Classical and tunneling pathways for crossing from the potential-energy surface of a reactant R to a product P. In the classical case crossing can take place only at the point of intersection C, and this requires an activation energy E_a from the lowest vibrational level v = 0. In the quantum-mechanical model nuclear tunneling (t) provides an alternative pathway which requires no activation energy. The tunnel width, which determines the rate of tunneling, decreases for higher vibrational levels (v = 1, v = 2), but such levels can be populated only by thermal activation. In this figure r represents the nuclear coordinate involved in the reaction

The current evidence concerning the details of the temperature dependence of e.t. processes, cannot be considered as decisive one way or the other; it may be surmised that this will become an area of very active research in the future. The reported observations of a bell-shaped temperature dependence of some e.t. processes at least is, however, not in favour of the radiationless transition model, because this does not predict the existence of any "normal" region in the rate constant vs energy function. On the other hand, the observation of temperature independent e.t. processes of widely different exergonicities (Ref. 101) would certainly speak against the Marcus model, which according to Eq. 3 predicts that the exponential factor which contains the $\Delta G^{\#}/RT$ term becomes constant only in the special condition $\Delta G^0 \simeq \lambda$.

Acknowledgement. This work is part of project No. 20-28.842.90 of the Swiss National Foundation for scientific research.

6 References

1. a. Dogonadze RR (1972) Theory of molecular electrode kinetics. In: Hush, NS (ed) Reactions of Molecules at Electrodes Wiley, New York
   b. Levich VG, (1970) Kinetics of reactions with charge transport. In: Eyrich H (ed) Physical chemistry Vol. 9B, Academic Press, New York
   c. Miller JR, Beitz JV, Huddleston RK (1984) J Am Chem Soc 106: 5057
   d. Marcus RA (1960) Discuss Faraday Soc 29: 21
   e. Leonhardt H and Weller A (1963) Ber Bunsenges Phys Chem 67: 791
   f. Hush NS (1961) Trans Faraday Soc 57: 557
   g. For a review of applications of electron transfer reactions, see: Julliard M and Chanon M (1983) Chem Rev 83: 425 (1983)
2. Hoffmann R and Woodward RB (1968) Woodward, Acc Chem Res 1: 17
3. Oevering H, Verhoeven JW, Paddon-Row MN and Warman J (1989) Tetrahedron 45: 4751
4. Parker CA (1968) Photoluminescence of solutions, Elsevier, Amsterdam
5. Marcus RA (1964) Ann Rev Phys Chem 15: 155
6. Hupp JT and Weaver MJ (1984) J Phys Chem 88: 1463
7. Cannon RD (1980) Electron transfer reactions, Butterworth, London
8. Marcus RA (1956) J Chem Phys 24: 966
9. Brunschwig BS, Ehrenson S and Sutin N (1987) J Phys Chem 91: 4714
10. Kim D, Lee ISH and Kreevoy MM (1990) J Am Chem Soc 112: 1889
11. Closs GL, Johnson MD, Miller JR and Piotrowiak P (1989) J Am Chem Soc 111: 3751
12. Pasman P, Rob F and Verhoeven JW (1982) J Am Chem Soc 104: 5127
13. Hupp JT and Weaver MJ (1985) J Phys Chem 89: 1601
14. Meyer TJ (1978) Acc Chem Res 11: 94
15. Eberson L and Shaik SS (1990) J Am Soc 112: 4484
16. Jortner J and Bixon M (1988) J Chem Phys 88: 167
17. Kahlow MA, Kang TJ and Barbara PF (1987) J Phys Chem 91: 6452
18. Weaver MJ and Gennett T (1985) Chem Phys Lett 113: 213
19. Sutin N (1986) J Phys Chem 90: 3465
20. Rehm D and Weller A (1970) Israel J Chem 8: 259
21. Indelli MT, Ballardini R and Scandola F (1984) J Phys Chem 88: 2547
22. Shida T, Haselbach E and Bally T (1984) Acc Chem Res 17: 180
23. Siders P and Marcus RA (1981) J Am Chem Soc 103: 748
24. Miller JR, Calcaterra LT and Closs GL (1984) J Am Chem Soc 106: 3047
25. a. Ohno T, Yoshimura A and Mataga N (1986) J Phys Chem 90: 3295
    b. Wasielewski MR, Niemczyk MP, Svec WA and Pewitt EB (1985) J Am Chem Soc 107: 1080
    c. Vauthey E, Suppan P, Haselbach E (1988) Helv Chim Acta 71: 93
26. Rehm D and Weller A Ber Bunsenges (1969) 73: 834
27. Goodman JL and Peters KS (1986) J Am Chem Soc 108: 1700
28. a. Rossbroich G, Garcia NA and Braslavski SE (1985) J Photochem 31: 37
    b. Poston P and Harris J (1990) J Am Chem Soc 112: 644
29. Levin PP, Pluzhnikov PF and Kuzmin VA (1988) Chem Phys Lett 147: 283
30. Lewis FD, Dykstra RE, Gould IR and Farid S (1988) J Phys Chem 92: 7042
31. Glasstone S, Laidler K and Eyring H (1941) Theory of rate processes, McGraw-Hill, New York p 519
32. Iwa P, Steiner UE, Vogelmann E and Kramer H (1982) J Phys Chem 86: 1277
33. Kubota T, Uno B, Kano K, Kawakita T and Goto M (1990) Bull Chem Soc Japan 63: 516
34. Born M (1920) Z Physik 1: 45
35. Bucher M and Porter TL (1986) J Phys Chem 90: 3406
36. Woodward CE and Nordholm S (1988) J Phys Chem 92: 497
37. Suppan P and Vauthey E (1989) J Photochem Photobiol A 49: 239

38. Porter G (1983) Photochemical conversions, Presses Polytechniques Romandes, Lausanne
39. Chibisov AK (1984) Progr React Kinetics 13: 1
40. Suppan P (1986) J Chem Soc Faraday Trans 1, 82: 509
41. Rudzki JE, Goodman JL and Peters KS (1985) J Am Chem Soc 107: 7849
42. Rosencwaig A, Photoacoustics and photoacoustic spectroscopy (Chemical Analysis Vol. 57) Wiley Interscience, New York (1980).
43. a. Komorowski SJ, Grabowski ZR and Zielenkiewicz W (1985) J Photochem 30: 141
    b. Terazima M and Azumi T (1990) Bull Chem Soc Jap 63: 741
44. Tam AC and Confal H (1984) Appl Physics Letters 42: 33
45. Isak SJ, Komorowski SJ, Merrow CN, Poston PE and Eyring EM (1989) Appl Spectroscopy 43: 419
46. Kaupp G and Jostkleigrewe E (1982) Angw Chem Suppl 1089
47. de Schijver FC, Boens N and Put J (1977) Adv Photochem 10: 359
48. Shida T (1983) Molecular ions: spectroscopy, structure and chemistry, ed Miller TA and Bondybey VE, North Holland, Amsterdam
49. Gould IR, Eges D, Mattes SL and Farid S (1987) J Am Chem Soc 109: 3794
50. Gschwind R and Haselbach E (1979) Helv Chim Acta 62: 941
51. McClelland BJ (1983) J Chem Soc Faraday Trans 2, 79: 1233
52. Guerry-Butty E, Haselbach E, Pasquier C, Suppan P and Phillips D (1985) Helv Chim Acta 68: 912
53. Kavarnos GJ and Turro NJ (1986) Chem Rev 86: 401
54. Devadoss C and Fessenden RW (1990) J Phys Chem 94: 4540
55. Eigen M (1954) Z Phys Chem (Wiesbaden) 1: 176
56. Heitele H, Pöllinger F, Weeren S and Michel-Beyerle ME (1990) Chem Phys Letters 168: 598
57. Grampp G, Harrer W and Jaenicke W (1987) J Chem Soc Faraday Trans 1, 83: 161
58. Hush NS (1958) J Chem Phys 28: 962
59. Simon JD and Su S-G (1988) J Phys Chem 92: 2395
60. Chandler D and Kuharski RA (1988) Faraday Disc Chem Soc 85: 329
61. Strauch S, McLendon G, McGuire M and Guarr T (1983) J Phys Chem 87: 3579
62. Domingue RP and Fayer MD (1985) J Chem Phys 83: 2242
63. Milosavljevic BH and Thomas JK (1983) J Phys Chem 87: 616
64. Perrin F (1924) CR Acad Sciences 178: 1978
65. Miller JR (1972) J Chem Phys 56: 5173
66. Zamaraev KI, Kairutdinov RF and Miller JR (1978) Chem Phys Letters 57: 311
67. Kakitani T and Mataga N (1988) J Phys Chem 92: 5059
68. Conklin KT and McLendon G (1988) J Am Chem Soc 110: 3346
69. Heitele H, Michel-Beyerle ME and Finckh P (1987) Chem Phys Letters 138: 237
70. Scherrer POJ and Fischer SF (1987) Chem Phys Letters 141: 179
71. Jortner J (1976) J Chem Phys 69: 4860
72. Vauthey E and Suppan P (1989) Chem Phys 139: 381
73. Rips I and Jortner J (1988) J Chem Phys 88: 818
74. Rettig W (1986) Angew. Chem. 98: 969
75. McManis GE, Golovin MN and Weaver MJ (1986) J Phys Chem 90: 6563
76. Truong TB (1984) J Phys Chem 88: 3906
77. Schmickler W (1976) J Chem Soc Faraday Trans 2, 72: 307
78. Ballardini R, Varani G, Indelli MT, Scandola F and Balzani V (1978) J Am Chem Soc 100: 7219
79. Liang N, Miller JR and Closs GL (1989) J Am Chem Soc 111: 8740
80. Siders P and Marcus RA (1981) J Am Chem Soc 103: 741
81. Pasman P, Mes GF, Koper NW and Verhoeven JW (1985) J Am Chem Soc 107: 5839
82. a. Miller JR and Beitz JV (1981) J Chem Phys 74: 6746
    b. Gould IR, Moser JE, Armitage B and Farid S, Goodman JI and Herman MS (1989) J Am Chem Soc 111: 1917

83. Kim H-B, Kitamura N, Kawanishi Y and Tazuke S (1987) J Am Chem Soc 109: 2506
84. Balzani V, Bolletta F, Gandolfi MT and Maestri M (1978) Top Curr Chem 75: 1
85. Shida T (1988) Electronic absorption spectra of radical ions, Elsevier, Amsterdam
86. a. Kakitani T and Mataga N (1985) Chem Phys 93: 381
    b. Kakitani T and Mataga N (1986) J Phys Chem 90: 993
    c. Kakitani T and Mataga N (1987) J Phys Chem 91: 6277
87. Hatano Y, Saito M, Kakitani T and Mataga N (1988) J Phys Chem 92: 1008
88. Heinziger K and Palinkas G (1985) Stud Phys Theor Chem 38: 313
89. a. Tachiya M (1989) Chem Phys Lett 159: 505
    b. Tachiya M (1989) J Phys Chem 93: 7050
90. Greiner G, Pasquini P, Weiland R, Orthwein H and Rau H (1990) J Photochem Photobiol A 51: 179
91. Lee EKC and Loper GL (1980) Radiationless transitions, p 7 ed. Lin SH, Acad Press, New York
92. Formosinho SJ (1976) J Chem Soc Faraday Trans 2, 72: 1313
93. Miller JR, Hartman KW and Abrash S (1982) J Am Chem Soc 104: 4296
94. Guarr T, McGuire ME and McLendon G (1985) J Am Chem Soc 107: 5104
95. Beitz JV and Miller JR (1979) J Chem Phys 71: 4579
96. Beitz JV and Miller JR (1981) J Chem Phys 74: 6746
97. Miller JR, Peeples JA, Schmitt MJ and Closs GL (1982) J Am Chem Soc 104: 6488
98. Brickenstein EK, Ivanov GK, Kozhushner MA and Khairutdinov RF (1984) Chem Phys 91: 133
99. Mataga N, Asahi T, Kanda Y, Okada T and Kakitani T (1988) Chem Phys 127: 249
100. Eberson L (1984) Acta Chem Scand B 38: 439
101. Gunner MR and Dutton PL, (1989) J Am Chem Soc 111: 3400
102. Gould IR, Ege D, Moser JE and Farid S (1990) J Am Chem Soc 112: 4290
103. Chen P, Duesing R, Tapolsky G and Meyer TJ (1989) J Am Chem Soc 111: 8305

Structure and Reactivity of Organic Radical Cations

Heinz D. Roth

Rutgers University, Wright-Rieman Laboratories, New Brunswick, N.J. 08855, USA

Table of Contents

Topics in Current Chemistry, Vol. 163

Photoinduced electron transfer has attracted much attention because of its central role in the chemistry of life and as a method for generating radical ion pairs in solution. Radical cations so generated can be characterized by many spectroscopic techniques, and their reactions can be studied conveniently. Photoinduced electron transfer is compared with alternative methods of radical ion generation; the application of these methods is discussed in conjunction with suitable techniques of observation.

The presence of both spin and charge in radical cations allows them to undergo elementary reactions typical for either free radicals or cations: unimolecular reactions such as rearrangements, isomerizations, cycloreversions, or fragmentations; or bimolecular reactions involving neutral or ionic substrates, free radicals, or radical ions of like or opposite charge and including additions, cycloadditions, hole transfer, complex formation, nucleophilic capture, spin labeling, disproportionation or dimerization, reverse electron transfer, proton transfer, atom or group transfer; or coupling.

Radical cation structure types can be classified according to the nature of the donor molecules, viz., π-, n-, or σ-donors, from which they are generated. Radical cations derived from typical π-donors may be closely related to the structure of their precursors, whereas substantial differences may be observed between the structures of radical cation and precursor for σ-donors. The potential surfaces of radical cations and their parents may differ in three features: reaction barriers may be reduced, free energy differences between isomers may be reduced or reversed; and energy minima on the radical cation surface may have geometries corresponding to transition structures on the parent potential surface. The pursuit of such novel structure types has given new direction to radical cation chemistry. Representative radical cation structures are discussed to document their rich variety and to illustrate the molecular features that determine their structures.

1 Introduction

Photoinduced electron transfer is one of the very few fundamental reaction types known in excited-state organic chemistry; rearrangements, fragmentations, additions, substitutions, insertions, and abstraction reactions being the others. Light induced electron transfer occurs in the solid state, in solution, as well as in the gas phase, including the Earth's atmosphere and outer space. Depending on the nature of the substrate and on the medium in which this reaction either occurs or is carried out, the products can be macromolecular "charge-separated" entities (in the photosynthesis of the green plant) [1], zwitterions, ion pairs, or essentially "free" radical ions, carrying a positive or negative charge in addition to an unpaired spin. This chapter deals with the structures and reactions of the positively charged variety of these species.

Organic radical cations, their structures and their reactions have been attracting ever increasing attention for over the last decade [2–11]. They are important intermediates in a wide variety of chemical and biological processes, encompassing photosynthesis [1] and interstellar chemistry [12] and ranging from conducting polymers [13] to synthetically useful reactions, such as electron transfer induced anti-Markovnikov additions [14], cation radical Diels-Alder reactions [10], or nucleophilic substitution reactions [15]. In addition, they command attention because of a multitude of unusual structure types and a variety of interesting reactions. The pertinent literature includes an impressive volume of work on donor-acceptor photochemistry in general, but also numerous papers specifically aimed at establishing the reactivity of these species. An equally impressive variety of spectroscopic techniques has been applied to establish their structural features.

In this chapter we will discuss various aspects of radical cation chemistry, including: 1) experimental methods to generate these intermediates; 2) spectroscopic techniques employed to observe radical cations and to characterize their properties and structural features; 3) an overview of selected interesting structure types; and 4) a cross section of representative radical cation reactions. Organic radical ions undergo a plethora of intramolecular transformations as well as intermolecular reactions with molecules, free radicals, and ionic substrates of like or opposite charge. The full range of reaction types goes far beyond the scope of this article. We will delineate and illustrate only selected examples. We begin with a discussion of the various methods available for the generation of radical cations.

2 Methods of Generation

2.1 Chemical Oxidation

A rich variety of chemical oxidizing reagents have been applied for the generation of radical cations. The principal reagent types include: Brønsted and Lewis acids; the halogens; certain peroxide anions or radical anions; numerous metal ions or oxides; nitrosonium and dioxygenyl ions; stable organic (aminium) radical cations;

semiconductor surfaces; and certain zeolites. The wide variety of reagents with a wide range of reduction potentials may make it possible to choose a reagent with a one-electron reduction potential sufficient for the desired oxidation and a two-electron reduction potential insufficient for further oxidation of the radical cation.

Among the Brønsted acids sulfuric acid was employed as early as 1835, when Laurent investigated a residue obtained from the oil of bitter almonds [16]. Treatment of one component, possibly diphenylamine, with sulfuric acid gave rise to a blue solution. Eighty years later Kehrmann applied this reagent in the one- and two-electron oxidation of phenothiazine and characterized the "semiquinoid" and "holoquinoid" forms by UV and visible spectroscopy [17, 18]. Later, colored solutions of aromatic hydrocarbons (perylene, anthracene) in sulfuric acid were found to be paramagnetic [19] and, shortly thereafter, their radical cations were postulated based on both optical [20] and ESR spectroscopic data [21]. While these oxidations are certainly interesting, details of their mechanism are still in question. We do not expect this oxidant to be of much value in generating radical cations of strained ring or cage compounds.

Lewis acids such as $AlCl_3$, $SbCl_5$, or PF_5 have been used successfully to generate a variety of radical cations. Antimony pentachloride was first used with hydrocarbons such as benzene or anthracene [22, 23]. Salts obtained from aromatic amines with this reagent were found to be paramagnetic [24]; eventually, well resolved ESR spectra identified the formation of radical cations [25, 26]. Although an electron transfer mechanism must be involved, the fate of the complementary radical anions and details of their decay are poorly understood. Once again, it appears doubtful that Lewis acids are suitable oxidants for the study of the sometimes delicate substrates discussed in this review.

As for halogens as oxidizing reagents, bromine has proved more useful than its homologs chlorine and iodine. It was employed as early as 1879 on di- and tetramethyl-*p*-phenylenediamine [27–29] and early in this century, Wieland used it to generate the aminium salts of triarylamines and tetraarylhydrazines [30, 31]. Since bromine adds readily to unsaturated as well as to some strained ring compounds, it is not expected to be very useful in the context of the radical cations discussed here.

A particularly attractive method of generation involves the use of nitrosonium and dioxygenyl ions, as the reduction products of these agents are gases and, thus, potentially easy to remove [32, 33]. Even at cryogenic temperatures NO and O_2 may diffuse away from the product cation. Dioxygenyl ion is the stronger oxidant of the two and its reduction product is less likely to interact with either the substrate or the resulting radical cation. At sufficiently low temperatures, both reagents may be quite useful for the preparation of radical ion salts [33].

Recently, Ramamurthy and colleagues demonstrated that certain zeolites, including Na-ZSM-5, spontaneously oxidize a variety of olefinic or aromatic substrates (Fig. 1) [34, 35]. Zeolites have been utilized frequently as supporting matrix materials [36–38]. These materials contain host cavities of well defined geometries and allow molecules of appropriate shapes to be incorporated. Typically, the host contained in the zeolite is oxidized by exposure to ionizing radiation (vide infra), and the resulting radical cation is protected against ion

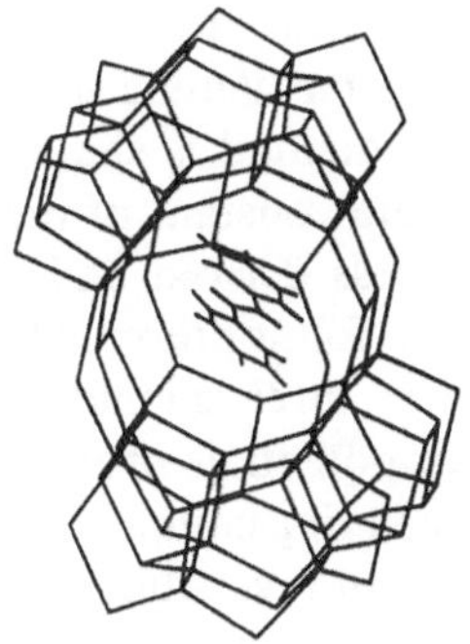

Fig. 1. Model showing the structure of *trans*-stilbene included in the channels of ZSM-5. The view is along the direction of the straight channel (crystallographic b axis) [35] (Reprinted with permission)

molecule or other bimolecular scavenging reactions [36–38]. The particular Na-ZSM-5 zeolites have nearly cylindrical cavities which incorporate rod-like (oblong) molecules. In addition, they contain metal ions within the cavities, which are capable of oxidizing a variety of substrates. The "reduction potentials" of these materials are reasonably well defined; they were determined by incorporating into the solid a series of 4-substituted *trans*-stilbenes with appropriate oxidation potentials and following the extent of oxidation; these experiments suggest a redox potential near 1.65 V. The authors consider this method general in scope and simple in practice.

2.2 Electrochemical Oxidation

A variety of electrochemical techniques have been developed for the generation and study of organic radical cations. In principle they fall into two classes, bulk electrolysis (coulometric) techniques and micro-electrode (voltammetric) techniques [39, 40]. Among these methods, cyclic voltammetry (CV) is by far the most frequently applied technique. This method utilizes a small stationary electrode, typically of platinum, in an unstirred solution. The potential of the electrode versus a saturated calomel electrode (SCE) is increased linearly with time over an appropriate range and then linearly decreased to the starting value. This experiment leaves the bulk of the electrolyte solution essentially unchanged, since the oxidation products are formed only in the immediate vicinity of the anode. During the experiment, the current is recorded as a function of the applied potential. The resulting curve (cyclic voltammogram) may provide important details about the oxidation product, e.g. whether it is formed by one- or two-electron oxidation and whether it is stable for the duration of forward and reverse sweep. If the oxidation product is stable under the conditions of the experiment, the reaction is reversible, and the oxidation potential for the reaction

$$S \rightarrow S^{+\cdot} + e$$

can be determined from the anodic peak potential. Thus, electro-chemical studies can provide evidence for the existence of organic radical cations, for their degree of stability, and for the energetics of their formation.

Unfortunately, this method is not generally applicable. In cases where the product of one-electron transfer ($S^{+\bullet}$) is unstable or has a shorter lifetime than required for a reversible electrode reaction, the anodic peak potential is shifted toward less positive values, depending on the scan rate, the rate constant of the decay reaction, and the reaction order. The oxidation potentials derived under these conditions are not reliable. On occasion, the peak potentials for nonreversible reactions are quoted in the literature, sometimes to as many as three significant figures. The critical reader will realize that nonreversible peak potentials provide a basis only for qualitative comparisons. Since few, if any, strained ring compounds have stable one-electron oxidation products, their oxidation is not reversible, and reliable oxidation potentials cannot be determined by the CV method.

Controlled potential electrolysis or coulometry can be used to generate radical ions in quantities sufficient for study by appropriate techniques such as optical or EPR spectroscopy. This method is routinely applied to characterize radical anions and has also been used extensively for studying radical cations. However, the application of coulometric techniques to the study of strained ring compounds is severely limited, even more than the application of cyclic voltammetry, by the limited stability of their one-electron oxidation products.

2.3 Radiolytical Generation of Radical Cations

The application of radiolysis for the generation of organic radical anions or cations was pioneered by W. H. Hamill [41–43]. This method utilizes the fact that high-energy photons such as X- or γ-rays cause ejection of highly energetic electrons from appropriate target (matrix) materials. The high-energy electrons, in turn, excite and ionize molecules present in the same system. Although the chemical effects of ionizing radiation are highly complex, this method can be used for the selective generation of radical cations chosen for study. A variety of halogen containing molecules can be used as matrix materials, among them tetrachloromethane, the chlorobutanes, and several freon mixtures. High-energy photons interact with these materials by ejecting electrons, while at the same time generating unstable radical cations (Eq. 1). Subsequently, appropriate solute molecules (S) at concentrations as low as 1 in 10^3 are oxidized, regenerating the matrix material (Eq. 2). The ejected electrons attach themselves to a halogen containing molecule (Eq. 3), causing its fragmentation to a halide ion and a free radical (Eq. 4). These processes are without direct consequence for the radical cation and are mentioned here only because they may interfere with the observation of the species chosen for study, or because the halide ions may scavenge the radical cations to form free radicals (Eq. 5).

$$\text{R–X} + \gamma \longrightarrow \text{RX}^{\bullet+} + e^- \quad (1)$$

$$\text{R–X}^{\bullet+} + \text{S} \longrightarrow \text{R–X} + \text{S}^{\bullet+} \quad (2)$$

$$\text{R–X} + e^- \longrightarrow \text{R–X}^{\bullet-} \quad (3)$$

$$\text{R–X}^{\bullet-} \longrightarrow \text{R}^{\bullet} + \text{X}^- \quad (4)$$

$$\text{S}^{\bullet+} + \text{X}^- \longrightarrow \text{S}^{\bullet}\text{–X} \quad (5)$$

Scheme 1

The application of this method has facilitated the generation and study of a substantial number of radical cations. However, the relatively high reduction potentials of the primary (matrix derived) radical cations combined with the relatively slow dissipation of excess energy in rigid matrices constitute a distinct drawback for the application of this method in the study of radical cations derived from strained molecules. It is generally recognized that the barriers to the rearrangements of radical cations are considerably lower than are those on the parent energy surfaces. As a result, some of the most interesting radical cations do not survive under the conditions of radiolytical generation.

2.4 Photoinduced Electron Transfer

A mild and versatile method for the generation of radical cation–radical anion pairs in solution is based on photoinduced electron transfer (PET; Scheme 2) [44–46]. This method utilizes the fact that the oxidative power of an acceptor and the reductive power of a donor are substantially enhanced by photoexcitation (Fig. 2). Thus, for donor–acceptor pairs with negligible or weak interactions in the ground state, electronic excitation of either reactant may lead to the generation of radical ion pairs via electron transfer. For the study of radical cations it is advantageous to excite the acceptor (Eq. 6). Depending on the nature of the acceptor and its lifetime, it will be quenched before or after intersystem crossing to the triplet state. In Scheme 2, the reaction is formulated for triplet quenching (Eq. 7), which generates radical ion pairs of triplet spin multiplicity. Even so, the resulting radical ions have limited lifetimes since the pairs readily undergo intersystem crossing (Eq. 8), followed by recombination of the singlet pairs (Eq. 9) or separation by diffusion (Eq. 10) to generate "free" radical ions.

$$A \longrightarrow {}^{1}A^{*} \longrightarrow {}^{3}A^{*} \tag{6}$$

$${}^{3}A^{*} + D \longrightarrow {}^{3}\overline{A^{\bullet-}D^{\bullet+}} \tag{7}$$

$${}^{3}\overline{A^{\bullet-}D^{\bullet+}} \longrightarrow {}^{1}\overline{A^{\bullet-}D^{\bullet+}} \tag{8}$$

$${}^{1}\overline{A^{\bullet-}D^{\bullet+}} \longrightarrow A + D \tag{9}$$

$${}^{3}\overline{A^{\bullet-}D^{\bullet+}} \longrightarrow {}^{2}A^{\bullet-} + D^{\bullet+} \tag{10}$$

Scheme 2

The most common triplet state electron acceptors are ketones and quinones, whereas aromatic hydrocarbons, often bearing one or more cyano groups, are the most frequently used singlet state electron acceptors. For the generation of radical cations from a given donor it is important that the exothermicity of the electron transfer reaction can be adjusted to fall within an appropriate range, typically

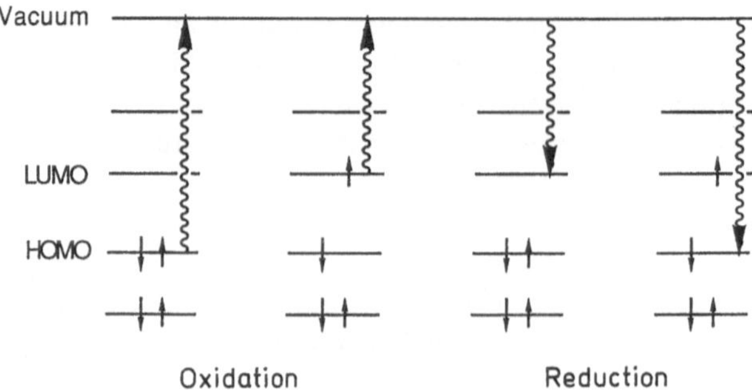

Fig. 2. A simple model for the enhanced redox reactivity upon photoexcitation

between 0.2 and 1.0 eV. The change in free energy (ΔG) for an electron transfer reaction is given by the Rehm–Weller equation (Eq. 11) [47]

$$\Delta G = -E_T - E_{red} + E_{ox} - e^2/\varepsilon a\,, \tag{11}$$

where E_T is the excited state energy (0–0 transition), E_{ox} is the one-electron oxidation potential of the donor, E_{red} is the one-electron reduction potential of the acceptor, and $e^2/\varepsilon a$ is the Coulomb term which accounts for ion pairing. Alternatively, one can define the reduction potential of the acceptor excited state:

$$^*E_{red} = -E_T + E_{red} \tag{12}$$

The application on PET is limited to the oxidation of substrates with oxidation potentials well below the threshold value defined by E_{red}. According to Eq. (11) the change in free energy of the reaction can be tuned by variation of the solvent (polarity) and of the acceptor (reduction potential, excited-state energy). For a given class of acceptors the excited state energies typically vary over a narrow range, whereas the reduction potentials can be altered substantially by the introduction of appropriate substituents (Table 1). Therefore, it does not generally present a problem to adjust the exothermicity to an appropriate range.

Two types of competing reactions pose potential drawbacks to the PET method. First, the principal types of electron acceptors are ketones and quinones, whose triplet states are known to abstract atoms with formation of neutral radicals. Second, many of the radical cations generated by PET are potential proton donors and the radical anions are comparably strong bases. Accordingly, proton transfer in the geminate radical ion pair may produce neutral radicals, and the potential involvement of two or more competing reactions may introduce mechanistic ambiguities. On the other hand, this feature has made it possible to study interesting electron transfer/proton transfer sequences. Both reaction types will be discussed in Section 4.3.

Table 1. Reduction potentials of selected electron acceptors

Acceptor	$E_{A/A}$ [a]	$E_{(0,0)}$ [b]	$^3E^*$ [c]	$^*E_{A/A}$ [d]
Singlet acceptors				
2,4,6-Triphenylpyrylium tetrafluoroborate	−0.29[48]	2.8[45]		2.5
2,6,9,10-Tetracyanoanthracene (TCA)	−0.45[45]	2.82[45]		2.35
1,2,4,5-Tetracyanobenzene (TCB)	−0.65[45]	3.83[45]		3.2
9,10-Dicyanoanthracene (DCA)	−0.89[45]	2.88[45]		2.0
1,4-Dicyanonaphthalene (DCN)	−1.28[45]	3.45[45]		2.15
9-Cyanoanthracene (CA)	−1.39[45]	2.96[45]		1.55
p-Dicyanobenzene (*p*-DCB)	−1.60[45]	4.29[49]		2.7
9-Cyanophenanthrene (CP)	−1.88[50]	3.42[45]		1.55
1-Cyanophthalene (1-CN)	−1.98[51]	3.75[45]		1.75
2-Cyanonaphthalene (2-CN)	−2.13[51]	3.68[45]		1.55
Naphthalene (N)	−2.50[52]	3.97[45]		1.45
Phenanthrene (P)	−2.45[52]	3.58[45]		1.15
Anthracene (A)	−1.96[52]	3.28[45]		1.3
Triplet acceptors				
Chloranil (CA)	+0.02[52]		2.7[53]	2.7
Benzoquinone	−0.54[52]		2.95[53]	2.4
Naphthoquinone	−0.60[52]			
Anthraquinone	−0.94[54]		2.72[54]	1.8
Benzil	−1.50[54]		2.36[54]	0.85
Benzophenone	−2.16[54]		2.95[54]	0.8

[a] Half wave reduction potential (V) vs SCE
[b] Singlet energy (eV) from the 0,0 transition of the fluorescence spectrum
[c] Triplet energy (eV)
[d] Excited state reduction potential

2.5 Electron Impact Ionization

Electron impact ionization is the method of choice for the generation of "molecular ions" to be analyzed in a mass spectrometer [55, 56]. This method is based on essentially the same principle as radiolytical generation, but it is used at considerably lower pressures, typically 10^{-6} torr. High energy electrons, tuned to energies between 10 and 70 eV, are ejected from a heated filament and impact on molecules contained in an evacuated ionization chamber. In the collision, energy is transferred to the substrate causing ionization.

Molecular ions and ions closely related to them can also be generated by an alternative technique, chemical ionization. This method is based on the reaction of the substrate molecule with a reactive ion, generated by electron bombardment of appropriate gases which are present at relatively high pressures. Substrates belonging to different structure types can be oxidized selectively with specific positive ions.

3 Methods of Detection

3.1 Mass Spectrometry

Without any doubt, mass spectrometry (MS) is by far the most generally applied technique dealing with organic radical cations. This technique identifies positive ions, generated in a high vacuum by an electron beam (typically 10–70 eV), and accelerated by electrostatic potentials of up to several thousand volts, by the different trajectories with which they travel through magnetic and/or electrostatic fields. These trajectories are determined by both the mass and the charge of the ions; accordingly the MS technique can identify the precise mass-to-charge ratio of an ion. This technique has substantial value as an analytical tool, because the family of positive ions generated from a given substrate upon electron impact (along with undetected neutral fragments and negative ions) has a mass distribution characteristic for the neutral precursor molecule [55, 56]. Unfortunately, however, this method usually provides little or no information revealing the structures of the radical cations that are being detected. In special cases, ions of a particular mass-to-charge ratio can be selected and subjected to ion molecule reactions. Under special circumstances structures may be inferred from the observed reactivity pattern.

3.2 Ultraviolet Photoelectron Spectroscopy

Among the methods that have found the most widespread application in the study of radical cations, ultraviolet photoelectron spectroscopy (UV-PES) has a special place, because it provides a wealth of detailed information concerning the orbital energies of organic molecules [57]. In this experiment, a substrate is ionized by ultraviolet radiation with photons of known energy ($E_{h\nu}$), e.g. the $He(I)_\alpha$ line (21.21 eV), and the kinetic energy (E_{kin}) of the emitted electrons is measured. The vertical ionization potential (I_v) can then be calculated from $E_{h\nu}$ and E_{kin} (Eq. 13).

$$I_v = E_{h\nu} - E_{kin} \tag{13}$$

Systematic comparisons between series of structurally related substrates and molecular orbital calculations have allowed the assignment of individual transitions to individual orbitals and have indicated how orbital energies change as a result of subtle structural changes. It is important to keep in mind that the information contained in a photoelectron spectrum reflects the structure of the molecule, but need not have direct bearing on the equilibrium (relaxed) structure of the radical cation derived from it. Nevertheless, photoelectron spectroscopic data are quite valuable in the context of radical cation structures, because they identify the highest occupied molecular orbitals (HOMOs) of the parent molecule and, thereby, identify the bond(s) most likely to be weakened during ionization. Thus, they provide an appropriate starting point for the structure(s) to be considered.

3.3 Optical Spectroscopy

It would appear that optical spectroscopy should be the most natural method of studying organic radical ions. After all, any species must have unoccupied orbitals into which electrons may be excited, providing an absorption spectrum characteristic for the energy differences between occupied and unoccupied orbitals and, particularly, for the HOMO–LUMO gap. In addition, many excited state species display emission spectra characteristic for the reverse transition. Furthermore, numerous methods have been delineated and sophisticated equipment has been developed for these experiments. Particularly, the time resolution of optical spectroscopy has been developing rapidly, from the first flash photolysis experiments with modest (millisecond) resolution [58, 59] to the current state of the art, where nanosecond resolution [60–63] must be considered routine and where picosecond [64–66] and even femtosecond time-resolved experiments [67] are being pursued in several groups. The high time resolution has made it possible to compare the decay of one transient with the rise of a subsequent one, often allowing the identification and correlation of several consecutive intermediates.

Accordingly, the application of optical spectroscopy has provided insight into many subtle problems. On the other hand, the general application of this technique is not without some problems, of which we will mention the following: First, the typical optical spectra observed in condensed media consist of broad bands without any identifying features. Hence, it usually provides structural information only by comparison with known species, or under the special conditions of high-resolution laser spectroscopy [68]. Second, the available equipment is generally limited to the spectral range of 250–800 nm, and often to a more narrow range. This limitation excludes many of the most interesting species from scrutiny. Further, the method of generation may limit the study of a species by optical spectroscopy, particularly when the absorption spectra of precursors and/or accompanying intermediates interfere with its unencumbered observation. Finally, optical methods are not suitable for the study of degenerate processes such as electron exchange between a radical cation and its neutral precursor.

3.4 Electron Spin Resonance

Among the techniques available to study free radicals or radical ions in solution, electron spin resonance (ESR) stands out as a technique with sufficient resolution to provide detailed information about the identity of the intermediate in question. In an external magnetic field the unpaired electron of such a species can adopt either of two spin orientations, parallel or antiparallel to the field H_0. The two orientations are of slightly different energies and transitions between them can be stimulated by applying radiation of a frequency satisfying the resonance condition in Eq. (14):

$$h\nu = g\mu H_0 \tag{14}$$

where h is the Planck constant, g is a parameter characteristic for the radical under scrutiny, μ is the Bohr magneton, and H_0 is the applied field strength.

Experimentally, resonance is approached by sweeping the field in the range near 3400 G and holding the frequency constant, typically at 9.6 GHz. A key contribution to the identification of the radical (ion) is due to the interaction of the unpaired spin with nearby magnetic nuclei, the so called hyperfine interaction. This gives rise to a pattern of signals characteristic for the radical (ion). The spacing of the signals and their relative intensities identify the magnitude of the interaction between the electron and a group of equivalent nuclei as well as the nuclei so coupled; this information is indicative of the spin density distribution in the intermediate under study [69, 70].

The hyperfine coupling constants of magnetic nuclei in organic radicals range from essentially 0 to 200 G (0 to 20 mT). The resulting spectra appear in absorption with intensities determined by the Boltzmann population of the states at thermal equilibrium. This feature limits the sensitivity of the ESR method and has restricted its application to relatively stable species with lifetimes greater than a few milliseconds. An additional limitation has its origin in the tendency of free radicals to dimerize, thereby annihilating the unpaired spins. Nevertheless, this method has been the method of choice for the study of numerous families of radicals and radical ions; it has proved invaluable for investigating their spin density distributions and has provided detailed insights into many structures.

Appropriate modifications of the ESR spectrometer and generation of free radicals by flash photolysis allow time-resolved (TR) ESR spectroscopy [71]. Spectra observed under these conditions are remarkable for their signal directions and intensities. They may be enhanced as much as one hundredfold and may appear in absorption, emission, or in a combination of both modes. These spectra indicate the intermediacy of radicals with substantial deviations from equilibrium populations. Significantly, the splitting pattern characteristic for the spin density distribution of the intermediate remains unaffected; thus, the CIDEP (chemically induced dynamic electron polarization) enhancement facilitates the detection of short-lived radicals at low concentrations.

Deviations from Boltzmann intensities were first noted in 1963 [72]; they are known as CIDEP [73, 74]. These effects can be caused by two different mechanisms, each giving rise to CIDEP spectra with different characteristic intensity patterns. CIDEP effects can have two sources; they may be transferred to the intermediates from their immediate precursor, typically a triplet state (triplet mechanism, TM), or generated early during their existence (radical pair mechanism, RPM) [75].

The TR-ESR experiment requires sophisticated instrumentation, as it involves the repetitive generation of radicals by short laser pulses and the subsequent time-resolved recording of the transients. An entire decay curve can be recorded at a given field, and the field advanced in discrete increments [76, 77]. Alternatively, a single data point can be measured by integrating the decay curve over a preselected time period [78–80].

3.5 Fluorescence Detected Magnetic Resonance

A related technique, fluorescence detected magnetic resonance (FDMR), is suitable for the observation of short-lived radical ion pairs with lifetime in the range of

10^{-7}–10^{-9} sec. This method requires relatively sophisticated equipment but its sensitivity exceeds that of the standard ESR method by several orders of magnitude. It also offers an improvement in the time resolution obtained by standard time-resolved ESR. The FDMR technique was first applied in the study of matrix isolated phosphorescent triplet states [81]. Microwave-induced transitions between the triplet sublevels were found to affect the phosphorescence intensity, because different sublevels have different rates of radiative decay. FDMR is also useful for examining the ESR spectra of short-lived radical ion pairs, if they can form reaction products in electronically excited states that decay by radiative pathways [82–84]. Strong ESR signals can be observed particularly, if intersystem crossing can be achieved as a result of the applied magnetic field. This is normally the case, when the precession period $(gH_1)^{-1}$ of an electron spin in a resonance microwave magnetic field H_1 is compatible with the pair lifetime. The FDMR method can be applied to radical ion pairs produced in nonpolar solvents by pulse radiolysis. It is useful to employ a perdeuterated aromatic electron acceptor (anthracene-d_{10}) to minimize the spectrum due to the radical anion.

3.6 Nuclear Magnetic Resonance Methods

Nuclear magnetic resonance has proved to be an invaluable technique for many applications and it has been used prominently in the characterization of organic radical ions. We mention three areas of application in this context: selective chemical shift differences; differential line broadening; and chemically induced dynamic nuclear polarization. Each of these methods has provided insights into the structures and reactivities of numerous radical ions.

The influence of nuclear spin–electron spin interactions on the position of ^{1}H resonance peaks was first observed in 1957 [85]. We mention particularly the nickel(II) chelates of *N,N'*-disubstituted aminotroponeimines (**1**) and *N*-alkyl-salicylaldimines (**2**), which exist as equilibria between square planar (diamagnetic) and tetrahedral (paramagnetic) conformers. Although the diamagnetic conformer is more stable, the free energy difference is sufficiently small to allow a significant fraction of the paramagnetic conformer to be present. The NMR resonance shifts reflect the unpaired spin density distribution over the ligand, R [86, 87].

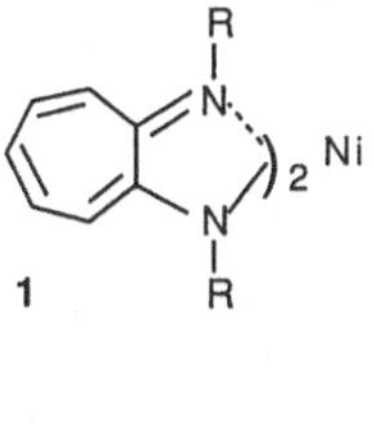

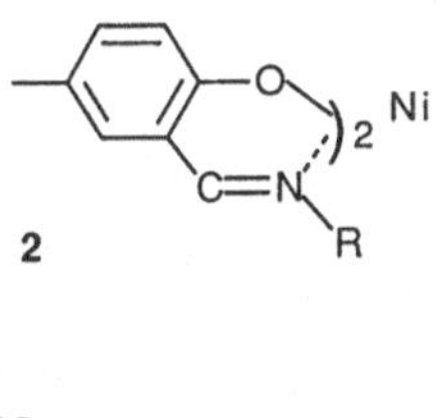

3.6.1 Differential Line Broadening

Interesting line broadening effects may be observed in solutions containing organic radical ions along with their neutral diamagnetic precursors [87]. In general, these effects can be explained by various chemical exchange mechanisms, involving, for example, degenerate exchange of hydrogen or iodine atoms, of triplet energy or, perhaps most frequently, degenerate electron exchange. For the exchange contribution to the NMR line width, ΔT_2^{-1}, two limiting conditions have been formulated: the strong-pulse (slow-exchange) limit (Eq. 15);

$$\Delta T_2^{-1} = k_e[P] \tag{15}$$

and the weak-pulse (fast-exchange) limit (Eq. 16);

$$\Delta T_2^{-1} = [P]\, a_n^2/([D]^2\, 4k_e) \tag{16}$$

where [D] and [P] are the concentrations of the diamagnetic species and its radical ion, respectively, k_e is the rate constant of electron transfer, and a_n is the hyperfine coupling constant (in radians s^{-1}) of the nucleus whose spectrum is being recorded. These equations offer three criteria to differentiate between fast and slow exchange. Since k_e must be expected to increase with temperature, the broadening should show a negative temperature coefficient in the fast exchange limit ($\Delta T_2^{-1} \propto 1/k_e$) whereas a positive temperature coefficient is expected for slow exchange ($\Delta T_2^{-1} \propto k_e$). In addition, the concentration of the diamagnetic precursor may have a marked influence on the degree of line broadening: in the fast exchange limit it should be inversely proportional to the square of [D], whereas it is expected to be independent of [D] in the case of slow exchange. However, applying [D] as a criterion can be problematic, if the radical ions are generated by photoinduced electron transfer. For example, if the broadened spectrum is that of the substrate used as quencher, the quantum yield of quenching and, accordingly, [P] show a complex dependence upon [D]. Therefore, examination of the temperature dependence appears to be the most obvious experiment to examine the kinetic limit of a particular line broadening effect. If it can be established that the fast-exchange limit is involved, the degree of broadening can be used to derive the relative magnitude of the hyperfine coupling constants and, possibly, the electron spin density distribution of the radical ion under investigation.

We illustrate the temperature dependence and the effect of quencher concentration with the broadened signals recorded during the electron transfer quenching of photoexcited decafluorobenzophenone by dimethoxybenzene (Fig. 3). Clearly, the degree of broadening shows a negative temperature coefficient and increases with decreasing quencher concentration. These findings are compatible with rapid degenerate electron exchange.

3.6.2 Chemically Induced Dynamic Nuclear Polarization

The technique most recently introduced to study radical cations is chemically induced dynamic polarization (CIDNP), a nuclear resonance method based on the observation of transient signals, substantially enhanced in either absorption or emission. These effects are induced as a result of magnetic interactions in radical or radical ion pairs on the nanosecond timescale. This method requires acquisition of an NMR spectrum during (or within a few seconds of) the generation of the radical ion pairs. The CIDNP technique is applied in solution, typically at room

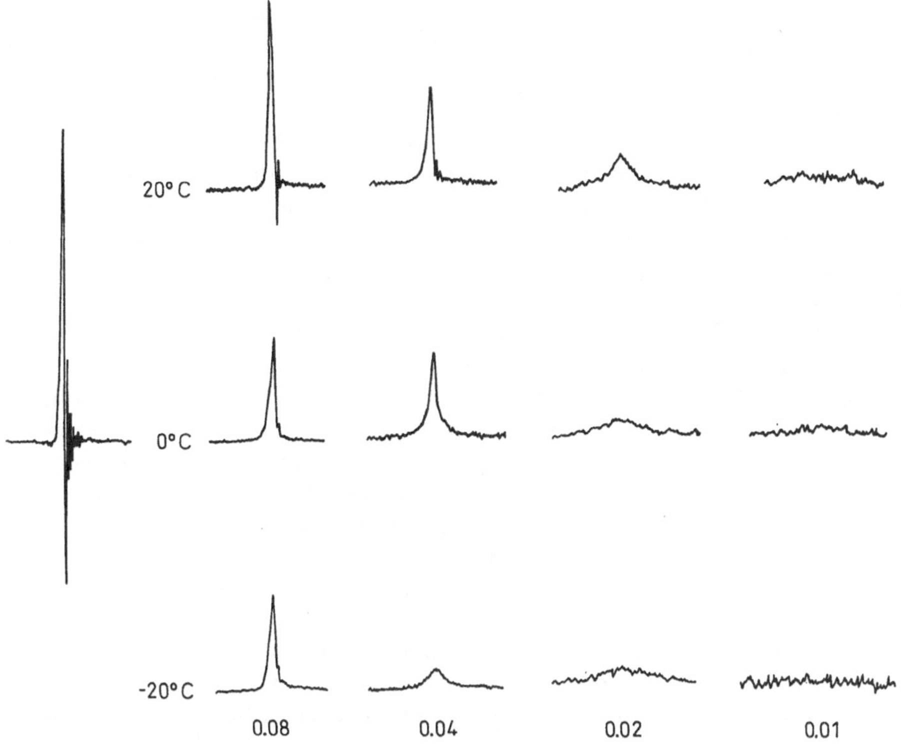

Fig. 3. Temperature and concentration dependence of the line broadening observed for the methoxy signal of 1,4-dimethoxybenzene (DMB) during the electron transfer quenching of triplet decafluorobenzophenone by DMB in acetonitrile-d_3 [178]

temperature, and lends itself to modest time resolution. The first CIDNP effects were reported in 1967, and their potential as a mechanistic tool for radical pair reactions was soon recognized [88, 89]. Nuclear spin polarization effects were discovered in reactions of neutral radicals. However, experiments in our laboratory established that similar effects could be induced in radical ions as well [90–92].

The theory underlying this effect depends critically on two selection principles: the nuclear spin dependence of intersystem crossing in a radical pair, and the electron spin dependence of the rates of radical pair reactions. Combined, these selection principles cause a "sorting" of nuclear spin states into different products and result in characteristic nonequilibrium populations in the nuclear spin levels of geminate reaction products (whose formation is allowed for singlet pairs but spin forbidden for triplet pairs) and in complementary nonequilibrium populations in the spin levels of free-radicals ("escape") products (whose formation is electron spin independent). The transitions between these levels will be in the direction towards restoring the normal Boltzmann population; their intensities will depend on the extent of nonequilibrium population. The observed effects are optimal for radical pairs with lifetimes in the nanosecond range. On a shorter time scale, hyperfine induced intersystem crossing is negligible whereas on a longer timescale, the polarization decays due to spin lattice relaxation in the radicals.

The quantitative theory of CIDNP [93–96] is developed to a state where the intensity ratios of CIDNP spectra can be computed on the basis of reaction and relaxation rates and the characteristic parameters of the radical pair (initial spin multiplicity, μ); the individual radicals (electron g factors, hyperfine coupling constants, a); and the products (spin-spin coupling constants, J). On the other hand, the patterns of signal directions and intensities observed for different nuclei of a reaction product can be interpreted in terms of the hyperfine coupling constants of the same nuclei in the radical cation intermediate.

The ^1H hfcs are related to carbon spin densities by different mechanisms of interaction. For π radicals, there are two principal mechanisms involving either an exchange interaction or hyperconjunction. Protons attached directly to carbon atoms bearing positive spin density have negative hfcs because of the preferred exchange interaction between the unpaired π spin density and the carbon σ electron (Fig. 4a). Positive hfcs, on the other hand, are usually observed for protons which are one C–C bond removed from a carbon bearing positive spin density. The positive sign is due to a hyperconjugative interaction which delocalizes the π spin density on carbon into an H_n "group orbital" (see Fig. 4b).

In a well designed experiment, the pattern of CIDNP signal directions and intensities observed for a diamagnetic product reveal the relative magnitude and the absolute sign of the hyperfine coupling constants of the corresponding nuclei in the paramagnetic intermediate. The hfcs, in turn, can be interpreted in terms of carbon spin densities and these reveal important structural features of the intermediates. These results often are quite unambiguous, because NMR chemical shifts are usually well understood, and the identity of the coupled nuclei is clearly established. Combined with PET as a method of radical ion generation, the CIDNP technique has been the key to elucidating mechanistic details of important reactions and provided insight into many short-lived radical cations with unusual structures, many of which previously had eluded any other technique.

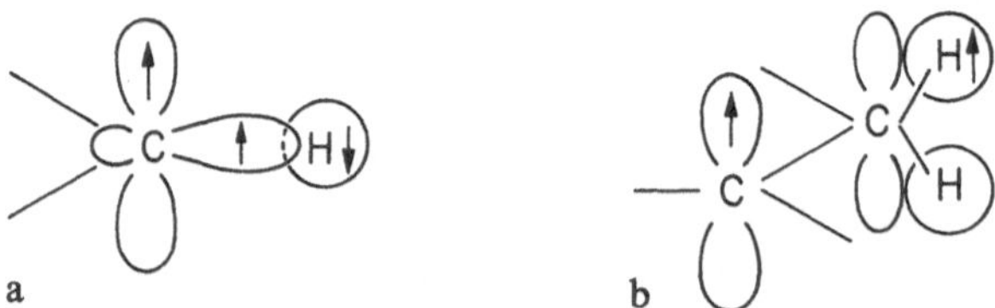

Fig. 4a Preferred configuration of electron spins in the σ orbital connecting a hydrogen atom to an sp^2-hybridized carbon atom bearing unpaired π spin density. **b** "Molecular π orbital" consisting of two carbon p_z orbitals and an H_2 "group orbital" generated by hyperconjugative interaction of an sp^2-hybridized C atom bearing unpaired spin with a CH_2-R group

3.7 Sources of Misinterpretation

In view of the significant conclusions that are derived on the basis of the experimental methods delineated above, it is appropriate to discuss the reliability of these methods as well as potential shortcomings and sources of misinterpretation. In this context, we will also comment on the significance of direct observation. We begin with potential problems with the interpretation of CIDNP results.

The CIDNP method is an indirect method, since the hyperfine pattern of a paramagnetic intermediate is derived from the unusual NMR intensities of a diamagnetic product derived from it. Accordingly, it appears necessary to discuss limitations and potential sources of misassignments. First, the CIDNP method does not provide any evidence for the nature of the paramagnetic intermediate, particularly there is no definitive evidence for the presence of the charge. This feature is derived typically on the basis of mechanistic considerations and from supporting secondary experiments.

Second, any CIDNP based assignments concerning the sign of hfcs are valid only if the radical pair mechanism (RPM) [93–96] is operative; they become invalid if the alternative triplet-Overhauser mechanism (TOM), based on electron nuclear cross relaxation [97–100] is the source of the observed effects. For effects induced via the TOM the signal directions depend on the mechanism of cross relaxation and the polarization intensities are proportional to the *square* of the hfc. Thus, they do not contain any information related to the signs of the hfcs. However, the TOM requires the precise timing of four consecutive reactions and, thus, is not very likely. In fact, this mechanism has been positively established in only two systems [98–100].

A third source of misassignment has its roots in the existence of nuclear–nuclear cross relaxation [101–104]. Again, depending on the mechanism of cross relaxation and on the polarization of the originally polarized nucleus, this may result in enhanced absorption or emission. It is of potential significance that this process induces nuclear spin polarization in nuclei without hfc, or alters the nuclear spin polarization of nuclei with weak hfcs. On the other hand, the magnitude of these effects may be quite small and fall below the threshold of chemical significance.

The final problem associated with CIDNP assignments is an artefact introduced by the manipulation of spectral data. Some CIDNP effects are weak compared to the steady state signals of substrates present in high concentrations. In such cases the CIDNP effect is evaluated as the difference between simultaneously processed free induction decays (FIDs) recorded in the dark and during irradiation, respectively. This method carries an inherent source of error, as any loss of starting material resulting from irradiation will manifest itself as net emission in the difference spectrum. This error is illustrated by effects observed during the irradiation of decafluorobenzophenone in the presence of *N*,*N*-diethyl-*p*-methylaniline. The light spectrum clearly shows selective line broadening without a superimposed CIDNP effect, but the difference spectrum seems to indicate CIDNP emission (Fig. 5).

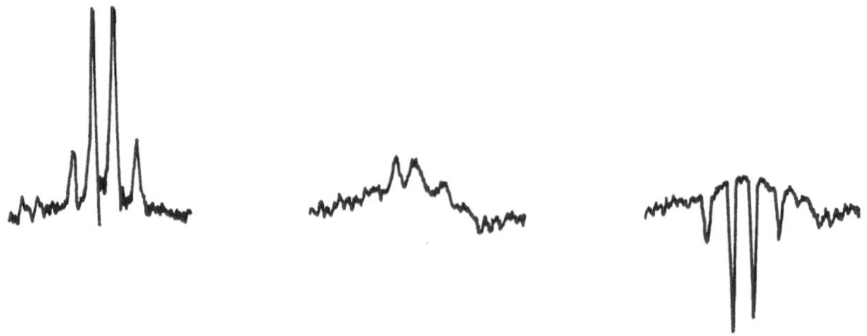

Fig. 5. PMR spectra (60 MHz) of the methylene quartet of *N,N*-diethyl-*p*-toluidine (0.02-M) in an acetonitrile-d_3 solution containing 0.02-M decafluorobenzophenone in the dark *(left)* and during UV irradiation *(center)*. The difference spectrum (light minus dark; *right*) seems to show emission signals, an artefact of the mathematical operation. The artefactual enhancement factor is near-1 [177]

In the ESR experiment transitions are recorded between the different levels of a "multi"-spin system consisting of one electron and several nuclear spins. The variations of radiofrequency intensities occuring in these experiments as a function of magnetic field are observed directly. Their interpretation in terms of a hyperfine coupling pattern and the assignment of the nuclei so coupled is secondary and subject to ambiguities. The typical ESR experiment reveals the absolute magnitude, but not the sign of the hyperfine couplings, and it identifies the number of nuclei so coupled, but not their identity. The occasional ambiguity can be exemplified by the ESR spectrum of the propene radical cation (Fig. 6) [105, 106].

The spectrum consists of 20 lines and the splitting pattern can be interpreted in terms of two different structures, containing either a freely rotating methyl group [106] or one whose rotation is substantially hindered [105]. This ambiguity was resolved by the ESR spectrum observed for the radical cation of selectively deuteriated propene-1,1-d_2; the three identical (methyl-) splittings found for this species clearly support a freely rotating methyl group [106].

As in the CIDNP experiment, the positive charge is not at all observed, but is an outgrowth of the chemical intuition of the investigator and is supported by appropriate secondary experiments. The potential problems with such an assignment will be illustrated for the secondary intermediate obtained upon warming a matrix containing the tetramethylcyclopropane radical cation (see Chap. 5).

Finally, misinterpretations can result from interpreting two overlapping spectra as a single spectrum. An example can be found in the recently presumed identification of the $^2B_2$ radical cation of hexamethyl (Dewar)benzene [107]; the supposedly observed septet turned out to be due to the superposition of a doublet of triplets and a single line of unassigned origin [108, 109]. The mistaken identification of the seven-line spectrum as a single species results in an unexplainably high electron g-factor [108, 109], an additional strong indication of the misassignment.

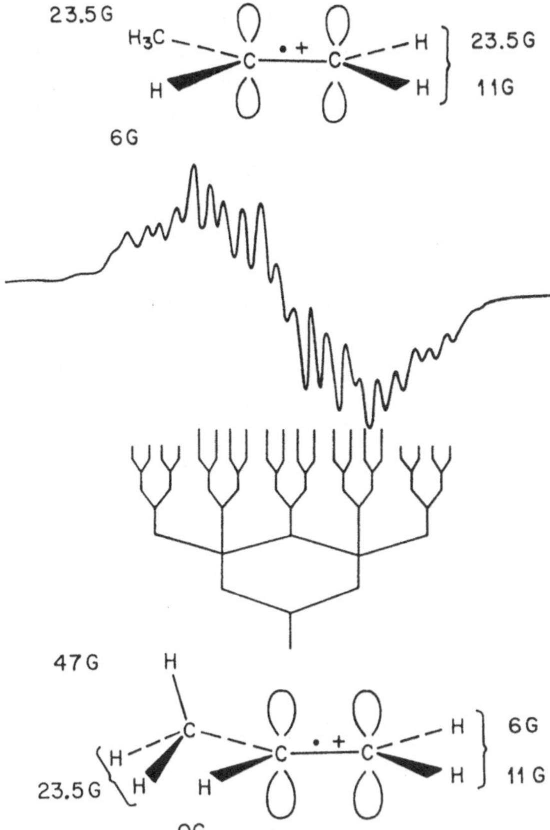

Fig. 6. ESR spectrum of propene radical cation generated by γ irradiation in a $CFCl_3$ matrix at 4.2 K and measured at 77 K *(center)*; analysis in terms of two different splitting patterns according to reference [105] *(bottom)* and [106] *(top)*, respectively

Concerning the ODMR experiment, the observed phenomenon is the luminescence from the decay of a scintillator excited state. While this luminescence is observed *directly*, its relation to the hyperfine coupling pattern of a radical cation is *highly indirect* and its connection to the assigned structure is even more remote. A second potential source of error lies in the presence of the strong background signal due to the deuterated scintillator anion. The effects due to the radical cation are often much weaker and occur as broad shoulders on that background signal. Due to this superposition of spectral features, a correlation of predicted and observed line intensities may be difficult. Furthermore, the weaker outside lines may not be observed, particularly if the number of equivalent protons is large. Finally, the assigned pattern is subject to the same ambiguities discussed for the ESR experiment.

The question of direct observation and its meaning for the validity of results or their interpretation seems to be misunderstood by some authors. Direct observation is used occasionally to support claims of significance and authenticity. Therefore, a clarification seems to be in order. Strictly, the term "direct observation" means nothing more than that some property of an intermediate is measured during its lifetime; in other words, the lifetime of the species exceeds the characteristic timescale of the method of observation. The simplest and oldest form of direct observation is that of absorption of light, which causes a transition to a higher lying nonbonding or antibonding state (Sect. 3.3), or of emission, indicating the return to the ground state. However, in most instances the absorption bands are broad and unspecific. In these cases, direct observation is without a particular benefit. For example, direct observation has no bearing on the ambiguity posed by the spectrum of the propene radical cation.

The problems inherent in the interpretation of direct observation are aptly illustrated by the many medieval accounts of unicorns or mermaids or, perhaps, by the post-Columbian accounts concerning the seven golden cities of Cevola. All of these reports were based on "direct observation", but none of them represent valid interpretations of the observed phenomena (results). To stay within the metaphor of a mermaid, the observation "of great charm and beauty" in a creature is necessary for establishing the existence of a mermaid, but it is not sufficient; one must observe the crucial fishtail.

3.8 Assignment of Structures Based on the Nature of Reaction Products

Both the general nature and the detailed structure of an intermediate is frequently inferred from the reaction conditions and from the type and structure of the reaction product. Thus, radical ions are frequently invoked to explain reactions occurring in polar solvents between pairs of substrates known to be electron donors and acceptors. The detailed examination of reaction products constitutes the classical approach to a mechanistic problem. In particular, the fate of chirality, stereochemistry, or of an isotopic label or a substituent is probed as a result of a chemical transformation.

However, this approach is not without its shortcomings. Typically, information about the products can be gathered only after the reaction is completed and the products are isolated. Moreover, the nature of the products cannot identify the number of intermediates, which may be involved either concurrently or consecutively. To be sure, the margin of error can be substantially reduced by judicious choice of an appropriate substrate and by careful evaluation of results obtained by a variety of complementary techniques. The literature is replete with elegant examples, in which the structures of intermediates or transition states were derived from the structures of reaction products. Nevertheless, a detailed mechanism, not unlike beauty, often lies in the eye of the beholder, if based on product studies alone.

3.9 Molecular Orbital Calculations

Molecular orbital calculations at different levels of sophistication have become an integral part in the characterization of organic radical cations in general, and those of strained molecules in particular. They are used to model the sequence of orbitals in the interpretation of PE spectra, to assign the electronic transitions observed in various spectra, and to determine detailed structures, including bond lengths and angles and, in favorable cases, the spin density distribution and even the hyperfine coupling pattern. The diverse molecular parameters of many molecules can be calculated by relatively simple methods. However, strained ring compounds often present challenging problems, whose solution requires the most sophisticated state of the art calculations.

The results of MO calculations depend on the nature of the basis set and on the extent to which polarization functions are evaluated. Open-shell systems, such as the radical cations discussed here, inherently present greater difficulties. Typically the structures are optimized at a somewhat lower level of theory; then the extrema so identified are subjected to single point calculations at higher levels of theory, including polarization functions. To confirm the identity of an extremum as a minimum, an analysis of the vibrational frequencies is carried out at an advanced level of theory. A single imaginary frequency typically disqualifies an extremum as a minimum.

4 Radical Cation Reactions

Having discussed methods for generating and observing organic radical cations we turn to the results obtained by applying these techniques in a variety of systems. The many research groups working in this exciting area approach the problems of photoinduced electron transfer from a variety of angles. Some research groups are concerned primarily with the electron transfer process and the factors affecting the rate of this reaction; others are interested in the paramagnetic intermediates and in the parameters that determine their structures and reactivities; several groups pursue mainly the ultimate diamagnetic products resulting from the electron transfer reaction. In addition to the wide range of experimental approaches there are various theoretical ones with a variety of goals which mirror the range covered by experimentalists.

The author was tempted to divide the material discussed in the following rigorously into two sections, one dealing with the structural features of intermediates and a second dealing with their various reactions. However, this strict division proved problematic, because the structure of the intermediate determines its reactivity, while in many others the structure of an intermediate is assigned on the basis of its reaction products. In essence, while it may be possible to discuss radical cation structure without considering reactivity, reactivity cannot be discussed without structure. Therefore, a modified approach was chosen: the multitude of reaction types are summarized in one section whereas selected structure types as well as their reactions will be discussed in a separate section.

Organic radical cations undergo a plethora of reactions, including both unimolecular processes and bimolecular reactions. Among the unimolecular reactions are geometric isomerization, rearrangement, cycloreversion, as well as fragmentation and other bond cleavage reactions. In bimolecular reactions radical cations can react with a) neutral reagents, b) ionic substrates, c) free radicals, and d) radical ions of like or opposite charge. These reactions include: a) ion–molecule reactions, such as cycloadditions, hole transfer, or complex formation; b) nucleophilic capture; c) spin labeling; d) dimerization or disproportionation, and reverse electron transfer, proton transfer, atom or group transfer reactions, or coupling. These reactions (see Scheme 3) will be exemplified below.

A. Unimolecular Reactions

- Geometric Isomerization
- Valence Isomerization
- Rearrangement
- Cycloreversion
- Fragmentation

B. Bimolecular Reactions

B1a. With Radical Anions

- Electron Return
- Proton Transfer
- Coupling
- Oxygenation ($O_2^{\cdot -}$)

B1b. With Radical Cations

- Dimerization (→ dimeric dication)
- Disproportionation (→ neutral plus dication)

B2. With Neutral Molecules

- Electron Transfer
- Addition to Olefins
- Cycloaddition to Diolefins
- Complex Formation

B3. With Ions or Dipolar Substrates

- Nucleophilic Capture

B4. With Radicals

- Spin Labeling
- Oxygenation ($^3O_2$)

Scheme 3: Reactions of organic radical cations

Compared to the analogous reactions of the parent molecules, many radical cation reactions show a dramatic decrease in activation barriers, one of the most striking aspects of radical cation chemistry. Intuitively, this observation can be ascribed to the fact that the highest occupied molecular orbital (HOMO) of a radical cation is occupied by a single electron. As a result, the bond strength of one or more key bonds must be reduced and the bonds more easily decoupled. However, the barriers to some radical cation rearrangements appear to lie even lower than might be expected on the basis of this simple model.

Lowered energy barriers may also be rationalized on the basis of the following qualitative consideration. Since the HOMO of a transition state lies higher in energy than the HOMOs of the adjacent (local) minima, the ionization of the less stable structure is likely to require less energy. Accordingly, the energy difference between the corresponding points on the potential surface of the radical cation may be substantially reduced. These simple considerations do not explain, however, why in some cases the relative energies of two minima should be reversed nor why the radical cation resembling the geometry of the transition state should become lower in energy than the cations corresponding to energy minima on the parent energy surface (vide infra).

For a radical cation generated by photoinduced electron transfer in the immediate vicinity of a radical anion, reaction with the geminate counter ion must be considered a most natural event. Reactions in this category include reverse electron transfer, proton transfer, and coupling, and any of these can be exceedingly fast. Yet, both unimolecular and bimolecular reactions may compete efficiently with the ion pair reactions. This is due, in part, to the low reaction barriers, but also to the strict spin multiplicity requirement of most electron return and coupling reactions. Normally, only pairs of singlet spin multiplicity can recombine, whereas radical ion pairs generated from triplet precursors need to undergo intersystem crossing before recombination can occur. Competitive unimolecular reactions include geometric isomerizations, rearrangements, and fragmentations; among the competitive bimolecular reactions are additions and nucleophilic capture. The efficiency of unimolecular reactions is determined by the barrier height and the rate of intersystem crossing, whereas bimolecular reactions obviously can be controlled by the concentration of the appropriate reagent.

In view of the various types of competing reactions available to any radical cation type, the author has chosen not to dissect the multitude of radical ion reactions according to reaction type, but rather according to donor type. Depending on the type of orbital from which the electron is being removed, organic electron donors are conveniently classified as π-, σ-, or n-donors. Among the radical cations of π-donors, those derived from aromatic hydrocarbons have been characterized in the greatest detail, whereas those of olefins and diolefins arguably show the greatest variety of reactions. We begin with selected π-donors, including alkenes, alkynes, and conjugated as well as cumulated dialkenes.

4.1 Electron Transfer Induced Reactions of Alkenes

The lightinduced interactions between excited-state molecules and alkenes may involve a variety of mechanisms, including exciplex formation, triplet energy transfer, bond formation to generate biradicals, or electron transfer to yield radical ion pairs. Among the reactions induced by electron transfer are geometric and valence isomerizations, rearrangements, and addition reactions that lead to cyclobutane as well as more complex dimers. In addition, nucleophilic capture, coupling of donor and acceptor, and photooxygenation can be observed. We will briefly discuss representative examples of selected reactions, before treating selected geometric isomerizations in more detail.

Several intramolecular [2 + 2] cycloadditions with formation of cyclobutane rings have been observed. For example, the tetracyclic diolefin (**3**), in which two double bonds are held rigidly in close proximity, undergoes valence isomerization to the cage structure (**4**); a notable reversal of the more general cycloreversion of cyclobutane type olefin dimers (vide infra). This intramolecular cycloaddition occurs only in polar solvents and has a quantum yield greater than unity [110]. In analogy to several cycloreversions [111, 112] these results have been interpreted as evidence for a free radical cation chain mechanism.

3 → 4

The monocyclic 1,2,5,6-tetraphenylcycloocta-1,5-diene (**5**), on the other hand, undergoes a "cross" cycloaddition [113]; the dicyanoanthracene sensitized irradiation of this diolefin produces a tricyclic product (**6**), most likely via the bicyclic bifunctional radical cation as an intermediate.

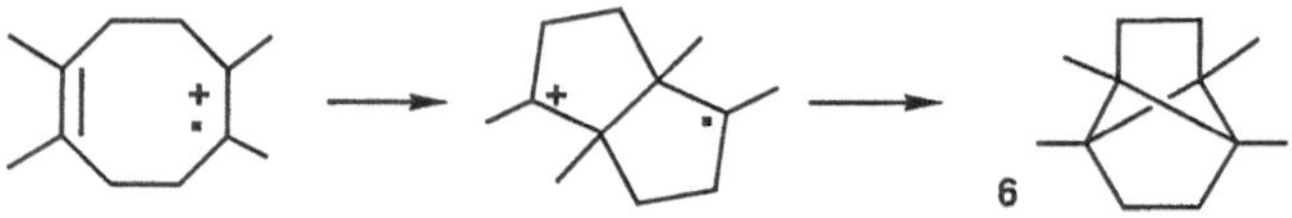

→ → 6

A large variety of unsymmetrically substituted olefins have been reported to form head-to-head dimers selectively. Among these we mention vinylcarbazole [44], assorted vinyl ethers [114–116], indenes [117, 118], and *p*-methoxystyrene [119]. The regiochemistry of the addition is compatible with a stepwise mechanism proceeding via a singly linked 1,4-bifunctional radical cation, in which spin and charge are located on two well-separated carbon centers and stabilized by two separate substituents.

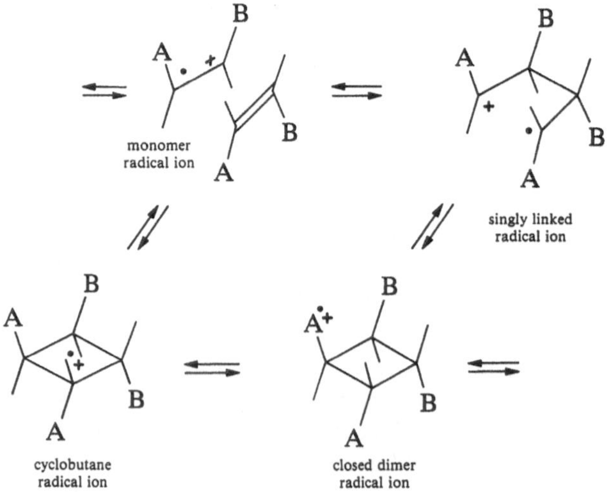

monomer radical ion

singly linked radical ion

cyclobutane radical ion

closed dimer radical ion

Scheme 4

Several of these dimerizations have quantum yields greater than unity, particularly when the oxidation potential of the dimer is either higher than that of the monomer or, at least, when the two oxidation potentials lie close to each other [114, 115, 120]. These observations, once again, support the involvement of radical cation chain processes, which seem to operate, whenever the electron transfer from the starting material to the product radical cation is efficient.

The 1,4-bifunctional intermediate is implicated further by the electron transfer sensitized isomerization of *cis*- to *trans*-1,2-diphenoxylcyclobutane [115], and by the unique CIDNP effects observed during the electron transfer sensitized cleavage of the anti-head-to-head dimer of 3,3-dimethylindene (vide infra) [121].

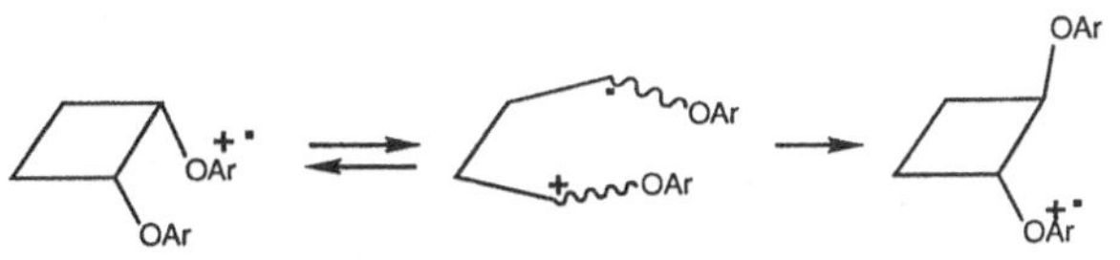

Dimerization products of alternative structure types have been observed for 1,1-diphenylethylene, where a 1,6-cyclization process occurs with participation of one phenyl group. The resulting ring-extended product (7) constitutes additional evidence for the 1,4-bifunctional intermediate [122].

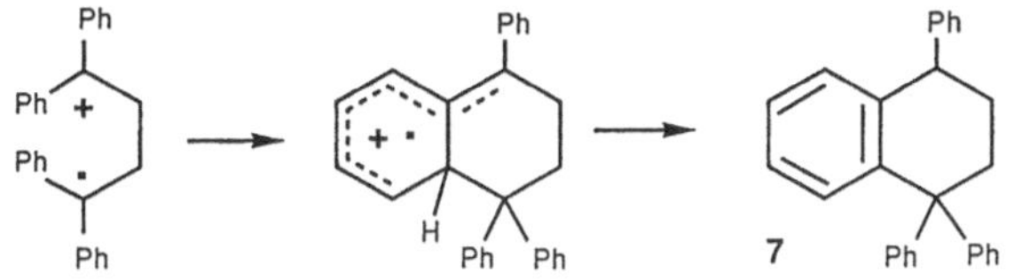

In a related electron transfer reaction, phenylacetylene reacts to form α-phenylnaphthalene, most likely by 1,6-coupling of yet another 1,4-bifunctional intermediate. In addition, formation of a pyridine derivative (**8**) is observed. Apparently, the 1,4-bifunctional intermediate can be trapped also by the solvent acetonitrile [123].

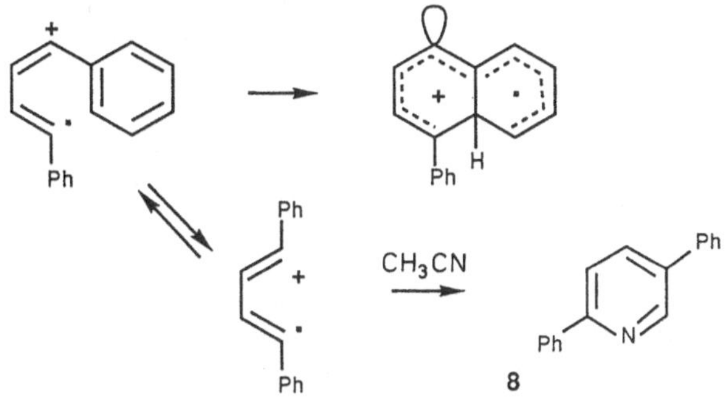

Although simple alkynes are marginal electron donors, they are oxidized readily upon γ-irradiation in frozen solutions. When butyne and several other acetylene derivatives are irradiated in rigid freon matrices at 77 K, the ESR spectra of the alkyne radical cations are observed. Subsequently, upon allowing the matrix to warm and soften, the ESR spectra clearly indicate the formation of cyclobutadiene radical cations, documenting an interesting cycloaddition [124–126]. Although no intermediate has been identified in any of these reactions, the net cycloaddition very likely is a stepwise process. In this context it is of interest that the tetra-*t*-butylcyclobutadiene radical cation can also be obtained by electron transfer induced rearrangement of the corresponding tetrahedrane [127].

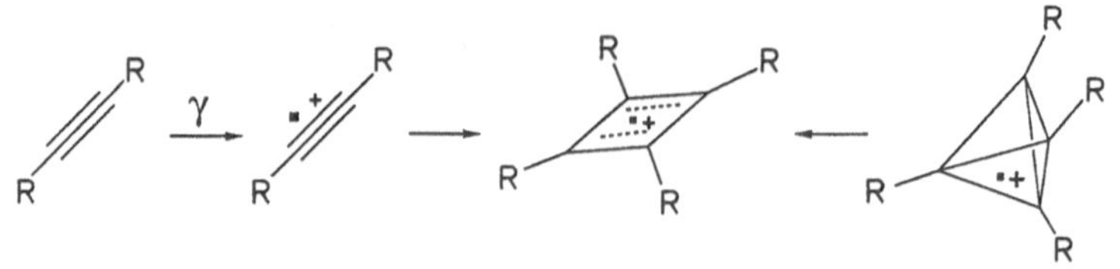

Diacetylenes also undergo interesting reactions upon radiolysis. Thus, deca-2,8-diyne (**9**) forms a cyclobutadiene radical cation at 77 K without requiring annealing at a higher temperature [126]. Hexa-1,5-diyne (**10**), on the other hand, gives rise to the hexa-1,2,4,5-tetraene radical cation, supporting a ready Cope rearrangement in the Freon matrix [128].

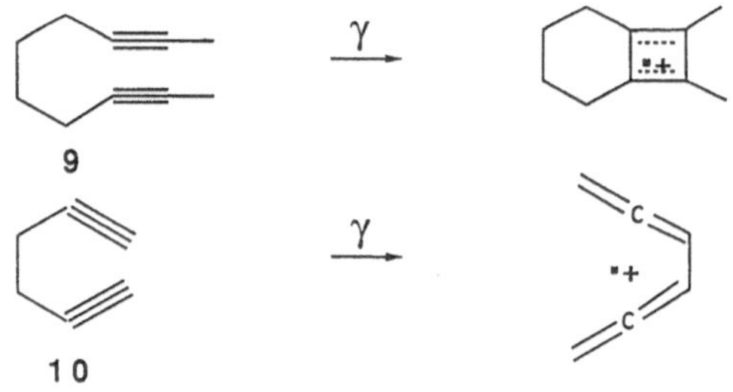

The electron transfer photosensitized reactions of diolefins results in the formation of [4+2]cycloadducts. For example, irradiation of octafluoronaphthalene [129] or dicyanoanthracene [130–132] in polar solvents containing cyclohexadiene leads to the formation of *endo*- and *exo*-dicyclohexadiene.

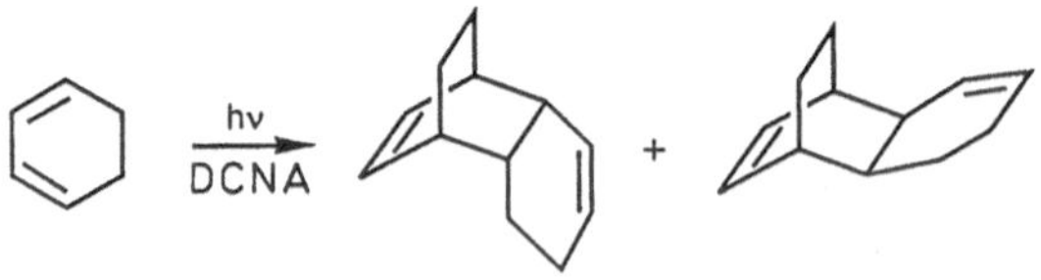

This type of reaction can be induced also by radiolysis [133, 134] or by chemical oxidation, particularly with tris-(*p*-bromophenyl)aminium salts (cation radical catalyzed Diels-Alder reaction) [10]. The scope of this reaction and its synthetic utility have been delineated in detail. The results unambiguously support a free radical cation chain mechanism [10].

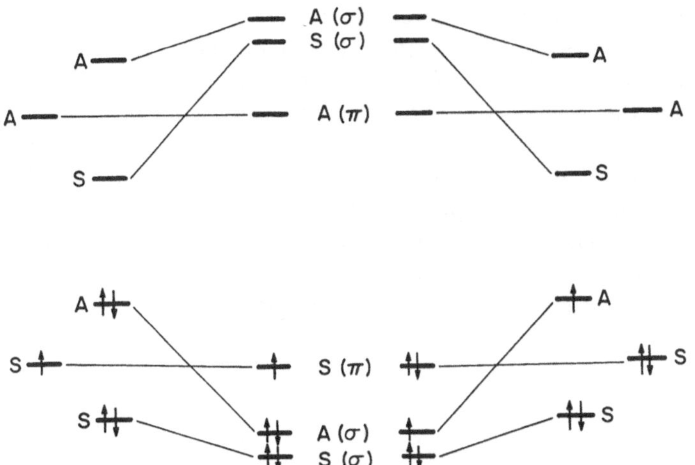

Fig. 7. Correlation diagram for the cycloaddition of ethene$^{\cdot +}$ onto butadiene *(left)* or of butadiene$^{\cdot +}$ onto ethene *(right)*. The former reaction is orbital symmetry allowed [10]

However, only limited experimental evidence is available concerning the key step of the dimerization, i.e. the addition of the radical cation to the parent olefin. Does this addition occur stepwise or in concerted fashion? Does the radical cation serve as a the "diene" component ([3+2]cycloaddition) or as dienophile ([4+1]cycloaddition)? The observed retention of dienophile stereochemistry and orbital symmetry arguments (Fig. 7) favor the [4+1]cycloaddition type. Although it is difficult to distinguish the [3+2] from the [4+1]addition type, a stepwise component for the cycloaddition and the complementary cycloreversion has been established in at least one system, viz., spiro[2.4]heptadiene.

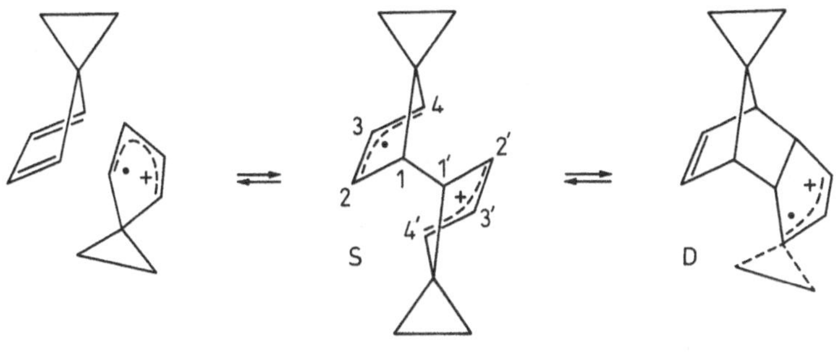

The electron transfer induced reaction of this diene system results in rapid [4+2]dimerization; conversely, the dimer rapidly undergoes cycloreversion upon electron transfer. Both reactions result in strong CIDNP effects. The monomer polarization supports a radical cation with a spin density distribution like those of the butadiene or fulvene radical cations. The dimer polarization identifies a dimer radical cation with appreciable spin density only on two carbons of the dienophile fragment; this species can only be the doubly linked radical cation **D** [135, 136]. Significantly, a second dimer radical cation is implicated in a pulsed laser experiment at high monomer concentrations. Under these conditions both monomer and dimer cations are rapidly quenched, the monomer cation by addition to monomer, the dimer cation by electron transfer from a monomer molecule. These conditions allow the polarization of a second radical cation of the dimer to be recognized, viz. the singly linked ion **S** (Fig. 8) [135, 136].

Electron transfer sensitized irradiation of acyclic dienes gives rise to mixtures of [4+2] and [2+2] dimers; for example, an 8:1 mixture was obtained from 2,4-dimethyl-1,3-pentadiene [137]. On the other hand, [4+2] dimers were obtained exclusively upon oxidation with aminium radical cation salts [138]. This difference has led some workers to question the key role of electron transfer in these reactions

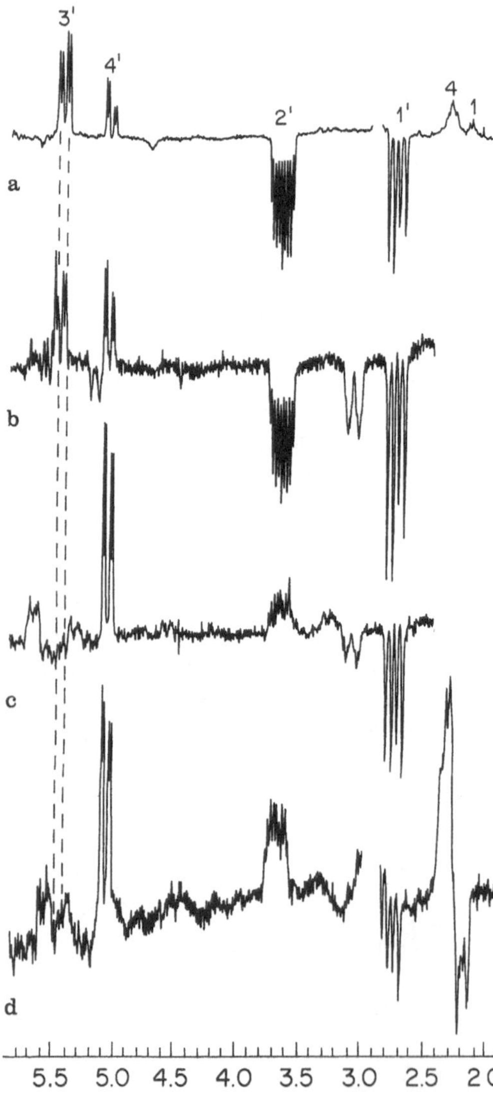

Fig. 8a–d. ^{1}H-NMR spectra (90 MHz) observed 100 µs after excitation of chloranil solutions (0.02 M) with the frequency tripled output (355 nm) of a Nd/YAG laser: (**a**) acetone-d_6 solution containing 0.01 M di(spiroheptadiene), 16 laser pulses; (**b**) acetonitrile-d_3 solution containing 0.005 M spiroheptadiene, 4 × 16 laser pulses; (**c**) acetonitrile-d_3 solution containing 0.1 M spiroheptadiene, 4 × 16 laser pulses; (**d**) acetone-d_6 solution containing 0.1 M spiroheptadiene, 4 × 16 laser pulses. The decrease of signal 3′, the increase of signal 4′, and the inversion of signal 2′ signify the altered nature of the intermediate [135, 136]

and propose protic acid catalyzed dimerization for the aminium salt catalyzed reaction [137]. However, although protic acid catalysis has been demonstrated [137], it appears that this mechanism is an occasional alternative rather than the rule [139].

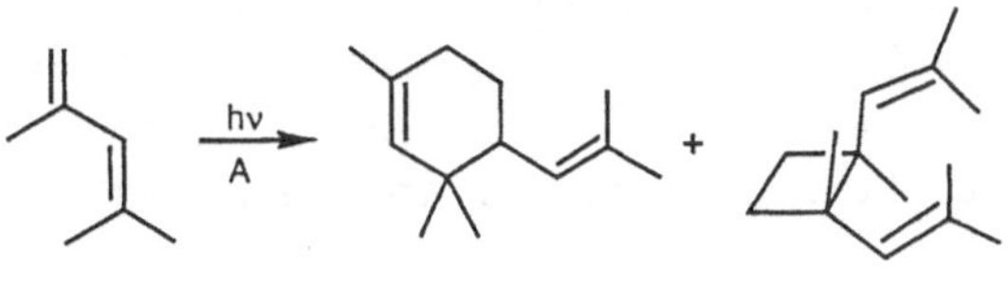

hν
A
+

In several photochemical electron transfer reactions, addition products are observed between the donor and acceptor molecules. However, the formation of these products does not necessarily involve direct coupling of the radical ion pair. Instead, many of these reactions proceed via proton transfer from the radical cation to the radical anion, followed by coupling of the donor derived radical with an acceptor derived intermediate. For example, 1,4-dicyanobenzene and various other cyanoaromatic acceptors react with 2,3-dimethylbutene to give aromatic substitution products, most likely formed via an addition–elimination sequence [140].

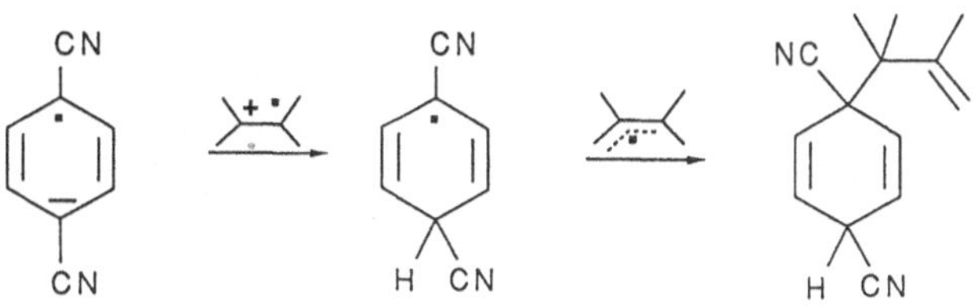

CN
CN
CN
H CN
NC
H CN

In some cases radical cations may undergo cycloadditions with an acceptor derived intermediate without prior proton transfer. This is observed especially for radical cations without sufficiently acidic protons, although it is not limited to such species. For example, the photoreaction of chloranil with 3,3-dimethylindene results in two types of cycloadducts [141]. In the early stages of the reaction a primary adduct is identified, in which the carbonyl oxygen is connected to the β-position of the indene (type B); in the later stages this adduct is consumed and replaced by an adduct of type A, in which the carbonyl oxygen is connected to the α-position. CIDNP effects observed during the photoreaction indicate that the type B adduct is formed from free indene radical cations, which have lost their spin correlation with the semiquinone anions.

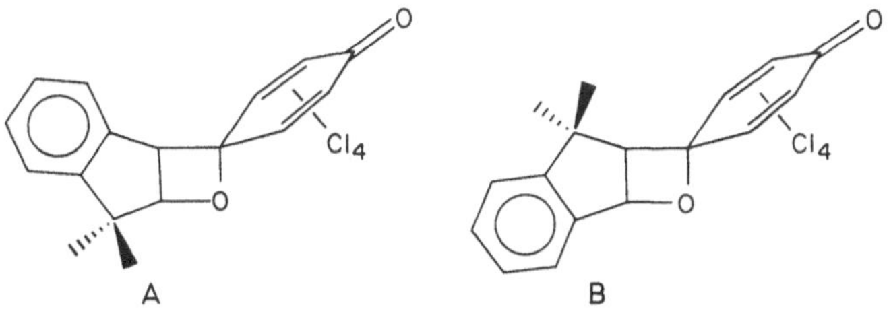

O
Cl_4
O
A
O
Cl_4
O
B

The dicyanonaphthalene sensitized irradiation of 1,3,3-triphenylpropyne (**11**) in the presence of methanol results in the formation of 1,1,3-triphenylallene (**12**) [142]. Although this conversion appears like a simple intramolecular hydrogen transfer, it cannot be formulated as such, since the corresponding allene radical cation is trapped efficiently by methanol resulting in the formation of the isomeric 2-methoxytriphenylpropenes (**13**, **14**) [142]. Accordingly the observed rearrangement must involve deprotonation, followed by electron transfer to the propargyl radical and protonation of the resulting anion. Apparently, methanol serves both as base and acid.

Ph Ph Ph Ph Ph Ph Ph Ph Ph

Ph Ph H Ph OCH_3 Ph Ph Ph **13** OCH_3 Ph Ph Ph **14**

When electron transfer reactions of olefins are carried out in nucleophilic solvents (alcohols) or in the presence of an ionic nucleophile (KCN/acetonitrile/2,2,2-trifluoroethanol), the major products formed are derived by *anti*-Markovnikov addition of the nucleophile to the olefin. In several cases, nucleophilic capture completely suppresses dimer formation [122, 143]. It is important to realize that the observed mode of addition reflects the formation of the more stable (allylic) intermediate and cannot be interpreted as evidence for the charge density distribution in the radical cation.

Ph OR R-OH Ph KCN Ph CN

In some cases the nucleophilic capture of a radical cation is followed by coupling with the radical anion (or possibly with the neutral acceptor), resulting ultimately in an aromatic substitution reaction. Thus, irradiation of 1,4-dicyanobenzene in acetonitrile–methanol (3:1) solution containing 2,3-dimethylbutene or several other olefins leads to capture of the olefin radical cation by methanol, followed by coupling of the resulting radical with the sensitizer radical anion. Loss of cyanide ion completes the net substitution reaction [144]. This "photochemical nucleophile olefin combination, aromatic substitution" (photo-NOCAS) reaction has shown synthetic utility (in spite of its awkward acronym).

CH_3OH OCH_3 NC OCH_3 CN

Photosensitized electron transfer reactions conducted in the presence of molecular oxygen occasionally yield oxygenated products. The mechanism proposed to account for many of these reactions [145–147] is initiated by electron transfer to the photo-excited acceptor. Subsequently, a secondary electron transfer from the acceptor anion to oxygen forms a superoxide anion, which couples with the donor radical cation. The key step, Eq. (18), is supported by spectroscopic evidence. The absorption [148] and ESR spectra [146] of *trans*-stilbene radical cation and 9-cyanophenanthrene radical anion have been observed upon optical irradiation and the anion spectrum was found to decay rapidly in the presence of oxygen.

$$^{1}A^{*} + D \longrightarrow {}^{1}\overline{A^{\bullet -}D^{\bullet +}} \qquad (17)$$

$$A^{\bullet -} + O_2 \longrightarrow A + O_2^{\bullet -} \qquad (18)$$

$$O_2^{\bullet -} + D^{\bullet +} \longrightarrow D{-}O_2 \qquad (19)$$

Scheme 5

The nature of the adduct formed according to Eq. (19) depends on the structure of the donor. When the donor is a monoolefin, the initial product, a dioxetane, may fragment into two ketones [143]. On the other hand, several substrates giving rise to bifunctional radical cations are trapped to form more or less stable adducts. Thus, tetraphenyloxirane gave rise to an ozonide (**15**) [149], several 1,4-bifunctional radical cations formed dioxanes (**16**) [150], and diolefins formed dioxenes (**17**) [151].

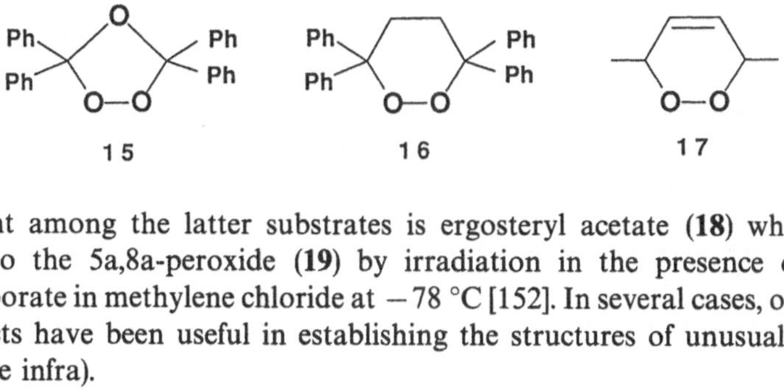

Prominent among the latter substrates is ergosteryl acetate (**18**) which was converted to the 5a,8a-peroxide (**19**) by irradiation in the presence of trityl tetrafluoroborate in methylene chloride at −78 °C [152]. In several cases, oxygenation products have been useful in establishing the structures of unusual radical cations (vide infra).

A series of hindered, α-branched tetraalkylolefins, such as biadamantylidene, have been converted to dioxetanes via an alternative oxygenation mechanism, involving a cation radical catalyzed chain (CRCC). This reaction can be initiated by chemical, electrochemical, or photosensitized oxidation [153, 154]. The following sequence of steps (Scheme 6) has been proposed [8]: the hindered radical cation reacts with triplet oxygen to form a bifunctional adduct (Eq. 20); subsequent ring closure generates a dioxetane radical cation with a three-electron π-bonded structure (Eq. 21); electron transfer from the olefin generates the dioxetane (Eq. 22); this step is efficient, because the dioxetane radical cation (E_{ox} = 2.3 V vs. SCE, −78 °C) is a better oxidant than the olefin radical cation (E_{ox} = 1.6 V vs. SCE, −78 °C).

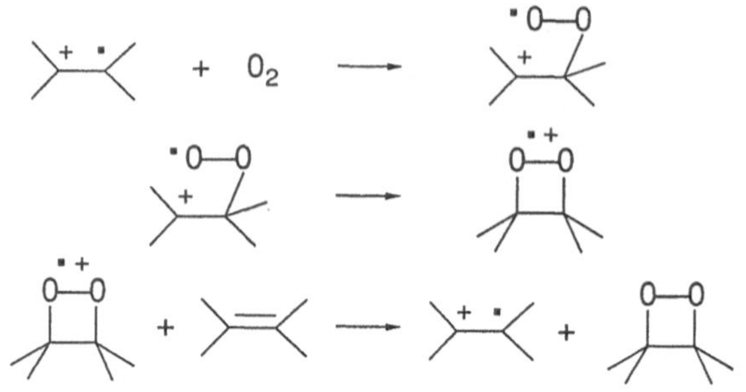

(20)

(21)

(22)

Scheme 6

A similar radical cation chain mechanism was suggested for the formation of 3,3,6,6-tetraphenyl-1,2-dioxane (**16**) upon chemical or photoinduced one-electron oxidation of 1,1-diphenylethylene [155, 156].

4.2 Electron Transfer Induced Geometric Isomerization of Alkenes

The topic of electron transfer induced geometric isomerization is treated in a separate section, because we wish to emphasize the existence of two fundamentally different mechanisms of isomerization. Although both mechanisms are initiated by an electron transfer step, the key intermediates involved in these isomerizations are fundamentally different. The interaction of acceptor sensitizers with donor olefins leads to the isomerization of alkene radical cations, whereas the reaction of donor sensitizers with acceptor olefins leads, eventually, to the population of alkene triplet states.

The isomerization of donor olefins is illustrated by the reaction of chloranil with a pair of geometric isomers, *cis*- and *trans*-1-phenylpropene. The irradiation of the quinone in polar solvents in the presence of either isomer results in nuclear spin polarization for both isomers. The key to understanding these effects lies in two observations (Fig. 9): (a) the polarization of the regenerated parent olefin is stronger than that of the rearranged olefin; (b) the reaction of the *cis*-isomer results in stronger overall effects than does that of the *trans*-isomer [157, 158].

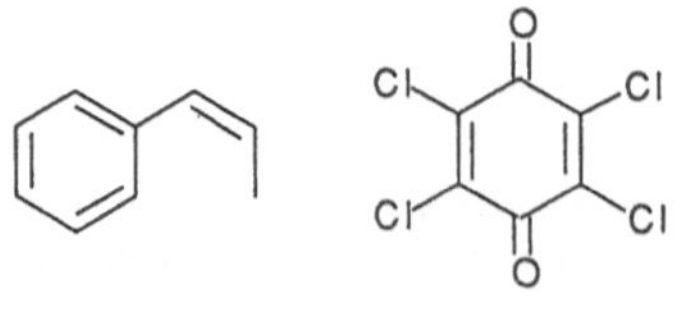

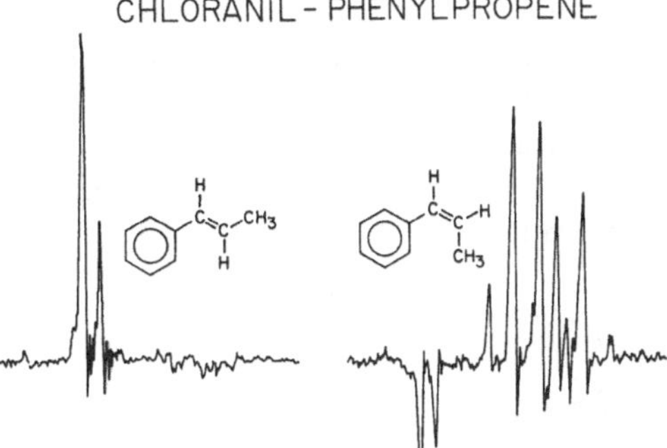

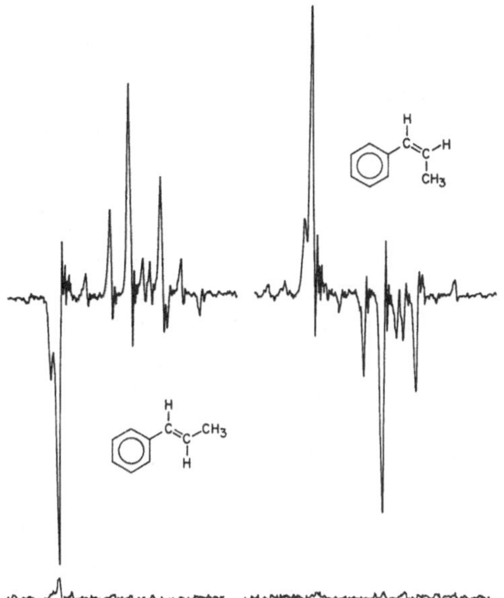

Fig. 9. [1]H CIDNP spectra during the photoreaction of choranil *(top)* and TMPD *(bottom)* with *trans (left)* and *cis*-1-phenylpropene *(right)*. In each spectrum, the multiplet on the right represents the β proton of the *cis* isomer whereas the multiplet on the left represents the corresponding proton of the *trans* isomer [157, 158]

The effects observed are ascribed to the radical ion pair generated by electron transfer from the olefin to the excited triplet state of the quinone. Irradiation of the quinone initially causes excitation to a short lived singlet state, which undergoes rapid intersystem crossing to the lowest triplet state. This state is quenched by interaction with an olefin molecule. The CIDNP signal direction of the reactant olefin (see Fig. 9, top) is consistent with reverse electron transfer in geminate radical ion pairs after intersystem crossing to the singlet state. The polarization of the rearranged olefin is compatible with a mechanism involving separation of the geminate ions by diffusion, geometric isomerization of free radical cations, and electron exchange of rearranged cations with the reactant olefin. The overall CIDNP intensity depends on the efficiency of the rearrangement since this step determines the efficiency of spin sorting. One of the factors affecting the rearrangement step is the energy surface of the radical cations. As in the olefin ground state, the radical ions have energy minima for structures allowing maximum conjugation, and maxima for nearly perpendicular arrangements. Because of steric repulsion, the energy of the *cis*-cation lies above that of the *trans*-cation and, therefore, closer to the (perpendicular) transition state.

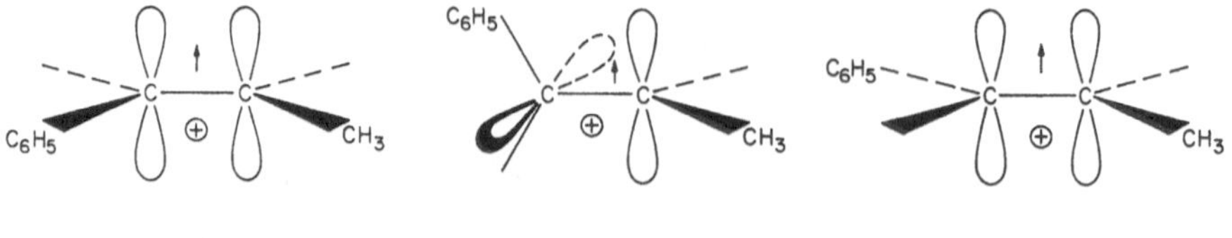

cis RADICAL CATION — TRANSITION STATE — *trans* RADICAL CATION

Accordingly the rearrangement of the *cis*-cation requires a lower activation energy, is faster, and results in more efficient spin sorting than is observed in the reaction of the *trans*-cation. In essence, the observed differences are characteristic for systems in which two cations of different energies are interconverted via a transition state.

The general features of the isomerization are compatible with a free radical cation chain mechanism, featuring electron transfer from unreacted olefin to rearranged radical cation. This chain mechanism was firmly established in several other isomerizations by the observation of quantum yields greater than unity. Thus, the dicyanoanthracene sensitized irradiation of *cis*-stilbene results in nearly quantitative isomerization (>98%) to the *trans*-isomer. In this system, the quantum yield increases with increased *cis*-stilbene concentration, solvent polarity, salt concentration, as well as decreasing light intensity [159].

The radical cations of fulvene systems are of interest, because steric and electronic factors might favor a perpendicular structure and because the energy difference between the respective *cis* and *trans* isomers are expected to be small. However, the chloranil photosensitized reaction resulted in CIDNP effects, indicating planar or slightly twisted structures. The *Z*- and *E*-2-*tert*-butyl-6-(dimethylamino)fulvene [**20**, R = $-N(CH_3)_2$] radical cations rearrange readily whereas di-*tert*-butylfulvene [**20**, R = $-C(CH_3)_3$] showed no interconversion under comparable experimental conditions [160].

E–20 ⇌ **Z–20**

A substantially different isomerization mechanism is operative in the reaction of strong electron donors with the isomeric phenylpropenes, in which the olefin serves as the electron acceptor. These interactions may also lead to geometric isomerization. The CIDNP effects observed during these reactions, for example, during the irradiation of tetramethyl-p-phenylenediamine (TMPD) in the presence of the phenylpropenes, show well balanced relative intensities for the *cis* and *trans* olefins (Fig. 9, bottom) regardless which isomer is the reactant [158].

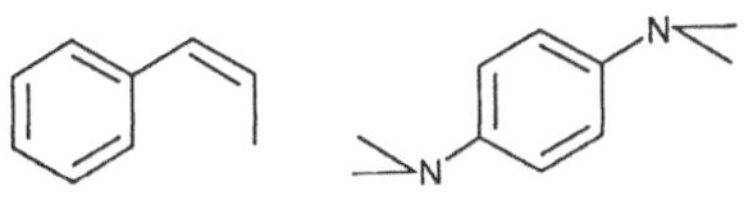

These effects are interpreted as evidence for an alternative isomerization mechanism. One significant difference between the two systems lies in the energies of the radical ion pairs relative to the olefin triplet state. In the chloranil reaction, the pair energy lies well below the triplet energy of either reactant. Therefore, reverse electron transfer can yield only ground state starting materials. However, the very negative reduction potential of the phenylpropenes ($E_{A^-/A} \sim 2.7$ V vs. SCE) causes the free energy of the pair in the TMPD reaction ($E_{D/D^+} = 0.16$ V vs. SCE; ΔG 2.7 eV) to lie above the olefin triplet state (E_T 2.6 eV).

This feature allows a special spin sorting principle to operate: both singlet and triplet pairs can undergo reverse electron transfer. Singlet pairs regenerate the reactants in their ground states, whereas triplet pairs populate one reactant triplet state and the ground state of the other reactant. In contrast to the energy surfaces of ground states and of most radical ions, olefin triplet states have energy minima near a perpendicular configuration, because this orientation minimizes the repulsive interaction of the unpaired spins. This unique geometry allows the triplet state to decay to either the *cis*- or the *trans*-isomer of the ground state olefin, regardless of the geometry of the radical ion from which it is generated. In essence, the rearrangement occurs via a common intermediate, a perpendicular triplet, which can be formed from the radical ion of either geometry [157, 158].

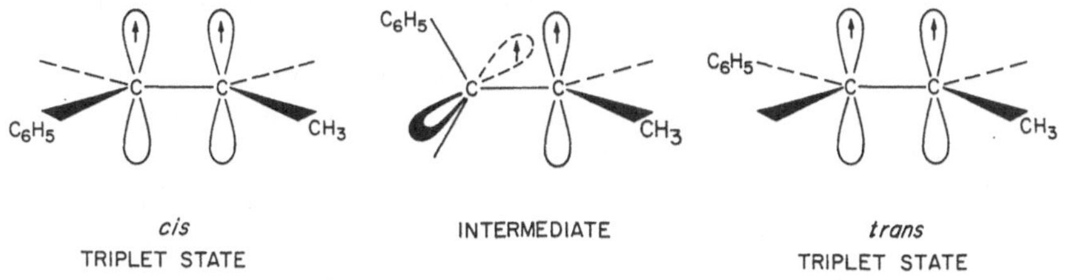

In this mechanistic scheme, the CIDNP intensities of reactant and product are determined by the competition of key steps at each stage of the reaction. For the system discussed here, the qualitative features of the observed polarization suggest that nuclear spin lattice relaxation during the lifetime of the olefin triplet state is negligible, that singlet and triplet pairs recombine with similar efficiencies, and that the triplet state decays to each of the isomers with equal efficiency.

The energy levels of the pertinent intermediates in the photoreactions of chloranil (acceptor) and tetramethylphenylenediamine (donor), respectively, with the "ambiphilic" phenylpropene isomers are compared in Fig. 10.

The involvement of an olefin triplet state in an electron transfer induced olefin isomerization was first recognized during the (electron transfer) quenching of aromatic hydrocarbon excited states by *Z*- or *E*-1,2-dicyanoethylene. Taylor

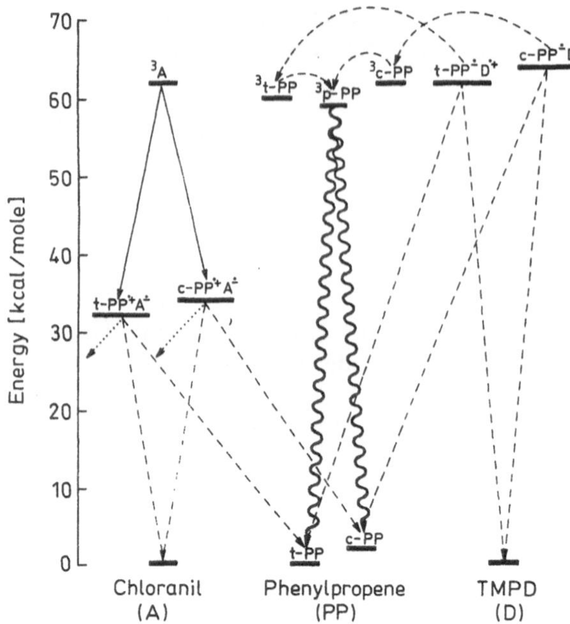

Fig. 10. Energy level diagram for the electron transfer reactions of chloranil and TMPD, respectively, with *cis*- and *trans*-1-phenylpropene [157, 158]. The generation of radical ion pairs is shown as solid lines, whereas singlet and triplet recombinations are depicted as *dashed* and *wavy* lines, respectively

observed characteristic CIDNP effects for the quencher and its geometric isomer and realized that the dicyanoethylene triplet state was populated by charge annihilation of triplet ion pairs [161]. Hyperfine induced singlet-triplet mixing, the key process underlying the induction of CIDNP effects, can cause intersystem crossing on a time scale of 1–10 nanoseconds. Very similar results were observed for the isomeric β-cyanostyrenes (cinnamonitriles) [157, 158] as well as for the *E*-, *Z*-pair of methyl cinnamates.

The importance of the relative energies of the reactant triplet states relative to the ion pair states is illustrated by photo-isomerizations of two stilbene derivatives. *E*-4,4′-dimethoxystilbene (**21**, R = $-OCH_3$) is isomerized in the presence of 9-cyanophenanthrene (**22**) as electron acceptor. In this system, the stilbene triplet state lies below both the pair energy and the triplet state of the sensitizer. The CIDNP intensities observed for the rearranged (*Z*-) isomer are stronger than those of the starting material, suggesting that the reverse electron transfer in triplet pairs is more efficient than in singlet pairs [158]. This behavior is in agreement with the Marcus theory of electron transfer. In contrast, the olefin triplet state is not formed efficiently, and little or no isomerization is observed in the electron transfer reaction between *E*-stilbene (**21**, R = H) and 9,10-dicyanoanthracene (**23**), because in this system the acceptor has the lower triplet energy [162].

Another extensively investigated system involves the interaction of two alkenes, each capable of geometric isomerization, viz., the system stilbene–dicyanoethylene, which also illustrates the involvement of ground-state charge-transfer complexes. Excitation of the ground-state complex results in efficient $Z \rightarrow E$ isomerization of the stilbene exclusively, because the stilbene triplet state lies below the radical ion pair, whereas the dicyanoethylene triplet state lies above it (Fig. 11) [163–166].

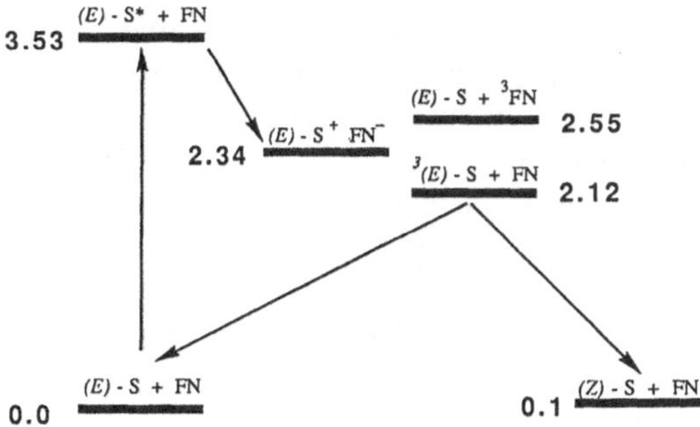

Fig. 11. Energy level diagram for the (*E*)-stilbene (S) excited singlet and triplet states, the (*E*)-stilbene–fumaronitrile (FN) exciples, and triplet fumaronitrile [165, 166]

In this system, the dynamics of pair decay (via intersystem crossing) and of pair separation by solvent molecules were determined by picosecond absorption spectroscopy [167, 168].

21 **22** **23**

In the context of the few $Z \rightarrow E$ isomerizations discussed above and the larger body of reactions which we had to pass over, we discuss in somewhat more detail the major schemes according to which geometric isomerizations occur. We find it useful to divide these reactions into three categories: 1) "thermal" rearrangements of radical cations, i.e. rearrangement under the conditions of their generation, when the available thermal energy is sufficient to overcome the barrier; 2) net rearrangements of the parent molecules due to a mismatch of energy sufaces; 3) rearrangements as a result of photo-excitation of the intermediates (photochemistry of radical cations).

The first few reactions discussed in this section clearly belong in the category of thermal rearrangements. In these systems the ground state and radical ion potential surfaces have minima at similar, though not necessarily identical, geometries. The isomerization occurs in the radical cation because of its lower barrier. Many ring-opening reactions of strained ring compounds (Sect. 4.4) belong to the same general reaction type. This isomerization scheme requires a specific temperature dependence. Since the key step is an activated process, the isomerization yield should fall off with decreasing temperature. This behavior has been confirmed in several cases, including the electron transfer induced isomerization of quadricyclane, benzvalene, and naphthvalene (vide infra).

The second group of isomerizations, featuring triplet recombination as a key step, owe their rearrangement to a mismatch of the energy surfaces of the ground and triplet state potential surfaces. They typically require the involvement of two consecutive intermediates, either two radical cations or a radical cation and a triplet state (if the latter is generated by triplet recombination). The rearrangement occurs because there is no unique correspondence between minima on the potential surfaces of an earlier intermediate and on those of a later intermediate or the ground state. The geometry of minimum energy in the intermediate corresponds to (or lies near) a saddle point on the potential surface of the subsequent state, so that two different minima can be populated.

The temperature dependence for this type of mechanism is quite different from the dependence discussed above. Here, the key step for the isomerization is a decay process, which should be temperature independent, at least over an appropriate temperature range. We have found CIDNP evidence for this behavior in several systems, including the rearrangement of bicyclo[6.1.0]nonatriene to cyclononatetraene (vide infra).

In this context it is useful to remember that the concept of the possible recombination of triplet radical ion pairs is not an ad hoc assumption to rationalize certain $Z \rightarrow E$ isomerizations, although the CIDNP effects observed during an isomerization reaction played a key role in understanding this mechanism. Triplet recombination has been accepted in several donor–acceptor systems as the mechanism for the generation of "fast" (optically detected) triplets [169–171], and invoked for several other reaction types [172]. The CIDNP technique is a sensitive tool for the identification of this mechanism, for example, in the geometric isomerization of *Z*- and *E*-1,2-diphenylcyclopropane and in the valence isomerization of norbornadiene (vide infra). Most of these systems have in common that the triplet state can decay to more than one minimum on the potential surface of the parent molecule.

The third mechanism of isomerization, photoinduced rearrangements of radical cations, has been pursued in a variety of systems. Matrix isolated radical cations have been noted to undergo some rigorous reorganizations as well as subtle ones. For example, the ring opening of cyclohexadiene to hexatriene radical cation and the interconversion of its different rotamers have been achieved by irradiation with UV or visible light [173–174].

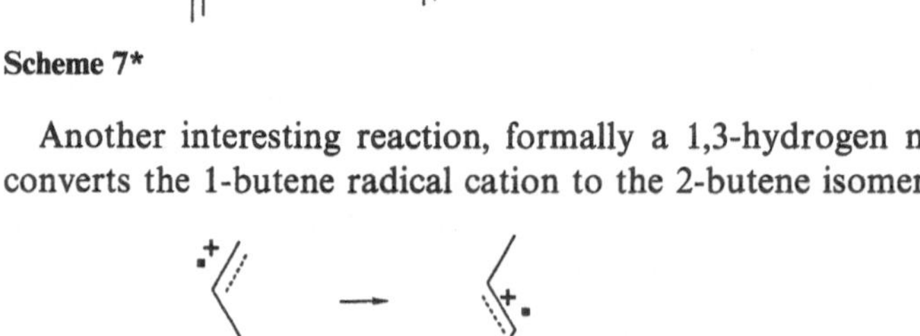

Scheme 7*

Another interesting reaction, formally a 1,3-hydrogen migration (vide infra), converts the 1-butene radical cation to the 2-butene isomer [75].

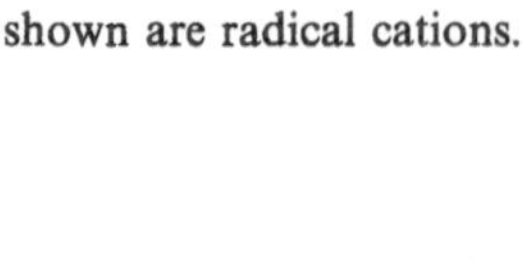

* All species shown are radical cations.

While solid matrices have been employed successfully, they may be less than ideal for controlled mechanistic studies. A more appropriate technique for "controlled" doublet photochemistry appears to be two-photon excitation in solution. In this experiment, the first photon is used to initiate radical ion formation, whereas the second photon, appropriately delayed to coincide with the maximum concentration of the radical cation so generated and tuned to its absorption maximum, serves to excite these intermediates. However, we hasten to add that the benefits of this technique have yet to be demonstrated. The photoinduced rearrangement of radical cations very likely will benefit substantially from a mismatch between (quartet vs. doublet) potential surfaces, much as triplet sensitized isomerizations can be ascribed to mismatches between triplet and ground state surfaces.

4.3 Radical Cations of n-Donors — Electron Transfer Reactions of Amines

Photoexcited acceptor molecules, such as aromatic hydrocarbons or carbonyl compounds, undergo fast reactions with amines. These may result in the reduction of the acceptor molecule or in the deactivation ("quenching") of the acceptor

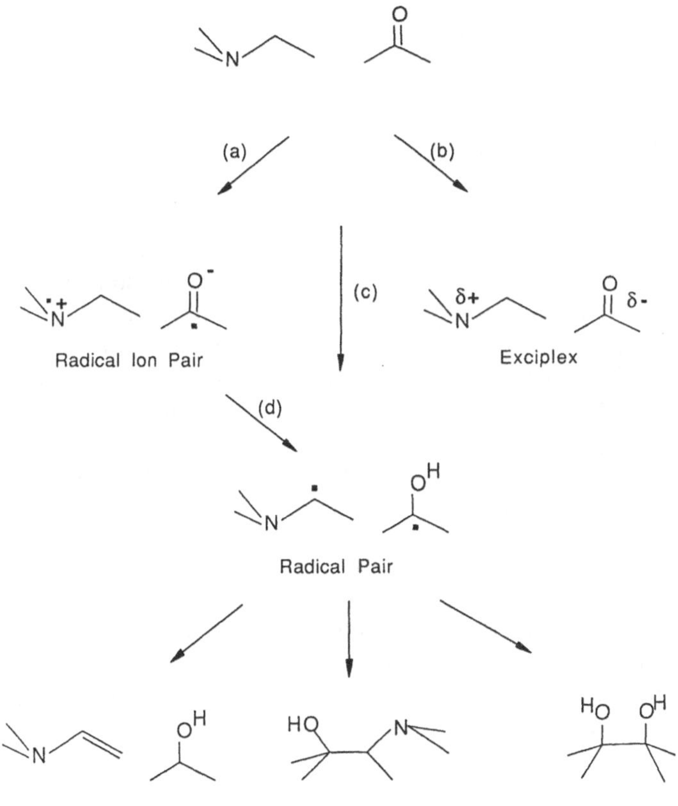

(a) (b) (c) (d)

Radical Ion Pair

Exciplex

Radical Pair

Scheme 8

excited state without any net chemical change. A large body of experimental results indicates that electron transfer from the amine to the acceptor must be considered in any mechanistic formulation [176]. The potential reactions of a carbonyl triplet state with a tertiary amine include: electron transfer, which should be favored in polar solvents; exciplex formation; or hydrogen abstraction (Scheme 8). The latter two pathways should be favored in nonpolar solvents. The neutral radicals produced by hydrogen abstraction are essential in accounting for any reaction products. However, they need not be formed directly; rather, their formation may involve a two-step sequence, electron transfer followed by proton transfer, yielding first a radical ion pair, which is then converted into a pair of neutral radicals.

Aminium radical cations and aminoalkyl radicals have substantially different spin density distributions and, therefore, substantially different hyperfine coupling (hfc) patterns. Aminium radical cations have appreciable proton hfcs only in the position adjacent to the nitrogen center whereas the neutral aminoalkyl radicals have sizable hfcs in both the α- and β-positions. CIDNP effects induced in these species are expected to reflect these differences.

Aminoalkyl radical	Aminium radical ion
$R_2N–\dot{C}H–CH_3$	$R_2\overset{\cdot+}{N}–CH_2–CH_3$
$a_{CH} = -13.6G$	$a_{CH_2} = +37G$
$a_{CH_3} = +19.6G$	$a_{CH_3} < 1G$

During the irradiation of ketones in the presence of amines, nuclear spin polarization effects are observed for two types of products. The reactant amine shows polarization only for the α-protons, suggesting that it is (re)generated from the aminium radical ion, most likely by electron return in a geminate radical ion pair. At the same time, a dehydrogenation product, a vinylamine, shows polarization for both α- and β-protons, in agreement with the involvement of neutral radicals [177]. Vinylamines are unstable under the conditions of these experiments and have not been isolated from these reactions. The observation of these unstable products illustrates another advantage of the CIDNP technique: a lifetime as short or shorter than a second is sufficient to observe a product using a conventional NMR spectrometer.

The different polarization patterns for the reactant amine and for the vinylamine demonstrate the involvement of two different types of intermediates in the photoreaction of carbonyl compounds with amines. The important question is whether these intermediates are formed independently, or whether the neutral radicals are secondary intermediates formed by deprotonation of the radical ions.

Evidence for the two-step alternative is provided by the effects observed during the reaction of several quinones with triethylamine [178]. The irradiation of anthraquinone in the presence of the amine gives rise to vinylamine polarization characteristic of an aminoalkyl radical intermediate. In contrast, the photoreaction of benzoquinone with the amine in solvents of different polarity gives rise to quite different effects (Fig. 12). These patterns are incompatible with either the neutral radical or the radical ion as the sole intermediate, but they can be explained as

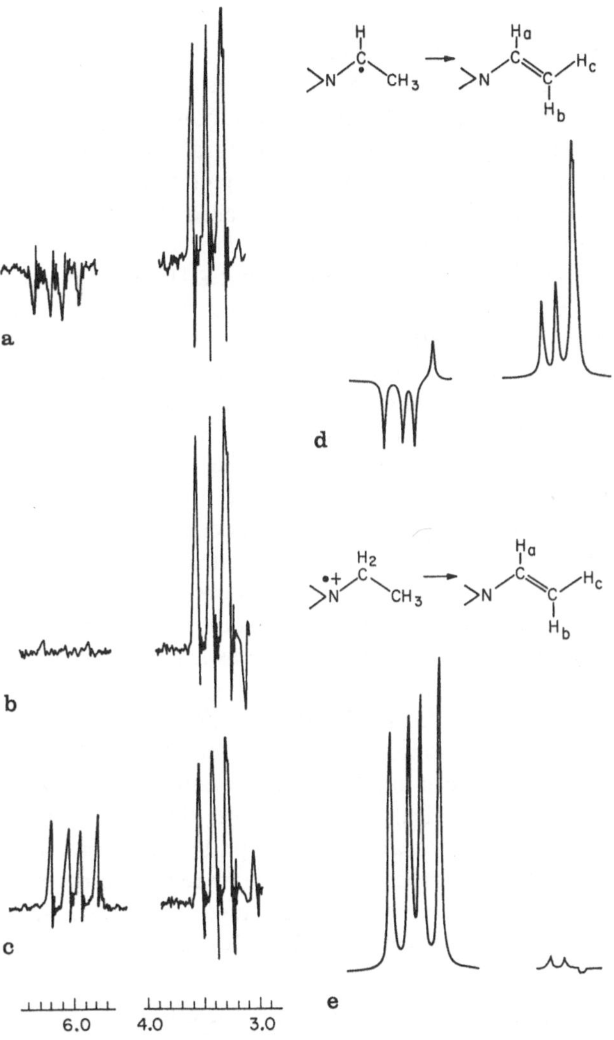

Fig. 12a–e. [1]H CIDNP spectra of diethylvinylamine observed during the reaction of (**a**) anthraquinone with triethylamine, (**b**) benzoquinone with triethylamine in acetone-d_6, (**c**) benzoquinone with triethylamine in acetonitrile-d_3. Traces (**d**) and (**e**) show theoretical spectra calculated for the exclusive involvement of the aminoalkyl radical and of the aminium radical ion, respectively. The change from emission (trace a) to enhanced absorption (trace c) for the doublets near 6.0 ppm indicates an increasing involvement of the aminium radical ion [178]

the result of two contributions due to the consecutive involvement of radical ion and neutral radical.

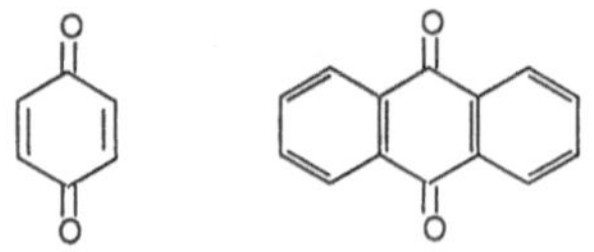

Since radical ion pairs are more stable in polar solvents, their contribution should be more significant in a more polar solvent. The results are fully consistent with this expectation: in acetonitrile ($\varepsilon = 38$) the contribution of the ion pair dominates the weak contribution of the neutral radicals (Fig. 12c), whereas in acetone ($\varepsilon = 21$) the two contributions are of comparable magnitude. As a result, the CIDNP effects in the α-position are (fortuitously) cancelled (Fig. 12b).

The polarization observed in the photoreaction of anthraquinone is also consistent with the concept of two consecutive intermediates [178]. Anthraquinone is a poorer oxidizing agent than benzoquinone so that the corresponding radical ion pair is less stable. Consequently, the aminium radical cation decays faster and does not contribute appreciably to the vinylamine polarization.

In summary, these results constitute strong evidence for the two-step reaction sequence. They require that the deprotonation of the aminium radical cation be competitive on the CIDNP timescale; i.e. surprisingly fast since it involves a carbon acid. The results delineate the fate of the amine derived intermediates with particular clarity, since they are observed directly for amine derived products. The conclusions based on the above CIDNP results were confirmed by time resolved optical spectroscopy in a variety of systems [179–182]. However, in essentially all these systems the reaction progress is monitored by following the "complementary" spectra of the acceptor derived radical intermediates, such as ketyl, semiquinone, stilbene, or thioindigo radical anions.

Tertiary amines have also been employed in electron transfer reactions with a variety of different acceptors, including enones, aromatic hydrocarbons, cyanoaromatics, and stilbene derivatives. These reactions also provide convincing evidence for the intermediacy of aminoalkyl radicals. For example, the photoinduced electron transfer reactions of aromatic hydrocarbons, viz. naphthalene, with tertiary amines result in the reduction of the hydrocarbon as well as reductive coupling [183, 184]. Vinyl-dialkylamines can be envisaged as the complementary dehydrogenation products; their formation was confirmed by CIDNP experiments [185].

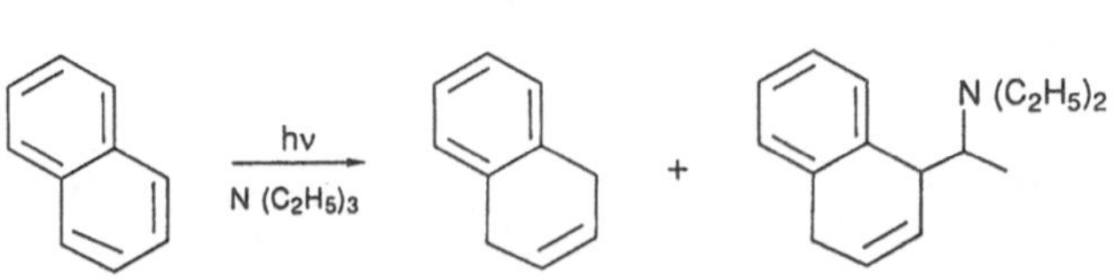

hv

$N(C_2H_5)_3$

+

$N(C_2H_5)_2$

The irradiation of chloro- or cyanosubstituted aromatic acceptors in the presence of tertiary amines give rise to (net) substitution products. In analogy to the coupling reactions of these acceptors with alkenes (Sect. 4.1) an addition–elimination mechanism is suggested. The initial electron transfer is followed by transfer of a proton to an *ipso*-position and coupling of the resulting neutral radicals in the 4-position. The reaction is completed by the 1,4-elimination of hydrogen cyanide [186, 187].

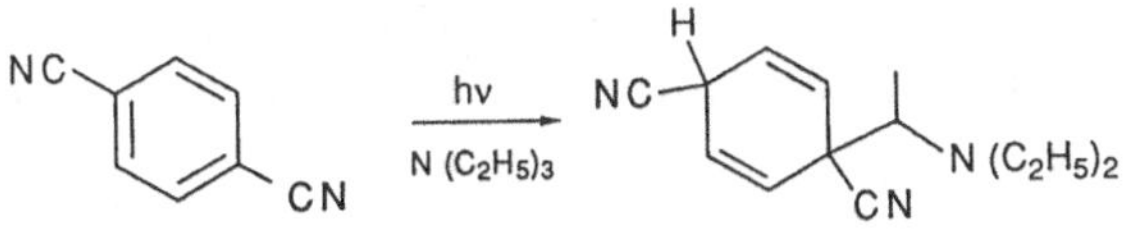

The photoreactions of the isomeric stilbenes with tertiary amines serve as additional examples of olefins functioning as electron acceptors. The irradiation of *trans*-stilbene in the presence of several amines results in the formation of assorted coupling and disproportionation products compatible with net hydrogen abstraction by the photoexcited olefin [188].

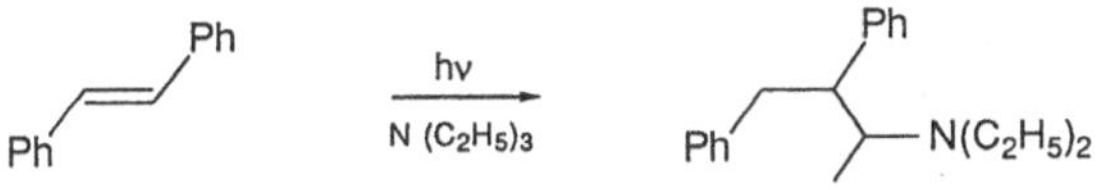

However, much evidence supports a two-step mechanism, involving consecutive electron and proton transfer steps (vide supra) [189]. Perhaps the most significant result is the observation of the Raman spectrum of *trans*-stilbene radical anion during the irradiation of *cis*-stilbene in the presence of ethyldiisopropylamine [190]. Additional observations supporting the electron transfer mechanism include the solvent polarity dependent quantum yield of product formation and the inverse relationship between the rate constants of fluorescence quenching and the oxidation potential of the amine quencher [191]. Tertiary amines containing primary as well as secondary alkyl substituents may give rise to two different α-aminoalkyl radicals. Interestingly, the irradiation of *trans*-stilbene in the presence of methyldiisopropylamine results exclusively in coupling to the primary carbon. This finding has been ascribed to a stereoelectronic effect [188, 192].

The deprotonation step, either by the sensitizer radical anion or by some adventitious base, is essential for the formation of any amine derived products. This step can be prevented if the α-hydrogens are arranged in a plane orthogonal to the singly occupied nitrogen *n*-orbital; a requirement which is met for the radical cation of 1,4-diazabicyclo[2.2.2]octane (DABCO). The low oxidation potential, due to the interaction of the pair of transannular nitrogens, makes this an excellent electron transfer quencher. Yet, no product formation is observed as a result of these interactions, with the possible exception of the zwitterionic adducts formed with highly electrophilic ketones [193].

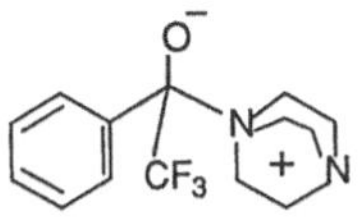

In addition to the various bond forming reactions discussed above, some electron transfer reactions of amines result in bond cleavage. The loss of an alkyl group, converting tertiary to secondary amines, has been noted in several of the systems discussed above [194–197]. The key to these reactions lies in the reducing nature of the α-aminoalkyl radicals [198]. Accordingly, they may undergo a (thermally induced) electron transfer reaction, generating iminium (cat) ions which, in the presence of water, can be hydrolyzed to yield a secondary amine and a carbonyl compound.

$$\text{R}_2\text{N}\dot{\text{C}}\text{H}_2\text{R} \longrightarrow \text{R}_2\overset{+}{\text{N}}{=}\text{CHR} \xrightarrow{H_2O} \text{R}_2\text{N{-}H} \quad \text{R{-}CHO}$$

Whether these radicals serve as electron donors (to yield iminium cations) or as hydrogen atom donors (to yield vinylamines) will depend on the nature of the reaction partner, particularly on its reduction potential. In any case, the quantum yield of C–N cleavage reactions are low, since they require several consecutive processes, i.e. successive transfer of an electron, a proton, and a second electron, or a mechanistic equivalent thereof.

More recently, a variety of C–C cleavage reactions were discovered in α,β-bifunctional substrates, viz., diamines or aminoalcohols, and their benzologs [198–203]. These reactions have high chemical yields (>90%), but their quantum efficiencies are often low (≦1%). They are formally analogous to the "thermal" two-electron fragmentations [204]; they also resemble various other C–C cleavage reactions of radical cations, particularly the ("mesolytic") cleavage of bibenzyl systems [205]. However, the reactions discussed here have special features that set them apart from either analog.

These reactions are initiated by electron transfer from the substrate to a photoexcited acceptor. Subsequently, the acceptor radical anion abstracts a proton from the functional heteroatom in the γ-position, leading to fragmentation. A moderate isotope effect on the fragmentation efficiency and a correlation with the radical anion basicity suggest that the proton is removed in the rate-determining step, possibly concerted with C–C cleavage, although not necessarily synchronous with it. The relative fragmentation rates of chiral aminoalcohols (*erythro* substrates undergo fragmentation more readily than *threo* compounds) support an *anti*-coplanar conformation as the optimum geometry for the fragmentation [199, 200].

$$\text{R}_2\text{N}^{\cdot+}\text{-C-C-O-H} \cdots \text{A}^{\cdot-} \longrightarrow \text{R}_2\text{N-}\dot{\text{C}} \quad \text{-C=O} \quad \text{H-A}^{\cdot}$$

The fragmentation leads to the simultaneous generation of two neutral radicals in close proximity, separated only by the carbonyl product. Depending on the nature of the substrate, these fragments may react by hydrogen atom transfer, either to the aminoalkyl (benzyl) radical or to the sensitizer derived radical. Accordingly, the acceptor can act either as a sensitizer or undergo two-electron reduction.

In general, the increased efficiency of C–C bond cleavage observed for these substrates can be ascribed to two features: a weakened C–C bond in the radical cation due to the presence of the second functional group; and the greater acidity of the N–H or O–H functions compared to the α-C–H acidity of the aminium radical cation.

These reactions are illustrated with the fragmentation of two isomeric aminoalcohols and one aminodiol, which have been observed upon thioindigo sensitized irradiation in benzene solution [203]. Under these conditions, the (*p*-*N*,*N*-dimethylaminobenzyl) benzyl alcohol (**24**) suffers fragmentation with apparent internal disproportionation; the benzyl alcohol function is oxidized, whereas the *p*-aminobenzyl function is reduced, giving rise to *N*,*N*-dimethyl-*p*-toluidine and benzaldehyde. The acceptor is involved only as a sensitizer and, hence, remains unchanged.

The benzyl *p*-dimethylaminobenzyl alcohol (**25**) is fragmented similarly with formation of *p*-(*N*,*N*-dimethylamino)benzaldehyde and toluence. Once again, the benzyl alcohol function is oxidized, the second benzyl function is reduced, and the acceptor remains unchanged; it is involved only as a sensitizer.

In the case of the diol **26**, the acceptor plays the dual role of sensitizer and hydrogen acceptor. In this case both benzyl alcohol functions are oxidized, whereas the acceptor is reduced to dihydrothioindigo.

In summary, the electron transfer reactions of amines offer an interesting variety of pathways, involving free radical or ionic intermediates and leading to two-electron oxidation as well as coupling and fragmentation reactions. In many systems, however, the most important reaction involves electron return from the acceptor radical anion, i.e. quenching of an excited state without net chemical change.

We have limited our discussion of *n*-donor derived radical cations to a single structure type. Other *n*-donors include hydrazines, polycyclic amines, ethers, phosphines, and thioethers. Several dioxetane radical cations were mentioned in Section 4.1. A detailed or even cursory discussion of these donors exceeds the scope of this article.

4.4 Reactions of Strained Ring Systems

The third class of organic donor molecules are σ-donors, viz., alkanes and cycloalkanes. These substrates have inherently high ionization and oxidation potentials. Therefore, their radical cations are not readily available by photoinduced electron transfer, but typically require radiolysis and electron impact in the condensed phases or the gas phase, respectively. Thus, radical cations of simple alkanes (methane [206], ethane [207]) or unstrained cycloalkanes (cyclopentane, cyclohexane) [208] were identified and characterized following radiolysis in frozen matrices. In contrast, strained ring compounds have significantly lower oxidation potentials so that the radical cations of appropriate derivatives can be generated by photoinduced electron transfer.

The radical cations of hydrocarbons containing strained-ring moieties have been attracting ever growing attention over the last two decades. Perhaps the earliest indication of their special nature was provided by the efficiency with which their precursors quench the fluorescence of excited state electron acceptors [209, 210]. Although this observation was originally ascribed to vibrational energy transfer, the charge transfer nature of this interaction was soon recognized [211, 212]. Subsequently, the unusually low oxidation potentials of strained rings were noted [213, 214]. The correspondence between the efficiency of fluorescence quenching and the vertical ionization potential of quenchers had been documented for π-donors such as aromatic [215] or olefinic quenchers [212, 216]. A similar relationship was confirmed for strained ring compounds [212]. Subsequently, the correlation between quenching efficiencies of strained ring compounds and their oxidation potentials was documented [217]. The lower oxidation potentials of strained ring compounds compared to less strained isomers can be interpreted as evidence that a significant fraction of the strain energy is released upon oxidation. In the following section selected electron transfer induced reactions of strained ring systems will be discussed. Interestingly, the principal reaction types of these substrates are quite similar to those of the olefins.

Among the typical electron transfer induced reactions of strained polycyclic ring compounds are efficient cyclobutane cycloreversions. For example, quadricyclane (**27**), a system containing three adjoining ("cumulated") strained rings, can be converted to norbornadiene (**28**). In this system, the driving force for the cyclobutane to diolefin conversion is increased by the breaking of two cyclopropane bonds, releasing their strain energy [218]. Since **27** is accessible by photo-sensitized conversion of **28** using sunlight, this system may hold some promise for solar energy storage and its subsequent electron transfer induced release. The conversion of the strained hydrocarbon can be achieved also by catalytic means [219, 220], or initiated electrochemically [221]. Nevertheless, the promise of solar energy utilization via the norbornadiene-quadricyclane pair has not come to fruition as yet. The chief problem lies in the fact that the substantial chemical energy of quadricyclane is converted to thermal energy, and that this process is carried out typically in dilute solutions. Under these conditions the utilization of thermal energy presents serious as yet unsolved challenges.

Several cyclopropane derivatives have been found to undergo geometric isomerization upon reaction with photo-excited singlet acceptors. This reaction type is exemplified by the electron transfer induced interconversion of *cis*- and *trans*-1,2-diphenylcyclopropane (**29**) in the presence of singlet sensitizers [222, 223]. Although CIDNP studies have clearly demonstrated the steric integrity of the two isomeric radical cations during the radical ion-pair lifetime (when they are generated by reaction with triplet sensitizers) [223], *cis-trans* isomerization results from the electron transfer reaction with several singlet sensitizers. This conversion has been ascribed to reverse electron transfer in pairs of triplet spin multiplicity, a process which populates a (ring-opened) triplet state with two orthogonal *p*-orbitals. The structure of this intermediate lies between the geometries of *cis*- and *trans*-isomer and can decay to regenerate either isomer (Section 4.2).

C_6H_5 C_6H_5 cis-**29** ⇌ trans-**29** C_6H_5 C_6H_5

An interesting variant of a geometric isomerization was observed for the 7,7-dimethylbicyclo[4.1.0]hept-2-ene system. The electron transfer reaction of the highly strained *trans*-fused isomer (**30**) with 1-cyanonaphthalene rapidly converted it to the *cis*-fused system (**31**) [224]. The observed rearrangement requires inversion at one of the tertiary cyclopropane carbons. This can be accomplished either by removal of a hydrogen (proton) or by cleavage of a cyclopropane or an allylic bond.

30 → **31**

Several interesting rearrangements (stereomutations) were observed for a series of 1-anisyl-2-vinylcyclopropanes (**32**, Ar = $C_6H_4-OCH_3$). These reorganizations have been elaborated in greater detail than most other radical cation reactions. For this reason, we will consider them in somewhat greater detail, even though they were initiated primarily by chemical, rather than photochemical electron transfer. Treatment of the *cis*-isomer (*cis*-**32**) with aminium or dioxygenyl salts at temperature between -90 and -40 °C produced the *trans*-isomer in high yield [225]. Compared to the thermal rearrangement of the parent molecule, a rate enhancement of 10^{24} was estimated for the aminium ion catalyzed reaction. The

facile isomerization of these substrates upon chemical oxidation stands in interesting contrast to the lack of isomerization observed for the diphenylcyclopropanes upon photoinduced electron transfer to triplet acceptors [222, 223].

A possible reconciliation of these seemingly conflicting results lies in the lifetimes of the individual radical cations under the different experimental conditions. In the PET experiment the lifetime is dictated by the rate of intersystem crossing, a hyperfine induced process, which often falls into the range 10^{-9} to 10^{-8} sec. The aminium salt catalyzed rearrangement is a free radical cation chain reaction. Under these conditions the radical cation lifetime is determined by the diffusion-limited encounter with a neutral molecule, which may be quite slow at the low temperatures of these experiments. Although any barrier to isomerization is larger at the lower temperatures, it is well-known that the barriers to many radical cation reactions are reduced drastically.

The different electronic nature of the two substituents may provide another reason why the unsymmetrically substituted cyclopropane (**32**) rearranges more readily. If the anisyl group preferentially delocalizes the positive charge, whereas the allyl moiety stabilizes the spin, the remaining bond between the two centers may be somewhat weaker than in the more symmetrically delocalized diphenylcyclopropane radical cations.

Some mechanistic aspects underlying the arylvinylcyclopropane rearrangement were examined by an elegantly designed doubly deuterium labeled derivative, *cis*-**33**, which resulted in a random (1 : 1) mixture of two *trans*-isotopomers. This observation was interpreted as evidence for "two-center rotation", i.e. ring opening to a bifunctional intermediate followed by random re-closure. Presumably this intermediate would be planar and generated by outward rotation of the two substituents [225].

cis-33 → trans-33a + trans-33b

We note, however, that this result can be explained also by the simultaneous involvement of two independent one-center inversions. This mechanism was specifically ruled out by the authors, because they thought it unlikely that two different reactions should have identical rates over a 50° temperature range. However this assumption does not appear so unreasonable, if the barriers between the respective minima are low, and the radical cation lifetime between electron exchanges is long. A detailed discussion of cyclopropane radical cation structures will be given in Section 5.1.

The failure to observe any ring enlargement to a cyclopentene derivative is noteworthy. It can be ascribed to the exclusive involvement of s-*trans* conformers (regardless of mechanism). That this failure is not due to an inherent inability of the s-*cis*-conformer to cyclize (for electronic or energetic reasons) is shown clearly by the ready ring enlargement of derivatives, e.g. **34** → **35**, in which the *cisoid* conformation is either enforced (by rigidly linking the cyclopropane and vinyl moieties) or less disfavored (due to "buttressing" at the junction between the two moieties) [226]; both the *cis*- and *trans*-isomer of the "buttressed" substrate were found to undergo ring expansion smoothly.

Ar
34
Ar
35

A further interesting facet of this rearrangement was revealed, when the second center engaged in bond formation also carried a label that would reveal the stereochemistry of bond formation [227]. The rearrangement of *cis*- and *trans*-vinylcyclopropanes gave the same 6:1 mixture of *cis*- and *trans*-substituted cyclopentenes. This result is best accommodated by a stepwise mechanism, involving a ring-opened bifunctional intermediate, which closes mainly to the (less hindered) *trans*-isomer, and to a minor degree to the *cis*-isomer (vide infra).

Lastly, it was noted that the product distribution obtained from appropriate substrates via thermal rearrangement on the one hand and electron transfer induced reorganization, on the other, may be quite different. For example, the thermal rearrangement of **37** involved ring opening with hydrogen migration (→**36**), whereas the electron transfer induced reaction proceeded with ring enlargement (→**38**) [227]. These differences can be explained readily by considering the reduced delocalization in the radical cation generated by the putative hydrogen migration. In general, while some bond breaking reactions of radical cations have substantially reduced barriers, others are facilitated to a minor degree and are therefore suppressed.

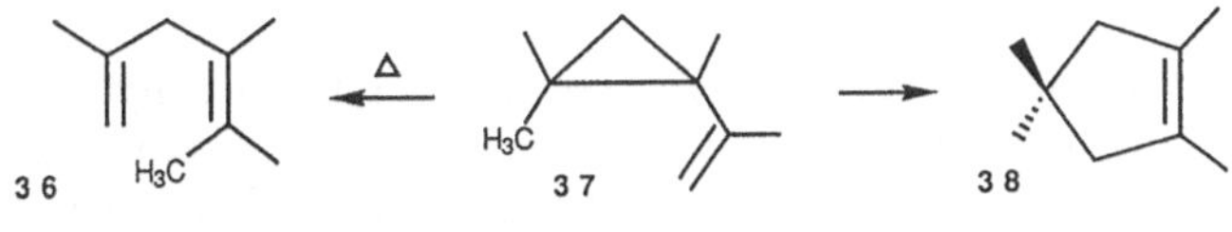

36 H_3C Δ H_3C 37 38

This change in product distribution can be rationalized in terms of substantial topological differences between the respective potential surfaces. The rearrangement of the neutral parent molecule involves hydrogen transfer in concert with ring opening. Accordingly, much strain is relieved in the transition state for hydrogen transfer. Considering the radical cation rearrangement, the ring strain is substantially reduced upon formation of the primary radical cation. Subsequent ring opening may be favored because of the extensive delocalization of spin and charge. Under these circumstances hydrogen transfer would only serve to limit delocalization and is, therefore, avoided.

Electron transfer sensitization of 1,1,2,2-tetraphenylcyclopropane (**40**) yields 1,1,3,3-tetraphenylpropene (**39**), a conversion readily explained via ring opening and subsequent hydrogen shift [228]. In contrast, irradiation of the charge transfer complex between **40** and tetracyanoethylene leads to a ring expansion product, 1,3,3-triphenylindene (**41**), along with reduction of the acceptor to tetracyanoethane. In analogy to the 1,6-coupling of formally 1,4-bifunctional intermediates (see Section 4.1), this reaction could be formulated as a 1,5-coupling of a formally 1,3-bifunctional (trimethylene) intermediate. The exact nature of the intermediate involved in the ring closure has not been established unambiguously. However, it is reasonable to rationalize the conversion via a three-step sequence, involving proton transfer to the sensitizer anion, followed by ring closure of the tetraphenylallyl free radical, and hydrogen atom transfer from the resulting cyclohexadienyl moiety to the counter radical.

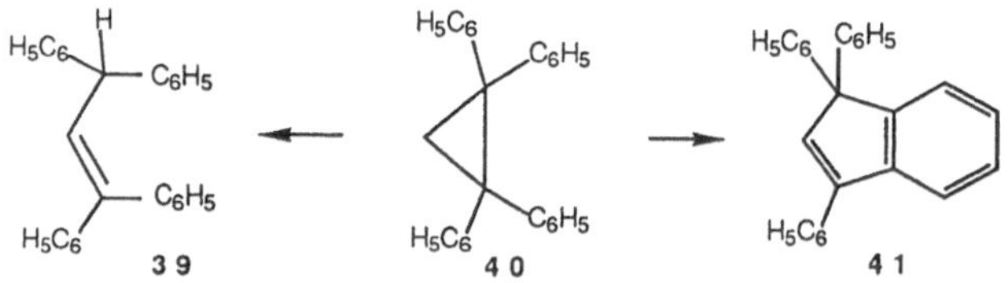

A related ring opening reaction was observed for benzonorcaradiene (**42**); the chloranil photo-sensitized reaction of this material leads to two polarized reaction products, 1- and 2-methylnaphthalene (**43, 44**) [229]. Apparently either bond linking the cyclopropane methylene group with the dihydronaphthalene skeleton may be broken. This reorganization also requires a net hydrogen migration from a tertiary to a secondary carbon. The driving force of the net conversion lies in the relief of ring strain and the formation of an aromatic system. However, mechanistic details have not been established.

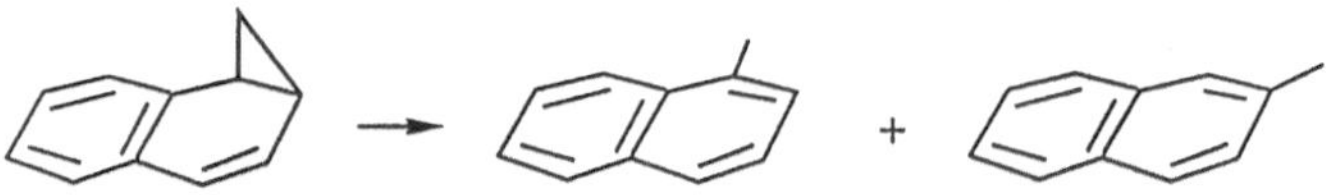

Several norcarene derivatives were shown to undergo dehydrogenation with ring opening to form the more extended π system of cycloheptatriene. Thus, irradiation of chloranil in the presence of tricyclo[4.4.1.0^{1,6}]undeca-3,8-diene (**45**) gave rise to bicyclo[4.4.1]undeca-1,3,6-8-tetraene (**46**), while the quinone was reduced to the corresponding hydroquinone [229, 230].

Some substituted cyclopropanes have been shown to undergo nucleophilic addition of suitable solvents (CH_3OH) [231]. For example, the electron transfer reaction of phenylcyclopropane (**47**, R = H) with *p*-dicyanobenzene resulted in a ring-opened ether (**48**), formed by *anti*-Markovnikov addition. More recently, the reaction of a 2,3-dimethyl derivative (**47**, R = CH_3) was shown to occur with essentially complete inversion of configuration at carbon, suggesting a nucleophilic cleavage of the "one-electron bond" [233]. This result is significant, since it requires an intermediate with the unperturbed stereochemistry of the parent molecule.

Interestingly, an aromatic substitution product (**49**, R = H) formed by coupling with the sensitizer anion was also isolated [231, 232]. This reaction is the cyclopropane analog of the photo-NOCAS reaction (Sect. 4.1) and preceeded it by almost a decade. No acronym was coined (photo-NCPCAS reaction?) and none is suggested.

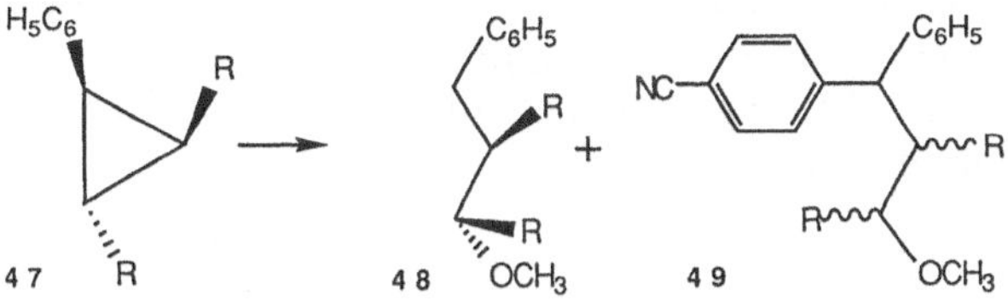

A related pair of products (**48**, **50**, R = H) was obtained upon photoinduced electron transfer reaction between phenylcyclopropane (**47**, R = H) and N-methylphthalimide [234]. In this system, as in the various coupling reactions discussed above, the formation of ethers indicates that the coupling reaction of the radical ion pair is slow, in spite of its geminate nature.

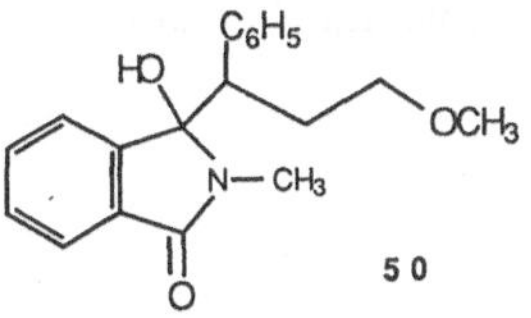

This conclusion raises the question, whether the radical anion is at all involved in the formation of coupling products, or whether these might not be formed by addition of adduct radicals onto neutral acceptor molecules. To the best of our knowledge no unambiguous evidence is available to decide this issue. However, CIDNP results obtained in several systems argue against geminate coupling. For example, CIDNP evidence suggests that free dimethylindene radical cations are involved in the formation of cycloadducts with chloranil (Sect. 4.1).

Two different 1:1 adducts were obtained from the reaction of 1,1-diphenylcyclopropane with tetracyanoethylene. A cyclic adduct (**52**) was accompanied by an acyclic addition product (**53**). Both adducts are compatible with a 1,5-bifunctional (zwitterionic or biradical) intermediate (**51**), which may react by ring closure or hydrogen (proton) transfer [235].

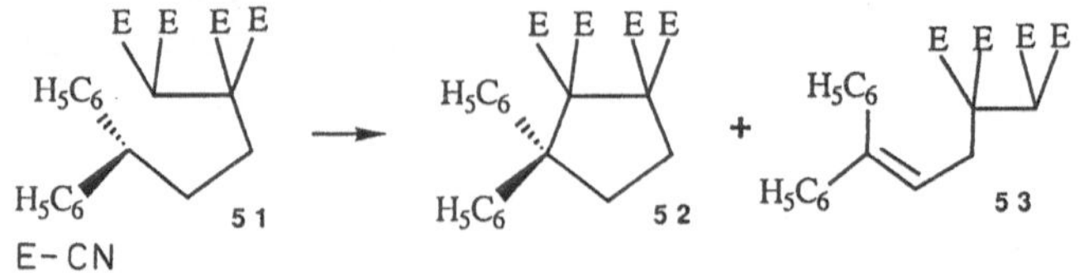

Cyclopropane derivatives of various structure types are involved in electron transfer induced oxygenation reactions. For example, the photoreaction of 9,10-dicyanoanthracene with 1,2-diarylcyclopropanes generates a mixture of *cis*- and *trans*-3,5-diaryl-1,2-dioxolanes [236]. Adducts of the same structure type could be observed upon irradiation of charge transfer complexes between tetracyanoethylene and tetraarylcyclopropane systems [237–239]. In some of these systems geometric isomerization was more efficient than oxygenation. The solvent polarity dependence of product yields suggests the involvement of radical cations which, depending on the reduction potential of the acceptor, will couple to superoxide or molecular oxygen.

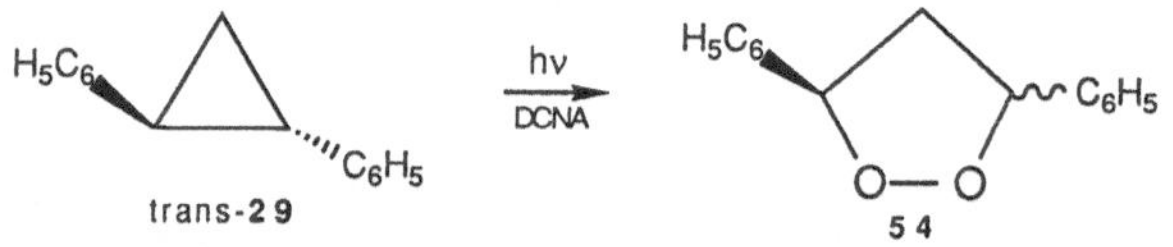

The chloranil sensitized isomerization of geminal diarylspiropentane systems (**55**) gives rise to a pair of isomeric methylenecyclobutanes (**56, 57**) [240]. When carried out in the presence of molecular oxygen, this reaction gives rise to two oxygenation products (**58, 59**). All four products are compatible with a single bifunctional intermediate; however, the available mechanistic data indicate a considerably more complex pathway.

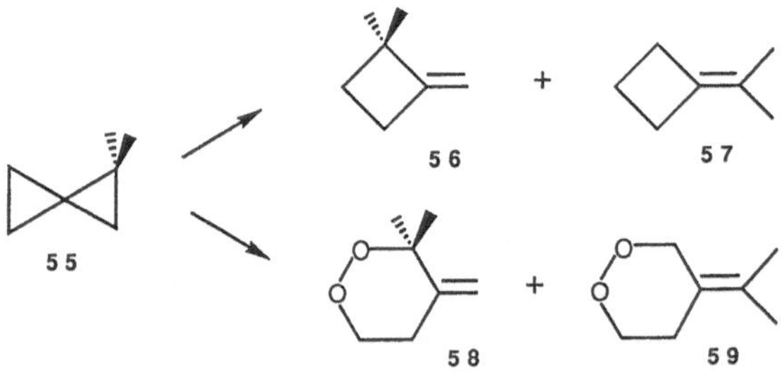

The electron transfer reaction of gem-diarylmethylenecyclopropanes (**60**) with singlet sensitizers results in the exchange of the *exo*-methylene and the secondary cyclopropane carbons [241]. The chloranil photo-sensitized reaction generates two unusual cycloaddition products (**61, 62**) [242], whereas the tetracyanoethylene sensitized oxygenation produces the respective dioxolanes [238]. These reactions are compatible with a ring-opened radical cation, and CIDNP experiments have revealed a specific bifunctional structure in which spin and charge stabilized in separate sections of the molecule. This structure will be discussed in more detail in Sect. 5.5 with related bifunctional systems.

H'2 Ar Ar 60 hν CA Ar Ar O 61 + Ar Ar O 62

Bicyclo[6.1.0]nonatriene (**63**) and its derivatives are among the most thoroughly investigated hydrocarbon systems. The electron transfer induced reactions of these compounds lead to several interesting rearrangements involving opening of the cyclopropane ring. Thus, the photo-sensitized reaction of the parent system with chloranil resulted in rearrangement to cyclononatetraene (**64**), a result ascribed to triplet recombination [243].

63 hν CA 64

On the other hand, 9,9′-disubstituted derivatives of **63** undergo more thorough rearrangements. For example, the reaction of singlet acceptors (9,10-dicyanoanthracene) with the 9-spirofluorene derivative (**63**, 9,9-o-o′-biphenylenyl) in polar solvents leads mainly to the corresponding barbaralane (**65**) system together with minor yields of 7-vinylcycloheptatriene (**67**) and fluorenespirobicyclo[4.2.1]nonatriene derivatives (**66**). In contrast, the analogous reaction of the 7,7-diphenyl system (**63**, 9,9-diphenyl) yielded mostly the cycloheptatriene derivative (**67**) [243, 244]. To account for the pronounced difference in reaction products, caused by a seemingly minor change in the substituent, two possible explanations were considered: the two derivatives may have different primary electron donating groups; or the intermediates may have different spin and charge density distributions. In addition, the solvent polarity dependence of product yields differs substantially. Thus, **65** (9,9-o,o′-biphenyl) is favored in polar solvents (CH_3CN), but its quantum yield decreased to one hundredth in benzene. In contrast, the formation of **66** and **67**, (9,9-o,o′-biphenylenyl) was not strongly affected by changes in solvent polarity. This observation is compatible with the involvement of an exciplex as a precursor for **66** and **67**, and a radical cation as a precursor for **65**. Obviously, this system is quite complex and a variety of different intermediates have to be considered.

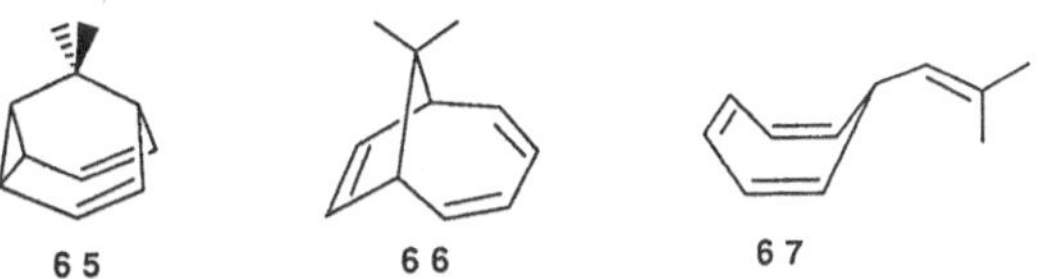

The electron transfer sensitized irradiation of allyl- or benzylcyclopropene systems (**68**) results in ring enlargement with dehydrogenation (→**70**) [245]. Formally, these structures can be considered as substituted hexadiene systems (vide infra). Although mechanistic details have not been established, the reaction very likely proceeds via bifunctional intermediates (**69**).

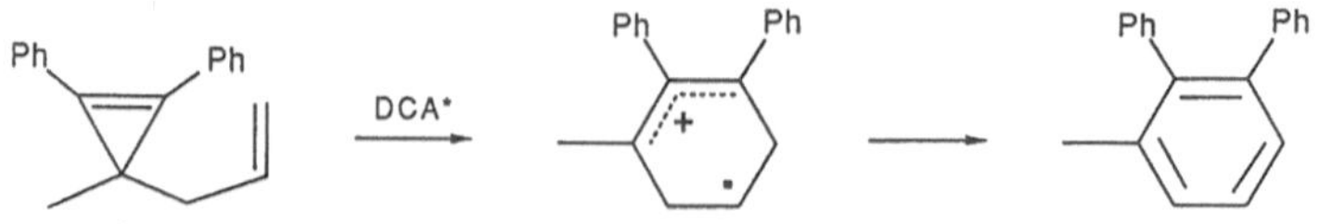

Ph Ph DCA* Ph Ph Ph Ph

68 69 70

Given the rich variety of electron transfer induced reactions observed for cyclopropane systems, it is not surprising to find a similarly rich reactivity for systems comprised of two fused cyclopropane rings. For bicyclobutane systems, such as the trimethyl derivative (**71**), nucleophilic capture leads to addition with cleavage of the transannular bond (→**72**). Alternatively, the initial capture and ring opening may be accompanied by dehydrogenation (→**73**) [246].

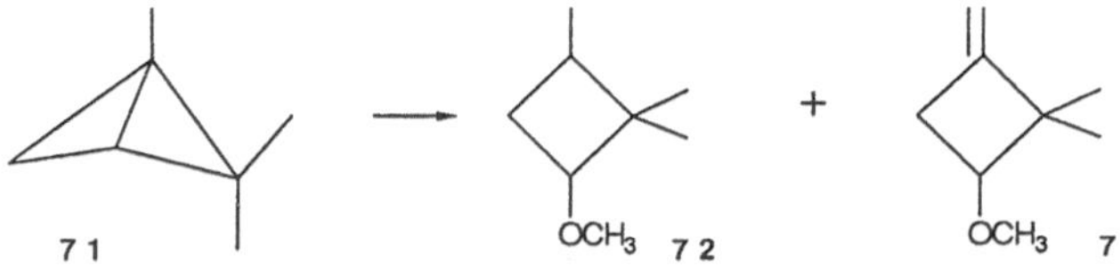

71 OCH_3 72 + OCH_3 73

Aside from quadricyclane the most thoroughly studied strained ring system is a bridged bicyclobutane derivative, tricyclo[4.1.0.0^{2,7}]heptane (**75**), possibly because of its ready availability. Irradiation in the presence of naphthalene resulted in rapid fluorescence quenching without rearrangement of the strained hydrocarbon. In contrast, either 1-cyanonaphthalene or 9,10-dicyanoanthracene caused the formation of a (dimeric) rearrangement product (**79**) [217, 247]. When the electron transfer reaction is carried out in nucleophilic solvents (CH_3OH, H_2O) or in the presence of an ionic nucleophile (CN^-), capture of the nucleophile leads to products formally derived by addition across the transannular bicyclobutane bond (**74**, **76**, **77**). Significantly, each product is derived by backside

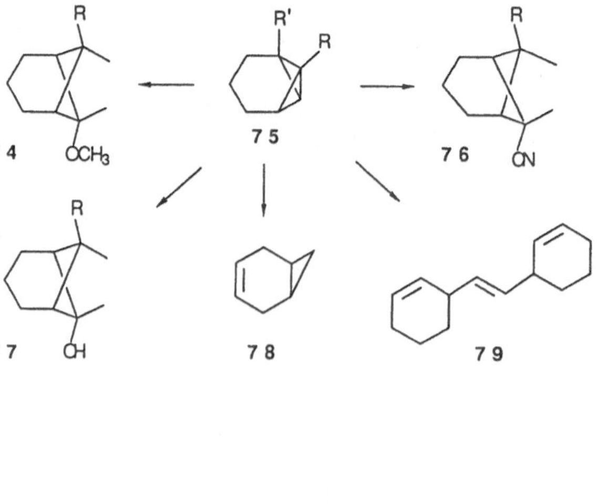

R R' R R

74 OCH_3 75 76 CN

R

77 OH 78 79

Scheme 9

attack, apparently in *anti*-Markovnikov fashion, as indicated clearly for the monomethyl derivative (**74**, **76**, R = CH_3) [217].

The bridged bicyclobutanes undergo several interesting rearrangements. While the above mentioned dienes can be explained without invoking hydrogen migration, several rearrangements of tricycloheptane to norcarene systems require either hydrogen (proton) migration or its intermolecular equivalents [248, 249]. The net conversion of the parent system (→**78**) is shown in Scheme 9, whereas the reaction of benzotricycloheptene (**80**) to benzonorcaradiene (**42**) is formulated below [248].

80 → 42

Two tricycloheptane derivatives carrying a bulky substituent in the 2-position [(**75**), R′ = $(CH_3)_3Si-CH_2-$, 1-CH_3-cyclo-C_3H_4-] undergo electron transfer induced conversions to bicyclo[3.2.0]hept-6-enes, whereas other similarly substituted derivative [(**75**), R′ = $(CH_3)_3C-$, $(CH_3)_3Si-$] gave only the unexceptional methanol adducts [250, 251]. This conversion was explained originally by a complex mechanism, involving four consecutive radical cations. We have proposed [9] that this reorganization can be interpreted, in analogy to the conversion of the neutral parent molecule [252], possibly as a conrotatory ring opening, **81**→**82**, followed by conrotatory ring closure, →**83**. Alternatively, the key intermediate **82** might be trapped by electron return, generating *cis,trans*-cycloheptadiene, a structure type which has been invoked as an intermediate in several bicyclobutane to cyclobutene rearrangements [253]. This revised explanation apparently has been accepted by the original authors [254]. Nevertheless, it remains to be established whether orbital symmetry plays a definitive role in radical cation reactions.

81 → 82 → 83

Among the electron transfer induced reactions of cyclobutane systems, cycloreversions are the most prominent. These reactions are the reverse of the cycloadditions discussed in Sect. 4.1. The reactivity of the corresponding radical cations depends on their substitution pattern. We have mentioned the fast "two-bond" cycloreversion of quadicyclane radical cation as well as the ready ring closure of a tetracyclic system (**3**, Sect. 4.1). A related fragmentation of *cis*-, *trans*-, *cis*-1,2,3,4-tetraphenylcyclobutane (**84**) can be induced by pulse radiolysis of 1,2-dichloroethane solutions. This reaction produces the known spectrum of *trans*-stilbene radical cation (**85**) without a detectable intermediate and with a high degree of

Ph Ph Ph Ph
84 —γ→ 85
Ph Ph

selectivity [255]. To our knowledge, this is the first case in which a stereoselective retro-cycloaddition has been demonstrated.

The ease of cyclobutane cleavage and the detailed mechanism can be affected by the nature of the substituents and the substitution pattern. While the above cyclobutanes are cleaved without a discernible intermediate, the *anti*-head-to-head dimer of dimethylindene shows a significantly different behavior. This substrate is cleaved in an apparent two-step process, involving a ring-closed radical cation (with spin and charge localized on one indan system) and a ring-opened 1,4-bifunctional radical cation. Apparently, the cleavage of the doubly benzylic cyclobutane bond is reversible. The involvement of more than one dimer radical cation is indicated by a unique polarization pattern (Fig. 13), which is incompatible with any one intermediate, but can be simulated on the basis of two successive radical cations (see Sect. 5.2) [256].

Another one-bond cleavage is indicated for the electron transfer induced reaction of α-pinene (**86**), a terpene containing a vinylcyclobutane moiety in a bisected conformation [257]. Irradiation of 1,4-dicyanobenzene in acetonitrile/methanol in the presence of α-pinene leads to the formation of a mixture of racemic 1:1:1 adducts (**87**) between **86**, methanol, and the acceptor. In contrast to the nucleophilic substitution of cyclopropane systems [233], the loss of chirality indicates the involvement of a symmetrical intermediate. This finding is compatible with ring-opening before nucleophilic capture (→**C**) [257], but it does not exclude a reaction path involving nucleophilic substitution at the quarternary carbon with ring opening (→**B**). Although this attack does not utilize the (kinetically favored) less hindered center, it does lead to the more stable intermediate. The same principle is followed also in several radical cation reactions resulting in *anti*-Markovnikov additions (vide supra).

The electron transfer induced cleavage of cyclobutane systems is of interest also in connection with photoreactivation, a repair mechanism for photo-damaged DNA. Upon exposure to UV light, DNA may be damaged by the formation of cyclobutane links between adjacent thymine units. The resulting dimers are cleaved, and DNA is restored, by a photoreactivating enzyme in the presence of near UV or visible light. The mechanism of reactivation may involve electron transfer, either to or from the pyrimidine dimer [258–260].

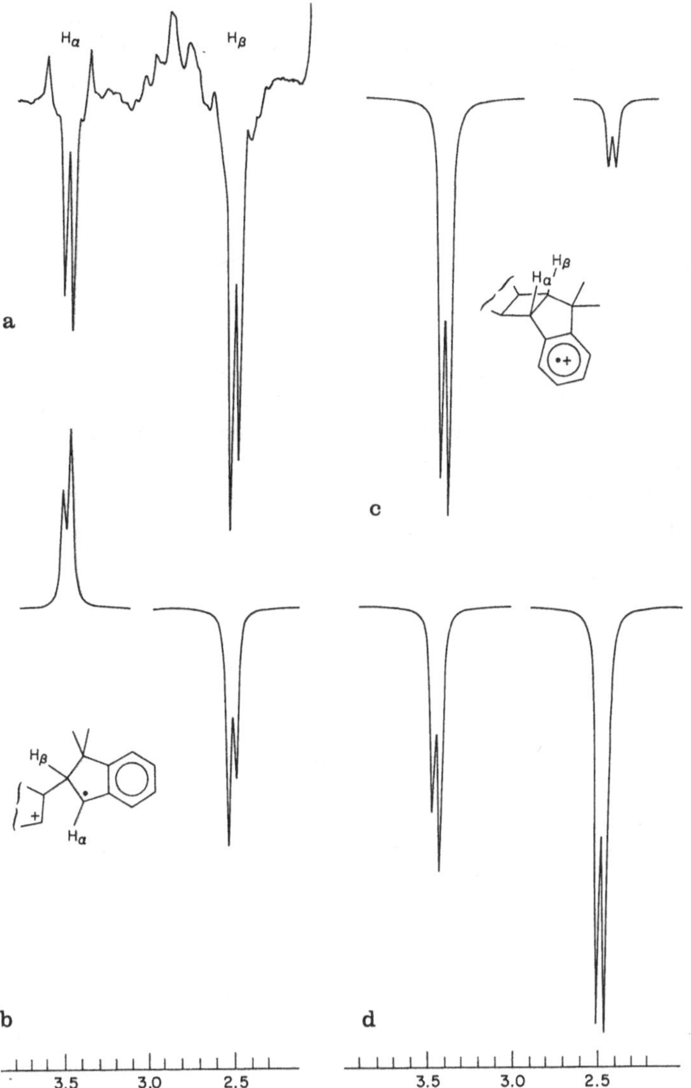

Fig. 13. CIDNP effects observed for the cyclobutane signals of the dimethylindene dimer during the photoinduced electron transfer reaction with chloranil (**a**), and simulated spectra based on the radical pair theory and assuming a ring-opened (extended) dimer radical cation (**b**), a ring-closed (localized) dimer radical cation (**c**) and the consecutive ('cooperative') involvement of open and closed radical cations (**d**) [256]

The mechanisms involved in dimer cleavage have been examined in simple model systems [90, 261–263], including bifunctional ones in which a sensitizer (e.g. indole) is linked to the pyrimidine dimer [264, 265]. For example, the *cis*-head-to-head dimer of dimethylthymine (**88**) was cleaved efficiently upon irradiation of anthraquinone-2-sulfonate; CIDNP effects observed during this reaction failed to reveal any evidence for the existence of the dimer radical cation [90]. Time-resolved

CIDNP studies limited the lifetime of the putative intermediate to a fraction of a μsec [262]. Most recently weak polarization was observed for a linked dimer (**89**), suggesting that the corresponding dimer radical cation is a discrete minimum, however short-lived it may be [263].

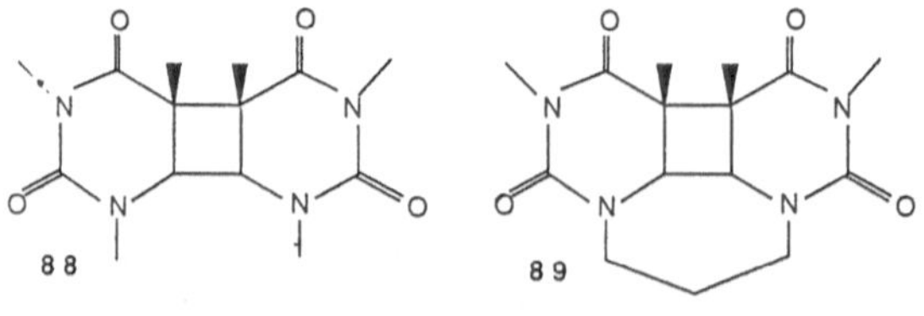

88 89

A variety of unsaturated ring systems undergo electron transfer induced electrocyclic reactions. Among these, the potential electrocyclic ring opening of cyclobutene systems have posed some interesting mechanistic problems. For example, Miyashi and colleagues observed the stereoselective conrotatory ring opening of *cis*-3,4-diphenylcyclobutene (**90**) upon electron transfer sensitization [266], and Kochi and collaborators observed analogous reactions for benzocyclobutene derivatives [267].

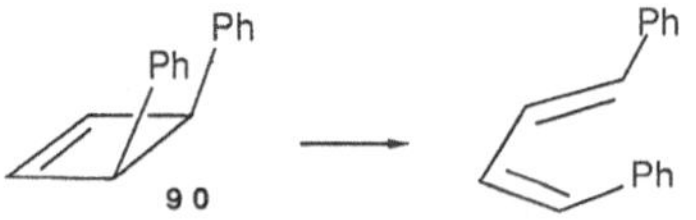

90

In contrast, the more highly substituted 1,2-diphenyl-3,3,4,4-tetramethylcyclobutene failed to show any evidence for "spontaneous" (thermal) ring opening [268]. This conversion (**92**) could be achieved only upon photo-excitation of the "stable" radical cation (**91**) [269].

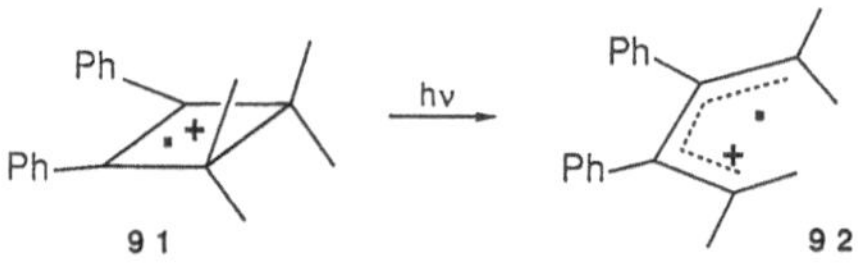

91 92

The observed ring opening, though not its stereoselectivity, can be explained by the release of ring strain and by the resonance stabilization of the extended conjugated system formed. The hexa-substituted system should also benefit from the release of ring strain, but the ring opened radical cation would be sterically crowded. In addition, it should show a minor degree of resonance stabilization, since the phenyl groups are attached to the centers of smaller spin density. In view of these arguments the failure to undergo a similar ring opening can be rationalized.

In this context it is also of interest that the parent cyclobutene radical cation can be observed at 77 K and, upon warming to 110 K, undergoes stereoselective ring opening to *transoid* butadiene radical cation [270]. A cyclopropylcarbindiyl radical cation has been proposed as an intermediate for this transformation [271].

Among four-membered ring compounds, the efficient conversion of hexamethyl Dewar benzene to hexamethylbenzene has attracted much attention as a free radical cation chain reaction [272, 273]. The structural feature(s) of the intermediate radical cation(s) have also attracted considerable attention. A detailed discussion will be presented in Sect. 5.3.

Additional cyclobutene ring openings have been documented for the 6′-chloro and 6′-methoxy-derivatives of 1,2,3,4,6-pentamethyl-5-methylenebicyclo[2.2.0]hex-2-ene (**94**). In the structures shown below, four methyl groups have been omitted for clarity. In these systems, the ring opening may be assisted by the presence of the *exo*-methylene group [274, 275]. Electron transfer sensitization of the 6′-chloro derivative (**94**, X = Cl) gave rise to pentamethylbenzyl chloride (**93**), suggesting cleavage of the transannular bond followed by chlorine migration. The methylenecyclohexadiene derivative (**95**), which might be an intermediate in this conversion, was observed in the analogous reaction of the methoxy derivative (**94**, X = OCH_3). This system proved to be particularly interesting, as an alternative ring opening, generating a doubly allylic radical cation, was found to be competitive [274]. This intermediate will be discussed in more detail in Sect. 5 with related bifunctional structures.

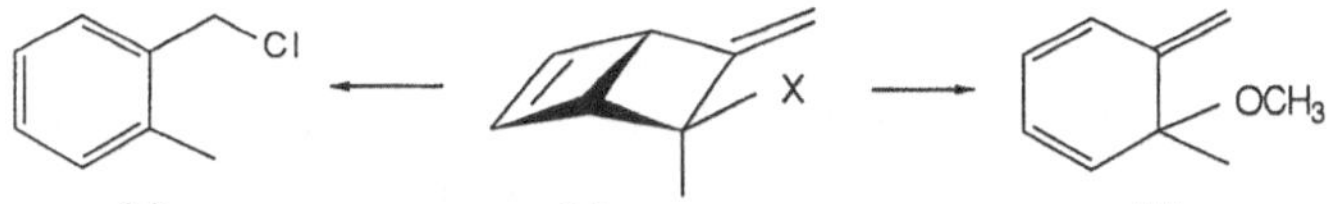

Cl ← X → OCH_3

93 94 95

In summary, substrates containing strained ring moieties undergo a range of electron transfer induced photoreactions. Most of their bimolecular reactions correspond closely to those of alkenes. However, the presence of strain manifests itself in various rearrangements, some of which do not occur in the reactions of their neutral diamagnetic precursors. Similarly, the "nucleophilic" ring opening observed for selected cyclopropane derivatives (but so far not for cyclobutanes) has at best limited precedent in alkane reactions. These unusual reactions suggest transition states as well as intermediates whose structures are without precedent on the potential surfaces of the parent molecules. These radical cation structures will be discussed in the following section.

5 Radical Cations with Unusual Structures

The majority of radical cations identified and characterized to date are relatively stable and their structures are closely related to those of the neutral diamagnetic precursors. In particular, a large number of species derived from aromatic hydrocarbons has been characterized by ESR [3] and optical spectroscopy [4]. The close structural similarity manifests itself in an interesting relationship between the UV spectra of selected radical cations and the UV photoelectron spectra of their parent molecules. Since both transitions lead to the same (excited) state of the radical cation, the excitation energies, ΔE, of the radical cation correspond to differences in ionization energies, ΔI, documented in the photoelectron spectroscopic data of the parent molecules [7, 276, 277].

However, the close correspondence between the geometries of radical cation and neutral parent, and the resulting relationship between the respective electronic structures can hardly be expected to be generally applicable. While the structure of a typical aromatic system need not be affected significantly upon one-electron oxidation, small ring compounds may undergo substantial structural changes. In fact, "exceptions" have been known for some time. Thus, for various stilbene derivatives several higher-lying radical cation excited states are without precedent in the photoelectron spectrum of the parent [278, 279]. More significantly, for the radical cations of systems **96–98** already the first excited state is without such precedent [278, 280]. Indeed, one might assume that even the ground states of these radical cations show significant differences. These states have been designated as "non-Koopmans" states.

96 97 98

We will approach radical cation structures according to the nature of the parent molecules, specifically according to the donor type, viz., π-, n-, or σ-donors, to which they belong. Among the radical cations derived from π-donors, those of aromatic hydrocarbons show the closest structural relationship to their parents. They also were the first class to be investigated in detail, because they are comparably stable and their spectra fall into a readily accessible range. This family shows the closest correlation between radical cation ΔEs, and parent ΔIs. On the other hand, cross-conjugated systems and alkenes may feature substantial differences between parent and radical cation electronic structures. Hence their tendency towards "non-Koopmans" type states.

Among radical cations of n-donors we mention briefly those of 1,4-diazabicyclo[2.2.2]-octane (**99**) and of the tricyclic tetraaza compound (**100**). For the bicyclic system a perfect correspondence has been reported between the ΔEs of the radical ion and the ΔIs of its precursor [276]. The radical cation of the tetracyclic system, on the other hand, is significantly distorted. While the parent system has D_{2d} symmetry and a b_2 HOMO, the radical cation is distorted towards two equivalent structures of C_{2v} symmetry (^2E), with a two-center three-electron N–N bond [281, 282]. The dioxetane radical cations (**101**), invoked as intermediates in oxygenations via oxygen capture (Scheme 6), and characterized by ESR spectroscopy [8] contain analogous three-electron O–O bonds.

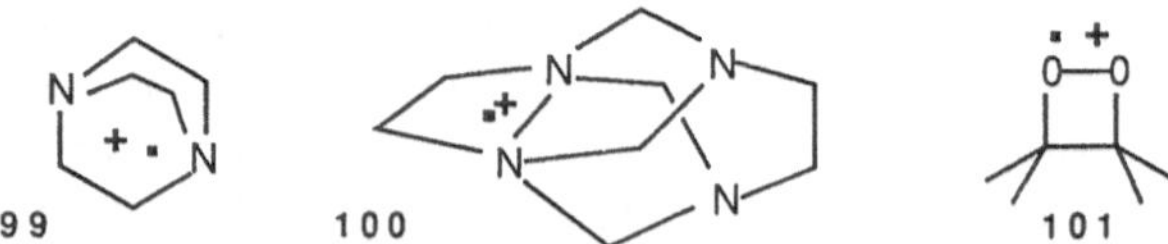

The most dramatic differences between radical cation structures and their parents is to be expected for σ-donors. We have noted above that the oxidation potentials of strained ring compounds may be lower than those of their unstrained isomers.

A plausible explanation for this fact involves the reduction of strain energy upon radical ion formation and corresponding noticeable changes in their structures. Accordingly, the radical cations of strained ring compounds are of major interest, particularly the distribution of spin and charge in response to the minimization of strain.

Recent investigations in the laboratories of the author and of several other groups have brought to light many interesting facts. In the following presentation we will compare the potential energy surfaces of radical cations with those of their neutral diamagnetic parents. An examination of the available information reveals three characteristic features, by which these potential surfaces may differ:

1) The barriers to radical cation rearrangements may be substantially reduced, and their reactions greatly accelerated; this feature appears to be generally recognized (Sect. 4.4)
2) The free energy difference between pairs of isomeric radical cations may be reduced because of the different oxidation potentials of the precursors (vide supra), or even reversed; the second case is particularly interesting, as it may provide access to otherwise inaccessible products (vide infra)
3) The energy minima on the radical cation surface may have geometries corresponding to transition structures on the parent surface; examples of these structure types appear particularly intriguing

Both we and others have established various radical cation structure types, which deviate in important features from the structures of their neutral diamagnetic precursors. The pursuit of these novel structure types has given new direction to radical cation chemistry. We have noted that some of these species resemble plausible transition structures for the thermal rearrangement of the parent molecules, i.e. saddle points on the corresponding potential surfaces. From a different point of view, they can be envisaged as one-electron oxidation products of biradicals or zwitterions. However, this relationship rarely serves as a practical approach to their generation, since the potential bifunctional precursors are often unstable and not readily accessible. These radical cations are usually generated from related hydrocarbons or cyclic azo compounds.

Before discussing some of the more interesting examples, we comment briefly on a suitable term to designate these unusual structures. As indicated above, radical cation excited states without precedent in the photoelectron spectrum have been called "non-Koopmans" states. Since the carbon skeletons underlying these states appear to be largely unaltered, species with significantly altered carbon skeletons lie beyond the "non-Koopmans" classification. In several cases, structures in which spin and charge are localized (or delocalized) in separate sections of the carbon skeleton have been called "distonic". We do not consider this a fortunate choice for the radical cations discussed here, because this term was coined for bifunctional species related by prototropy to a "vertical" oxidation product (e.g. Fig. 14) [283]. The geometries of such species correspond directly to minima on the parent potential surface (Fig. 14). More importantly, several of the unusual radical cations to be discussed are fully delocalized and do not meet the criterion for distonicity. We have designated the family of unusual radical cations in the following section as "non-vertical". This term does not refer to an unusual process by which they may be generated (non-vertical oxidation), but to

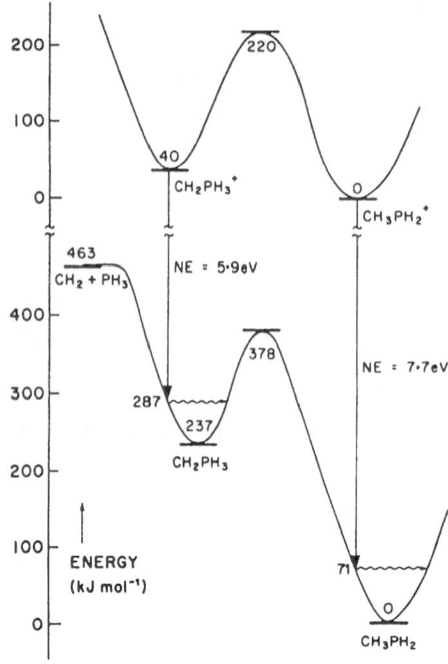

Fig. 14. Schematic potential energy profile showing the production of CH_2PH_3 by vertical neutralization of $CH_2PH_3^{+\cdot}$ and possible rearrangement and fragmentation reactions [283] (Reprinted by permission)

the complete lack of even a remotely vertical relationship between a radical cation and its precursor.

Several possible relationships between the geometries of radical cations and their precursors are shown schematically in Fig. 15. In addition to the purely

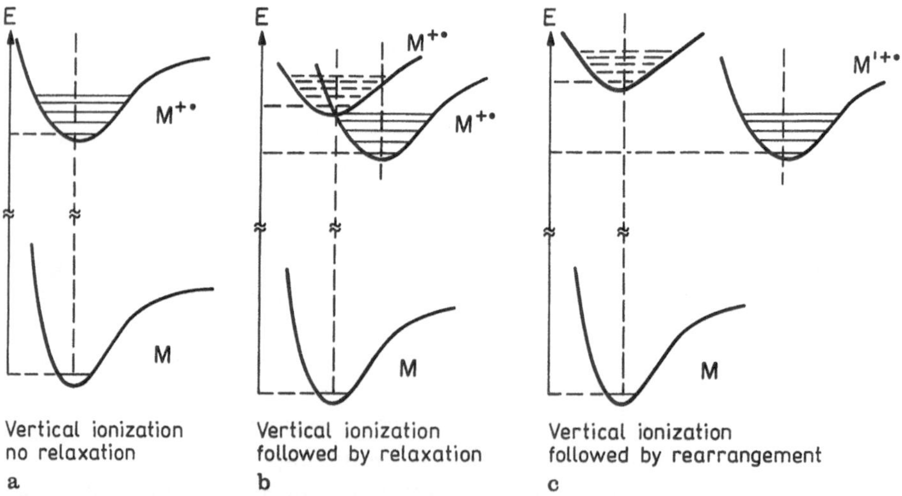

Fig. 15a–c. Schematic comparison between potential surfaces of radical cations and their neutral diamagnetic precursors. (**a**) vertical ionization without relaxation; (**b**) vertical ionization followed by relaxation; (**c**) vertical ionization followed by rearrangement

vertical relationship (Fig. 15a) minor geometry changes ("relaxation"; Fig. 15b) and major geometry changes ("rearrangement"; Fig. 15c) are illustrated. Rearranged radical cation structures may correspond to local minima or saddle points on the parent potential surface. Both cases will be illustrated, but we will focus mainly on the latter, more unusual one.

In the following sections selected radical cation structure types derived from σ-donors will be discussed. The selection is aimed to be illustrative of their rich variety with an emphasis on the molecular features that determine these structures.

5.1 The Cyclopropane System

Cyclopropane, as the simplest strained ring system, continues to be subject to intense and detailed scrutiny by a variety of techniques. Its photoelectron spectrum was investigated over twenty years ago. It shows Gaussian shaped bands without any fine structure, particularly a double peak centered near 11 eV and a broad signal near 13 eV. Molecular orbital calculations suggest that the first two bands be assigned to a $^2E'$ state. The large splitting of these bands, which is also observed for numerous derivatives, is ascribed to Jahn-Teller distortion [284]. A qualitative explanation of this phenomenon is shown in Fig. 16. The two states resulting from the Jahn-Teller split correspond to two different structures, $\mathbf{^2B_2}$ and $\mathbf{^2A_1}$, respectively, which pose interesting questions. Which structure is of lower energy? Can the $\mathbf{^2A_1}$ species undergo ring opening? How is the resulting trimethylene radical cation related to the propene radical cation? How do substituents affect the relative energies of the $\mathbf{^2B_2}$ and $\mathbf{^2A_1}$ structure types? These questions have been addressed in theoretical as well as experimental studies. We will discuss them

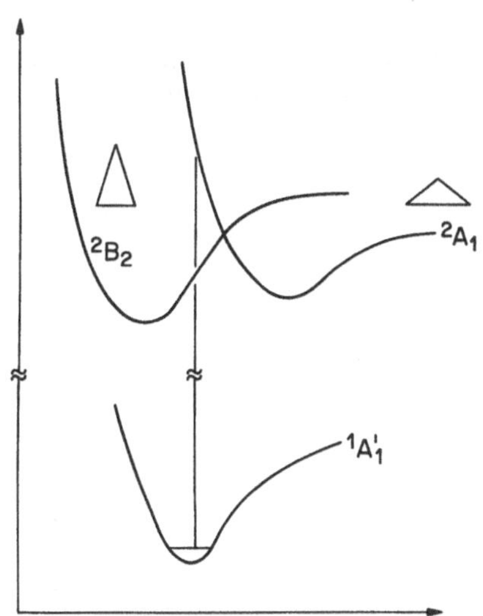

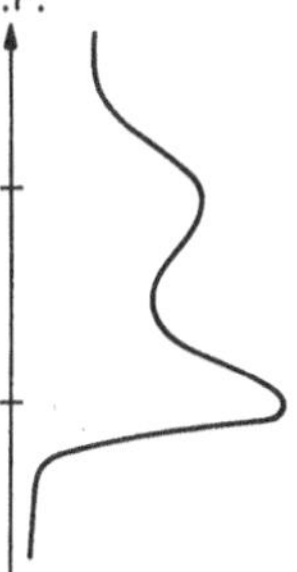

Fig. 16. Potential energy diagram for cyclopropane in its ground state ($^1A_1'$) and the two lowest states of its radical cation ($^2A_1$ and $^2B_2$). The vertical transition yields two bands (Jahn-Teller split) in the photoelectron spectrum

in moderate detail, from the vantage point of molecular orbital calculations and in the light of experimental results in both the condensed and gas phases.

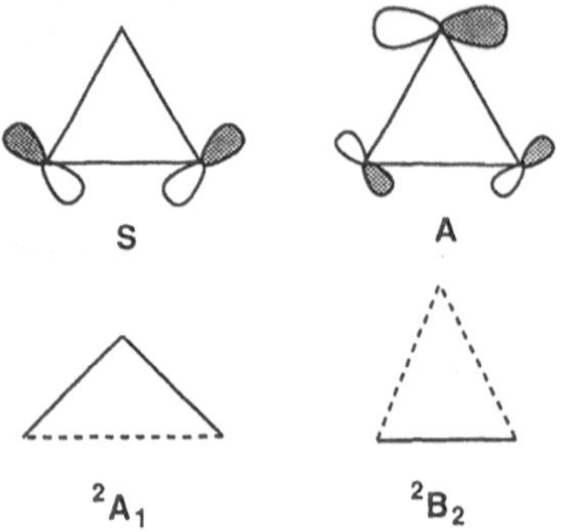

5.1.1 Molecular Orbital Calculations

The radical cations of cyclopropane have long been the target of theoretical investigations; for example, we refer to the important contributions by Haselbach [285], Rowland [286], Collins and Gallup [287], and Wayner and colleagues [288, 289]. However, we will discuss mainly the most recent thorough ab initio study by Borden and coworkers [290, 291], dealing with the cyclopropane radical cation, **102**$^{\cdot+}$, its potential ring opening to trimethylene radical cation, **103**$^{\cdot+}$, and the further rearrangement to propene cation radical **104**$^{\cdot+}$. In addition, we will briefly mention a recent study by Krogh-Jespersen and Roth [292], which deals with the existence of the ^{2}**B**$_2$ structure type and its potential stabilization by appropriate substituents.

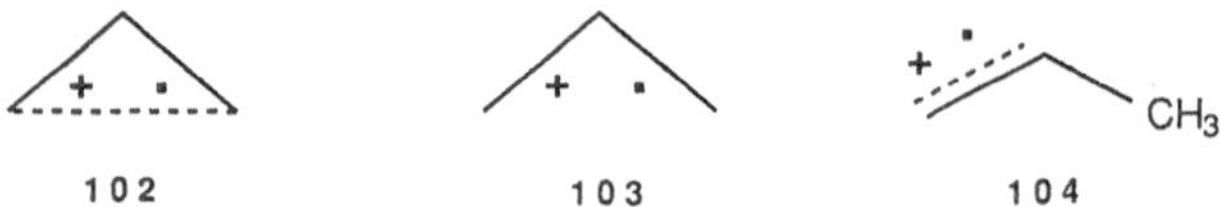

It is hardly surprising, that the propene radical cation, **104**$^{\cdot+}$, was found to be the intermediate of lowest energy on the $C_3H_6{\cdot}^+$ potential surface. At the unrestricted Hartree-Fock (UHF)/6-31G*//MP2 level (C_s symmetry), **104**$^{\cdot+}$ lies $\sim$10 kcal mol^{-1} below the radical cation, **102**$^{\cdot+}$, in which a single C–C bond is lengthened relative to the parent molecule. A vibrational analysis of structure **102**$^{\cdot+}$ shows only positive frequencies, thus identifying this species as a local minimum. In contrast, the ring-opened trimethylene radical cation, **103**$^{\cdot+}$, does not appear to be a minimum, since a 6-31G* vibrational analysis (C_{2v} symmetry) showed one imaginary and one low frequency. Additional geometries in which the terminal CH_2 groups were rotated between the plane of the three-carbon unit (orientation "0") and planes perpendicular to it (orientation "90") likewise failed to qualify as local minima (Table 2).

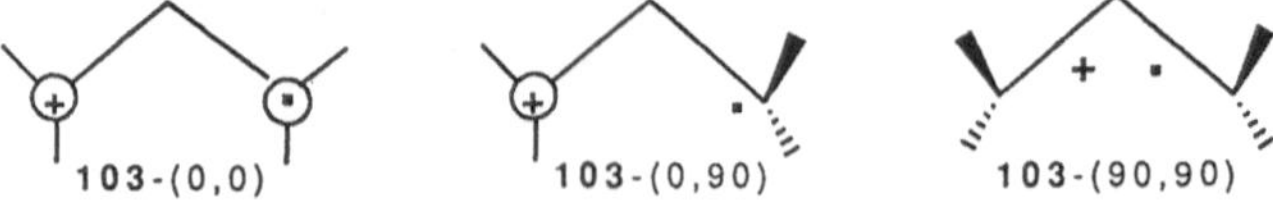

Table 2. Relative energies (kcal mol^{-1}) of C_3H_6 radical cations

Cation Radical	UHF/6-31G*	MP2/6-31G*
Propene$^{+\cdot}$	0[a]	0[b]
Cyclopropane$^{+\cdot}$ (90,90)[c]	15.3	9.8
Trimethylene$^{+\cdot}$		
(C_{2v}) (0,0)[d]	42.1	31.5
(C_s) (0,0)[d]	35.6	37.6
($^2A'$) (0,90)	35.5	39.1
($^2A''$) (0,90)	37.3	38.4
(C_s' (7,−7)	37.3	29.6

[a] E = 116.7734 hartrees [b] E = 117.1122 hartrees [c] dihedral angles between the terminal methylene planes and that of the three carbon unit [d] the most stable geometry of $2^{+\cdot}$ should have C_{2v} symmetry

Finally, the calculations revealed a substantial energy difference (~22 kcal mol^{-1}) between **103**$^{+\cdot}$ (C_{2v}, 0,0 geometry) and **102**$^{+\cdot}$ but failed to indicate a chemically significant barrier for the conversion of **103**$^{+\cdot}$ to **104**$^{+\cdot}$ (90,90 geometry). Both results argue against the opening of **102**$^{+\cdot}$ to **103**$^{+\cdot}$ reportedly occurring in $CF_2Cl-CFCl_2$ matrices, even at cryogenic temperatures, and supposedly irreversible. [293].

Recently Krogh-Jespersen and Roth evaluated the potential stabilization of the $^2\mathbf{B}_2$ structure type by methyl substitution. For both mono- and gem-dimethyl derivatives the $^2\mathbf{A}_1$ type structures are minima, whereas the $^2\mathbf{B}_2$ type structures are transition states. They undergo second-order Jahn-Teller distortion to unsymmetrical structures with one very long C−C bond.

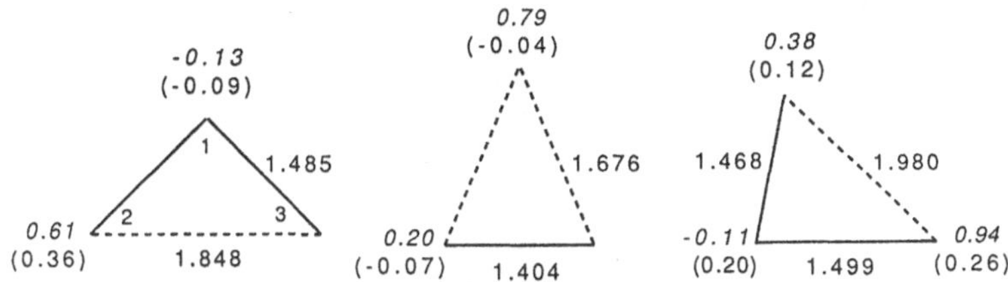

The bond lengths, spin densities (italics), and charge densities (in parentheses) of the respective 1,1-dimethyl derivatives are shown near the structural representations.

5.1.2 Cyclopropane Radical Cations Exemplified

The cyclopropane radical cation can be prepared by γ-radiolysis in rigid matrices; its ESR spectrum at 4.2 K shows evidence for static Jahn-Teller distortion, resulting in a structure of the trimethylene ($^2\mathbf{A}_1$) type [294, 295]. Irradiation of several methyl substituted derivatives at 77 K gave rise to a family of radical cations of the same structure type [293], which had been previously identified on the basis of CIDNP results [229, 230]. We begin with a discussion of the CIDNP investigations, since they preceded the ESR studies of all species but the prototype.

The first CIDNP results attributed to a cyclopropane radical cation were observed during the photo reaction between 1,4-dicyanonaphthalene and *cis*-1,2-diphenylcyclopropane [222]. However, the nature of the cyclopropane radical cation was characterized by CIDNP effects observed during the reaction of chloranil with *cis*- and *trans*-1,2-diphenylcyclopropane [223]. The pattern of benzylic and geminal polarization (Fig. 17) supports radical ions with spin density on the benzylic carbons. Strictly, the results do not differentiate a priori between a "closed" and an "open" radical cation. The "closed" structure was assigned because the reaction did not cause geometric isomerization, suggesting that the stereochemistry at the key carbons be preserved in the intermediate. These results establish local minima on the radical cation potential surface but have no bearing on the existence of additional minima with altered stereochemistry. Nor do they allow any conclusion concerning the global minimum on the radical ion energy surface.

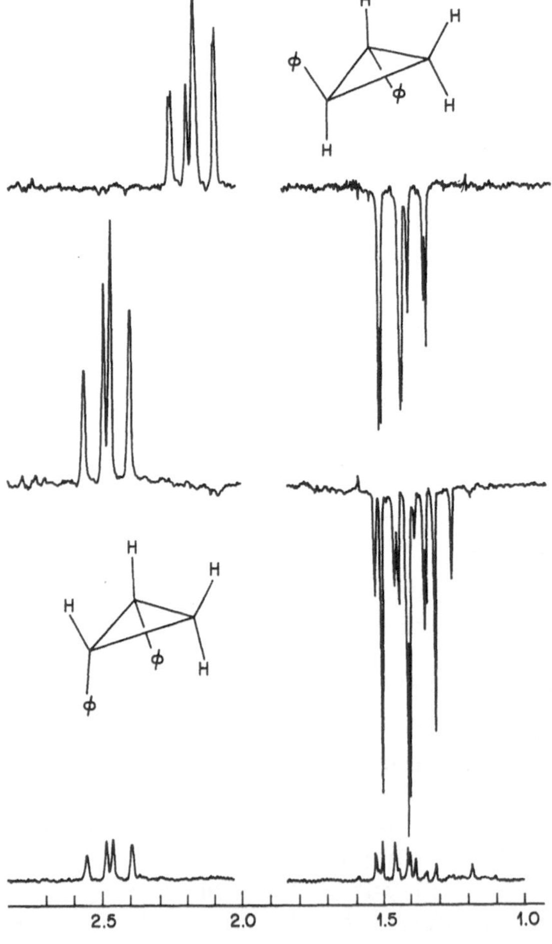

Fig. 17. PMR spectra (90 MHz) observed during the irradiation of chloranil (0.02 M) in acetonitrile-d_3 solutions containing 0.02 M *trans-(top)* or *cis*-diphenylcyclopropane *(bottom)* [223]

Radical cations of the same general structure type as those derived from *cis*- and *trans*-diphenylcyclopropane have been established for numerous cyclopropane derivatives, including the parent, 1,2-di-, 1,1,2-tri- and 1,1,2,2-tetramethylcyclopropane (Table 3). Two of these systems provide a direct comparison between the results of CIDNP and ESR experiments. In both cases, the ESR spectra observed by Williams and coworkers following pulse radiolysis in frozen solutions [293, 296, 297] show splitting patterns supporting the presence of spin density on two carbon centers, thus confirming the structure type (**102**) assigned on the basis of CIDNP results.

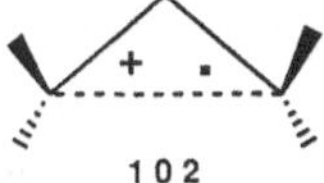

In this context we consider the potential effects of substituents on the structure of the cyclopropane radical cation. For molecules with a pair of degenerate HOMOs suitable substitution might be expected to lift the degeneracy and favor

Table 3. ^{1}H Hyperfine coupling constants [G] for the radical cations of selected cyclopropanes

	CIDNP	ESR
b, a	no result	a − 12.5 b + 21.0
c, b, a	no result	a (−) 10.4 b (+) 20.5 c (+) 20.5
c, b, a	no result	a (−) 11.9 b (+) 21.8 c (+) 21.8
d, b, c, a	a − b + c + d +	a (−) 9.8 b (+) 14.5 c (+) 20.6 d (+) 17.9
b, a	a + b +	a (+) 15.0 b (+) 18.7
b, Ar, Ar, a	a − b +	
b, Ar, a, Ar	a − b +	

one structure over the other. Qualitative predictions of the favored structure can be based on a frontier (F) MO/perturbational (P) MO approach [298, 299]. The substrates are dissected into molecular fragments and the potential interactions of the component FMOs are considered. According to PMO theory [298] the strength of the fragment perturbation is approximately proportional to $S^2/\Delta E$, where S is the overlap integral between the components and ΔE is the difference between the FMO orbital energies. For the S^2 term, three factors will be of primary importance: the FMO orbital symmetry (where present); the magnitude of the coefficients at the point(s) of union; and the orientation of the fragments relative to each other [299].

For cyclopropane, substituents at a single carbon might most effectively stabilize the antisymmetrical HOMO, whereas substitution at two carbons is expected to stabilize the symmetrical orbital. Since there is ample evidence for radical ions derived from the prototype of $^2A_1$ symmetry (vide supra), cyclopropane radical cations with the alternative, antisymmetrical singly occupied (SO) MO appeared be of particular interest. We have identified two substrates, benzonorcaradiene (**105**) [229] and spiro[cyclopropane-1,9′-fluorene] (**106**), whose radical cations belong to this category [299, 300]. Interestingly, there is, as yet, no theoretical support for such species.

The assignment of an antisymmetrical cyclopropane SOMO to the radical cation of **105** is based on a comparison of CIDNP effects (Fig. 18) with those for *cis*-1,2-diphenylcyclopropane. While the nuclei of the aromatic segments show identical or very similar polarization, the cyclopropane protons show characteristic differences. This suggests significantly different spin density distributions for the cyclopropane moieties of the two species and, thus, different structures [229]. The benzonorcaradiene radical cation should owe its structure to the symmetry of the fragment FMOs at the points of union. The styrene HOMO is antisymmetric at the positions of attachment, suggesting preferred interaction with the antisymmetric cyclopropane HOMO (as shown below).

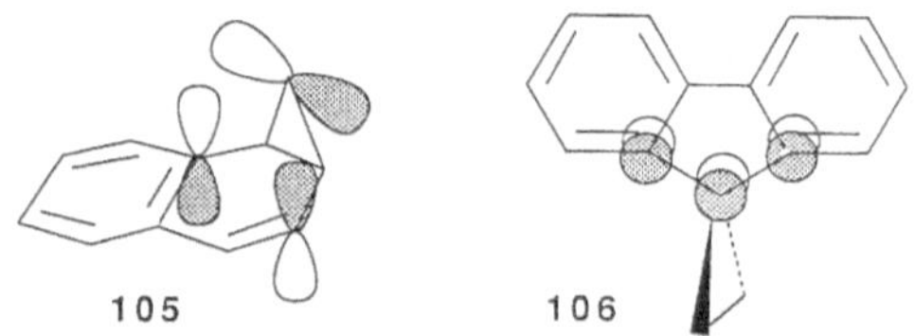

The radical cation of the spirofluorene (**106**) appears to be less stabilized, but quite interesting nevertheless; the orientation of the fragments allows only the *second* highest MO of the biphenyl moiety to interact with a cyclopropane FMO (the A HOMO shown above). The radical cations of **105** or **106** have not been examined by ESR or optical spectroscopic techniques. We suggest **105** as an attractive target for such studies, whereas the cyclopropane hyperfine couplings of **106** are expected to be weak because of a mismatch in orbital energies. In the context of providing guidelines for identifying the structure of cyclopropane radical cations we note that the stereochemistry of the diamagnetic products generated

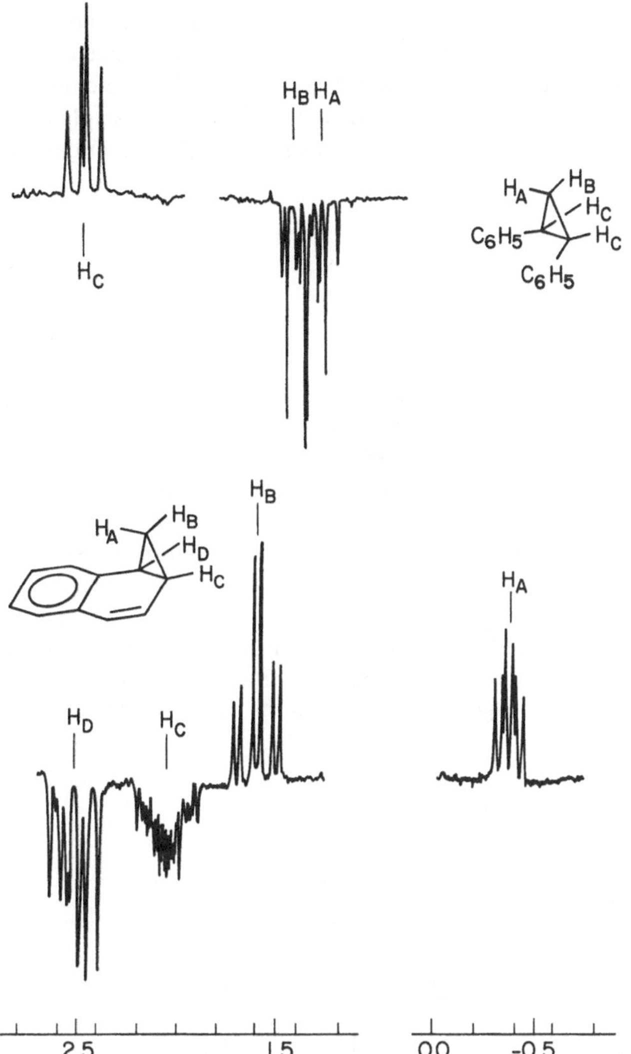

Fig. 18. ^{1}H CIDNP spectra (cyclopropane resonances) observed during the electron transfer photoreaction of chloranil with *cis*-1,2-diphenylcyclopropane *(top)* and benzonorcaradiene *(bottom)*. The opposite signal directions observed for analogous protons in the two compounds constitute evidence that the two radical cations belong to two different structure types [9]

by nucleophilic ring opening [233] would be the same for either structure type. Accordingly, this experiment clearly rules out the ring-opened structure, but is not suitable for differentiating between the alternative ring-closed radical cation structures.

The previously mentioned rearrangement of the *trans*-fused cyclopropane **30** can be rationalized also via the antisymmetrical cyclopropane SOMO. The rearrangement requires inversion at one of the tertiary cyclopropane carbons. The orientation of the olefinic bond relative to the ring may result in weakened external cyclopropane bonds. The isomerization can occur most readily by stretching and one-center inversion of one such bond. One potential drawback to this explanation is the distance between these orbitals (~3 Å). Nevertheless, this system deserves further scrutiny.

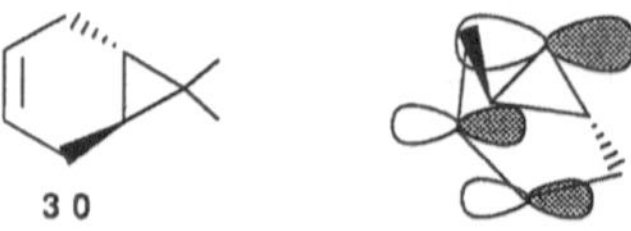

The interaction between a cyclopropane group and olefinic or aromatic systems conjugated to it depends on their relative orientation. While the tertiary–tertiary cyclopropane bond of benzonorcaradiene (**105**) lies in a plane perpendicular to the aromatic π system, [1:2, 9:10]bismethano[2.2]paracyclophane (**107**) features a parallel arrangement of these elements [301]. The photoreaction of this obviously strained compound with chloranil gives rise to CIDNP effects not unlike those observed for *cis*-diphenylcylopropane (*cis*-**29**). However, the relative signal intensities of **107** are noticeably distorted (Fig. 19). The (secondary) *endo* proton (1.2 ppm) is more strongly enhanced (indicating a larger hyperfine coupling) than both the *exo* (2.0 ppm) and the benzylic protons. This finding was interpreted in terms of different dihedral angles, Θ_B and Θ_C, between the singly occupied orbital

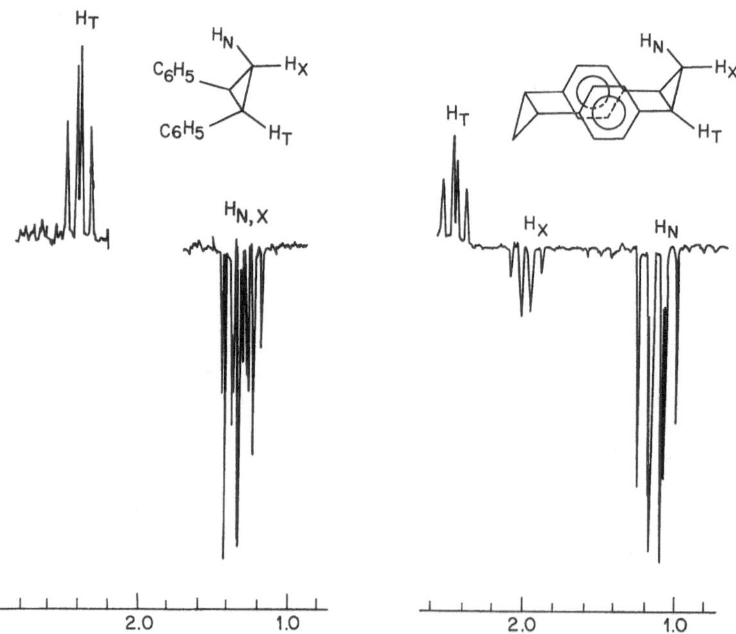

Fig. 19. A comparison of the CIDNP effects observed for the cyclopropane protons of *cis*-diphenylcyclopropane *(left)* and [1:2, 9:10]bismethano[2.2]paracyclophane *(right)* [229]

and the two geminal C–H bonds. Nonidentical dihedral angles for the secondary cyclopropane protons can result if the angle, Φ (Fig. 20), and the distance between the benzylic cyclopropane carbons is increased in the radical cation relative to the parent hydrocarbon. The *endo* proton, which in the diamagnetic molecule lies in the shielding cone of the aromatic moieties, has in the radical the smaller dihedral angle Θ with the benzylic "π" orbital and, therefore, the greater hyperfine coupling [302].

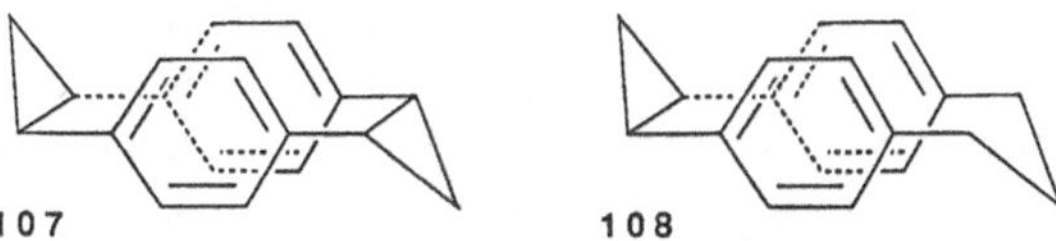

The comparably weak enhancement of the benzylic protons and the relatively strong ^{13}C polarization for the benzylic carbons can be explained if the benzylic carbon adopts a pyramidal structure. Although radicals containing pyramidal carbon are reasonably rare, the existing examples are derived from strained ring systems [303–306]. For the radical cation of **108**, the change in the angle, Φ, the stretching of the doubly benzylic bond, and the adoption of a pyramidal structure apparently relieve some of the strain in the carbon skeleton, yet maintain a reasonable degree of overlap between the benzylic carbons and the benzene rings [302]. The sum of these changes may well amount to a "ring-opened" cyclopropane radical cation (vide infra).

The [10:11] methano[3:2]paracyclophane system (**108**) [229] appears less strained than **107** yet more restricted than diphenylcyclopropane. The CIDNP effects observed are closer to that of *cis*-diphenylcyclopropane than to that of **107**. Obviously, a large fraction of the strain inherent in **107** is relieved by substituting a trimethylene bridge for one of the cyclopropane moieties [229].

This conclusion raises the question whether the radical cation of **107** may be localized on a single cyclopropane entity rather than involving both strained rings. The CIDNP method does not lend itself to decide this issue, because the two cyclopropane groups are magnetically equivalent in the substrate/product. The ESR method, on the other hand, is well suited to elucidate the precise nature of

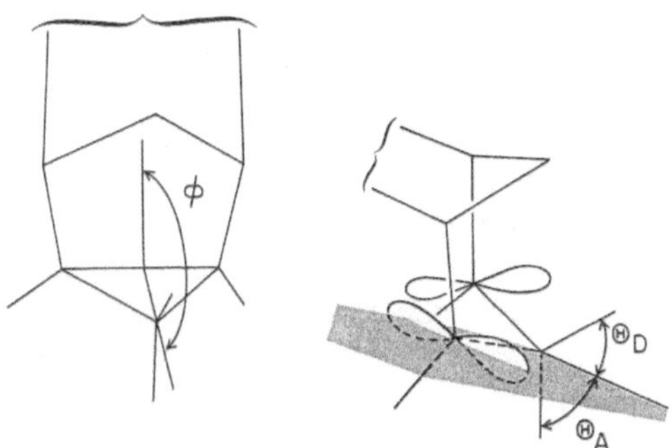

Fig. 20. Two partial views of [1:2, 9:10]bismethano[2.2]paracyclophane showing the angles Φ and Θ discussed in the text [229]

the intermediate, since a radical cation involving only one cyclopropane group will have fewer interacting nuclei. In the case of two interacting cyclopropane groups it is possible, in principle, to distinguish a fully delocalized system from rapidly equilibrating "localized" systems. This assignment requires the angle Θ and the hybridization of the radical centers; the latter parameter affects the proportionality constant relating spin density to hyperfine coupling [307, 308].

5.1.3 Ring-Opened (Trimethylene) Radical Cations

The final topic to be discussed in this section concerns the prospect of "opening" or "breaking" the weakened bond in the radical cations of the $^2A_1$ structure type. This topic has been the subject of some controversy. As mentioned above, ab initio calculations fail to support the existence of such a species. Yet, they have been postulated to rationalize some ESR observations and also to explain the geometric isomerizations of 1-aryl-2-vinylcyclopropanes upon reaction with aminium radical cations [225].

Thus, the primary ESR spectra decayed irreversibly at temperatures characteristic for the substrate, typically near 100 K; they were replaced by a second type of spectrum, in which the protons at one cyclopropane center no longer interact with the electron spin. This coupling pattern was interpreted as evidence for a ring-opened trimethylene species (**109**) in which one terminal carbon has rotated into an orthogonal orientation [293, 296, 297].

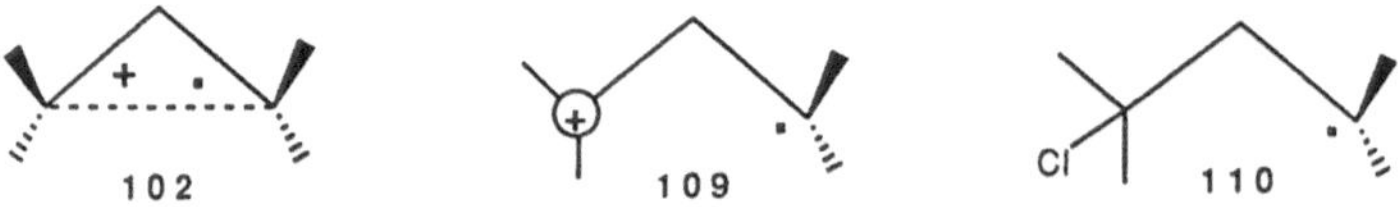

However, the putative rearrangement was found to be limited to two matrices, $CFCl_2CF_2Cl$ and CF_2ClCF_2Cl, posing an intriguing question concerning the role of the matrix in the "rearrangement". Another problem with the existence of species **109** is its failure to undergo a proton shift, generating the known and stable propene radical cation; proton shifts are known to occur under the conditions of these experiments (vide infra). Because of these problems, several alternative explanations were also advanced, including structures in which the cationic center of **102** or **109** has captured a chloride ion (→**110**) [288–291], interacts with a matrix molecule to form a chloronium substituted free radical [309], or interacts with several nearby matrix molecules [297]. If the actual species were a β-chloroalkyl radical (**110**) formed by nucleophilic ring opening by chloride ion, one possible role of the matrix may lie in preventing the approach of the ion. However, the ionic diameter of chloride ion, ~363 pm, would require a rather tight matrix, perhaps not a very likely assumption after hours of radiolysis.

A more detailed evaluation of the diverse structures proposed for the secondary species would go beyond the scope of this review. We merely emphasize, that the ESR results provide detailed evidence for the nature of the radical center, but offer little detail about the cationic center. Rather, its identity is left to secondary considerations or speculation [297]. We also note that any alternative structure has the virtue of not contradicting the ab initio calculations and that the potential capture of chloride ion has precedent in the nucleophilic substitution at a cyclopropane carbon (Sect. 4.4).

Ring-opened cyclopropane radical cations have also been postulated to account for the stereochemistry of the aminium radical cation catalyzed rearrangement of 1-aryl-2-vinylcyclopropanes [225]. These systems, of course, contain substituents that may veil the "true" nature of the cyclopropane radical cations by delocalizing spin and charge. In addition, we note that the experimental findings allow some latitude in their interpretations.

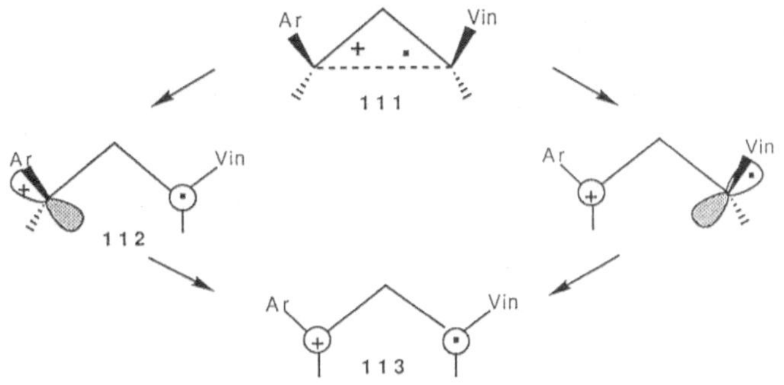

It is quite reasonable to expect that the potential surface of this radical cation contains several shallow minima separated by low barriers. If this premise is correct, the ultimate reaction products may offer little information about any one minimum. In fact, even ESR results, could they be obtained at −90 °C, might not differentiate unambigously between a planar ring-opened radical cation (**113**) and a rapidly equilibrating pair of orthogonal bifunctional intermediates (e.g. **112**). Perhaps the most likely explanation involves a pair of isomeric cyclopropane radical cations with one very weak bond, viz., **111**, which might allow either or both tertiary carbons to rotate. The bifunctional structures shown above, either orthogonal (e.g. **112**) or planar (**113**) may signify minima or transition structures.

The intriguing problem of a ring-opened cyclopropane radical cation has been investigated also in the gas phase. Structures of ions generated in a mass spectrometer are assigned on the basis of their reactivities in selective ion–molecule reactions. The ions of m/z 42^+ formed by electron impact on cyclopropane react with NH_3 by proton transfer ($\rightarrow NH_4^+$) and formation of CNH_4^+ and CNH_5^+ [310, 311]. The results show clearly that two different ions are involved, but leave some doubt as to their identity. Initially, the second species was assigned the structure of propene cation [310, 312]. However, more recent results show that a significant portion of the m/z 42^+ ions formed above 15 eV exhibit reactivities different from either propene or the cyclopropane ion; vibrationally excited cyclopropane cation or trimethylene cation are proposed as potential structures [313, 314].

5.1.4 The Radical Cations of the System Quadricyclane – Norbornadiene

No discussion about cyclopropane radical cations would be complete without reference to the quadricyclane radical cation ($\mathbf{Q}^{\cdot +}$) and its valence isomer, the norbornadiene radical cation ($N^{\cdot +}$). This pair has attracted considerable attention; one features two rigidly arranged adjacent cyclopropane rings, whereas the other contains two ethene π systems which are uniquely suited for the examination of through-space interactions [315–319]. In addition, the potential energy surface of these radical cations poses the interesting question whether two discrete minima exist, or whether there is a single minimum, accessible upon oxidation of either parent. While other methods had failed to provide evidence for more than one intermediate [319], the CIDNP technique furnished clear-cut evidence for the existence of two distinct species with characteristic spin density distributions (Fig. 21) [320, 321]. This result is compatible with molecular orbital considerations, which suggest the antisymmetric combination of two ethene π orbitals (**114**) or cyclopropane Walsh orbitals (**115**) as respective HOMOs of the two parent molecules. The SOMOs of the radical ions have different orbital as well as state symmetries.

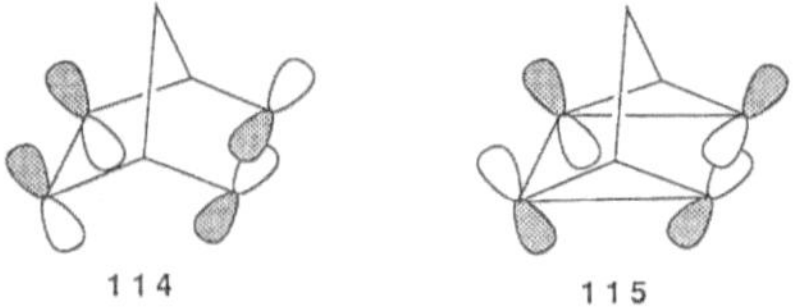

The structure of these ions rests on the following results (Table 4): detailed CIDNP spectra delineating the hyperfine patterns of both ions [320, 321]; ab initio calculations with a 6-31 G* basis set for both ions [322]; and ESR [323, 324] as well as ENDOR data [324] for the norbornadiene radical cation. The CIDNP results indicate the absolute signs and relative magnitude of the hyperfine coupling constants; a comparison with the calculated values (Fig. 22) shows satisfactory

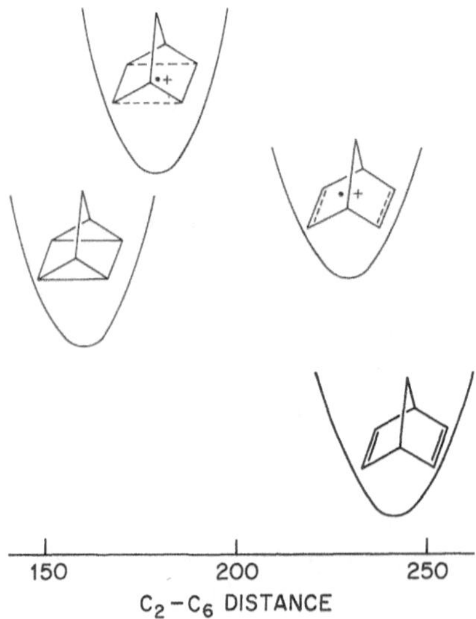

Fig. 21. Schematic energy diagram for norbornadiene, quadricyclane, and their radical cations. Each minimum on the radical cation potential energy surface corresponds to a minimum on the energy surface of the neutral diamagnetic precursor, with characteristic changes in bond lengths (ab initio, 6-31 G*) and angles [322]

Table 4. ^{1}H Hyperfine coupling constants [G] for radical cations of quadricyclane, norbornadiene, and their 7-methylene derivatives

	Quadricyclane			Norbornadiene			
	CIDNP[a]	Calc.[b]	ESR	CIDNP[a]	Calc.[b]	ESR[c]	ESR[d]
H_{ol}	− l	−10.5		− l	−13	−7.8	8
H_{bh}	+ l	+ 7.9		− vs	− 0.7	−0.49	
H_b	− m	− 4.0		+ s	+ 2.8	+3.04	3.3
	7-Methylenequadricyclane			7-Methylenenorbornadiene			
	CIDNP[e]			CIDNP[3]			
H_{ol}	− l			− l			
H_{bh}	+ l			+ vs			
H_{me}	+ m			− m			

[a] See [320]; l = large, m = medium, s = small, vs = very small; [b] See [322]; [c] See [324]; [d] See [323]; [e] See [325]

agreement [322]. The calculations at either the MINDO/3 [319] or the 6-31 G* level of ab initio theory also indicate that each radical cation is related uniquely to the geometry of one of the precursors.

For both species the unpaired spin density resides in four equivalent carbons; ^{1}H nuclei attached to these centers (N_0, Q_0) have sizeable hfcs due to the familiar π, σ spin polarization (cf. Fig. 4a). The CIDNP effects of the bridgehead (β) protons indicate major differences: a sizeable positive hfc of $\mathbf{Q}^{\cdot+}$ can be ascribed to hyperconjugation (π, σ spin delocalization; cf. Fig. 4b); in contrast, $\mathbf{N}^{\cdot+}$ shows only very weak and negative (!) hfcs. Since the β protons of $\mathbf{N}^{\cdot+}$ lie in the nodal plane of its SOMO, the hyperconjugative interaction is inefficient; the observed sign of the β protons was therefore ascribed to "residual" π, σ polarization [324]. This type of interaction is usually obscured by the typically much stronger hyperconjugative interaction.

The CIDNP effects observed for the protons of the methylene bridge again are quite different, suggesting a sizeable positive hfc for $\mathbf{N}^{\cdot+}$ and an even larger negative hfc for $\mathbf{Q}^{\cdot+}$. The positive sign for the γ hyperfine coupling of $\mathbf{N}^{\cdot+}$ can be ascribed reasonably to a "long-range" π, σ spin delocalization which is aided by an approximate W arrangement of the γ C−H bond relative to the p orbitals at the olefinic carbons. The bridgehead (γ) protons of $\mathbf{Q}^{\cdot+}$ lie in the nodal plane of the SOMO, rendering the π, σ spin delocalization mechanism inefficient. The relatively large negative hyperfine coupling indicated by the CIDNP results might suggest a σ, σ polarization mechanism operating between the bridgehead (β) and bridge (γ) protons. The dihedral angle ($H-C_\beta-C_\gamma-H \sim 60°$) is compatible with a sizeable interaction. An independent confirmation of this interesting assignment by an ESR and ENDOR study appears desirable, but so far $\mathbf{Q}^{\cdot+}$ has proved to be elusive.

The radical cations of the 7-methylene derivatives, **7-MQ**, and **7-MN**, also show interesting contrasts [325], illustrating the role of homoconjugation in delocalizing

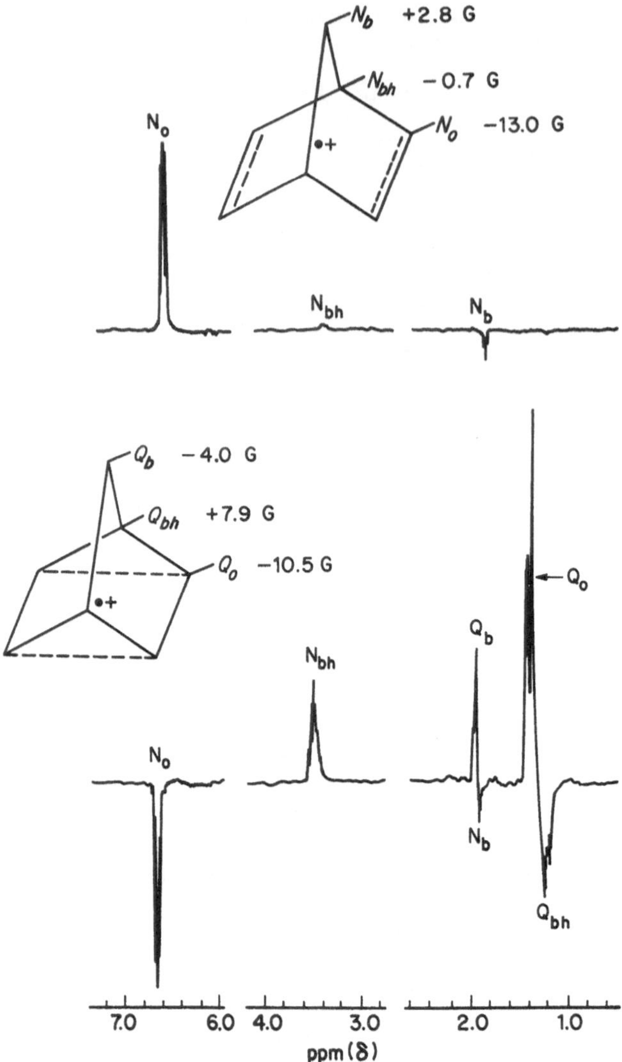

Fig. 22. ^1H CIDNP spectra observed during the photoreaction of chloranil with norbornadiene *(top)* and quadricyclane *(bottom)*. The insets in each spectrum show the hyperfine coupling constants calculated for the two radical cations [322]. For the comparison of the calculated couplings with observed signal intensities, it should be noted that there are twice as many nuclei in the position N_0 and Q_0 as in the positions N_{bh}, N_b and Q_{bh}, Q_b

spin and charge [299]. The symmetry plane of the 7-methylene *p* orbital coincides with that of the norbornadiene HOMO (cf. **116**), but is orthogonal to that of the quadricyclane HOMO (cf. **117**). In agreement with these considerations, the CIDNP effects for **7-MN** support negative hfcs for both sets of olefinic protons along with very weak positive hfcs for the bridgehead protons. These results suggest delocalization of spin and charge throughout both types of olefinic moieties and another case of π, σ polarization for a ^1H nucleus β to the unpaired spin.

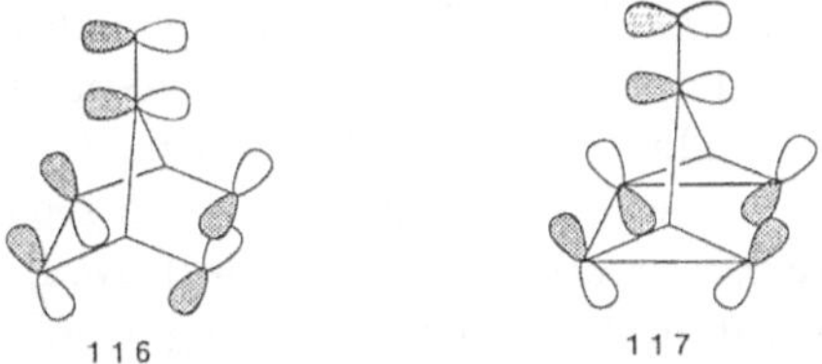

116 117

In contrast, the CIDNP effects for **7-MQ** support "cyclopropane" hfcs as for the parent system, but the *exo*-methylene polarization is compatible with positive hfcs, though somewhat smaller than those for the bridgehead positions (see Table 4). This observation suggests an interesting "long-range" interaction, through space or through the σ frame, between each *exo*-methylene proton and a pair of *p*-type orbitals bearing unpaired spin density [325]. Since there is, to our knowledge, little precedent for such an interaction, this system should make an attractive target for ab initio calculations as well as ESR/ENDOR studies. Preliminary results in the author's laboratory seem to indicate that **7-MQ**$\cdot^+$ may be somewhat less elusive than **Q**$\cdot^+$.

5.1.5 Radical Cations of Oxirane and Derivatives

Replacement of one of the methylene groups of cyclopropane by an oxygen changes the system considerably. Photoelectron results indicate that an oxygen *n*-orbital, b_1, and an a_1 ring orbital are the highest lying MO's [326]. Involvement of one of these orbitals would give rise to species **118** ($^2B_1$), in which spin and charge are localized on the oxygen atom, whereas the other would lead to radical cation **119**, ($^2A_1$) in which spin and charge reside on the two carbon atoms.

118 119 120

The oxirane radical cations have a means of stabilization which is not present in their alicyclic analoga: ring opening and rotation of the methylene groups may give rise to a resonance stabilized species (**120**), which is iso-π-electronic with the allyl radical. EPR spectra observed upon radiolysis of oxirane have been interpreted as evidence of either the ring-closed [327, 328] or the ring-opened structure [329–331]. Neither the g-factor of this species (2.0022–2.0024) [329] nor the magnitude of the hfcs of the β-hydrogens (16.2 G) are compatible with an oxygen centered radical (**118**), but tend to support the oxallyl alternative (**120**). Further disagreement centers on the interpretation of the ESR spectra obtained for the doubly ^{13}C-labeled radical cation [328, 330].

For a resolution to this controversy one may rely on ab-initio calculations [332–336] or on a comparison of the electronic absorption [337] as well as ESR spectra [338] of the parent oxirane radical cation with those of the tetramethyloxirane and 9,10-octalin oxide radical cations. These results leave little doubt that the radical cations of simple oxiranes assume ring-opened structures. The ring-closed structure of the octalin oxide radical cation can be considered yet another manifestation of Bredt's rule [339]. It remains to be seen whether the radical cations of $^2A_1$ structure can be established as (short-lived) intermediates in the formation of the ring-opened ones, or whether the ring opening occurs with a negligible activation energy. Similarly, the subtle question whether the ring-opened radical cations have single-minimum or double-minimum potential surfaces [333] may still await a more definitive answer.

5.1.6 Bicyclobutane Radical Cations

The unique bonding in the bicyclobutane system, which is reflected in its unusual chemistry, has also spurred an interest in its radical cation. For example, the photoinduced electron transfer chemistry of bicyclobutane derivatives has been investigated in detail (cf. Sect. 4.4). The product distribution obtained under a variety of reaction conditions suggests a radical cation structure in which the transannular bond is substantially weakened. This structure type has been confirmed by CIDNP results for several derivatives [248, 249] and by ESR results for the parent system [340].

The photoreactions of electron acceptors with several bridged bicyclobutane systems, e.g. tricyclo[4.1.0.0^{2,7}]heptane (**75**), give rise to CIDNP effects supporting negative hfcs for H_1 and H_7 and positive hfcs for H_2 and H_6 as well as $H_{3,3'}$ and $H_{5,5'}$. These results would place electron spin density at C_1 and C_7, in agreement with the involvement of the A_1 HOMO (**121**) [249, 284].

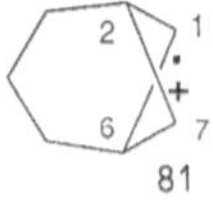

More recently, the radical cation of the parent system has been characterized by ESR/ENDOR spectroscopy [340]. These experiments revealed medium-sized hfcs for both the bridgehead (−11.4 G) and the equatorial protons (+11.4 G) and large positive hfcs (+77 G) for the axial protons (which in **75** are replaced by the trimethylene bridge). The large difference between the couplings of the axial and equatorial protons indicates a nonplanar, puckered geometry of the radical cation. No evidence for interconversion is apparent up to 160 K; this finding requires an inversion barrier of at least 12 kJ mol^{-1}.

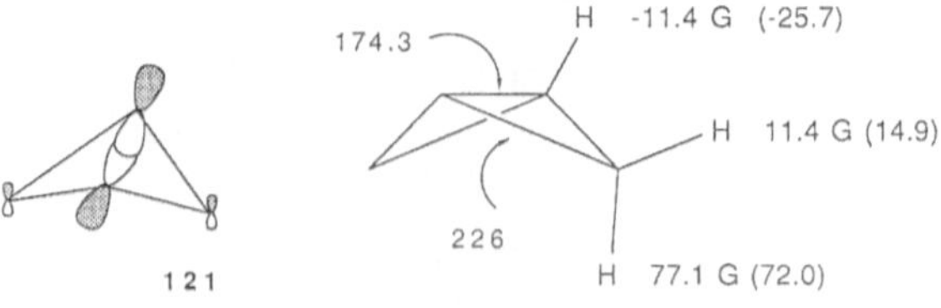

Both MNDO [340] and ab initio calculations [341] have been performed on the radical cation. The MNDO geometry optimization yields a flap angle of 132° and a C_1-C_3 bond distance of 178.6 pm; INDO calculations at this geometry give hfcs of 12.0, +18.3, and +78.0 G. The ab initio calculations at the 6.31 G level yield a slightly smaller bond distance (174.3 pm) and Fermi contact terms of −25.7, +14.7, and +72.0 G.

5.2 Cyclobutane

The formation and cleavage of cyclobutane systems have been discussed in Sect. 3.1 and 4.4. The structure of the intermediates is of major interest. The cyclobutane radical cation has been calculated by several groups. Bauld and coworkers [342] modeled the cycloaddition of ethene radical cation to ethene by the MNDO method. At this level of theory an unsymmetrical structure with one long one-electron C−C σ-bond is of lowest energy (Scheme 10, type C).

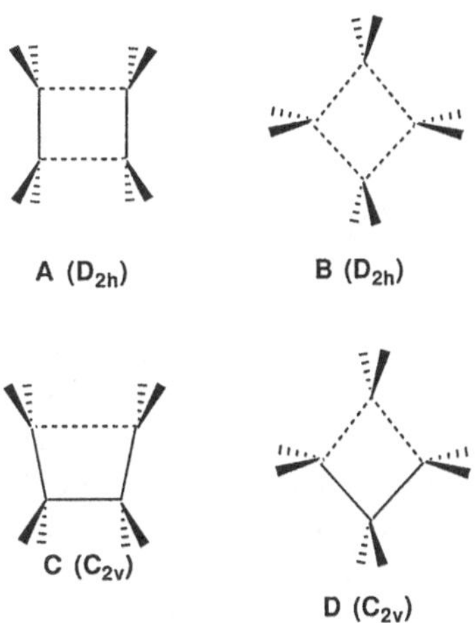

Scheme 10

Bouma et al. [343] carried out ab initio molecular orbital calculations. They found that loss of an electron from one of the e orbitals of cyclobutane leads to a Jahn-Teller unstable radical cation, giving rise to distorted structures of D_{2h} and C_{2v} symmetry. Among four local minima, a rectangular structure with two weakened C−C bonds (type A) and a rhomboidal structure with four weakened C−C bonds (type B) were found to be of equal energy, 16 kJ lower than a C_{2v} structure with one weakened C−C bond. In analogy to the cyclopropane radical cation one might consider ring opening, i.e. rotation of a methylene group into the ring plane, for structure type C. To our knowledge, such a structure has not been calculated in detail.

Ushida et al. [344] observed EPR spectra for X-irradiated cyclobutane in frozen $CFCl_3$ solution. At 4 K these spectra showed four strongly coupled protons (two pairs with splittings of 49 and 14 G, respectively) and four weakly coupled protons (5 G). These findings were interpreted in terms of a radical cation that had undergone static Jahn-Teller distortion to a rhomboidal structure of C_{2v} symmetry (type D). Above 77 K the Jahn-Teller distortion is averaged out and a nearly isotropic nine-line spectrum is observed. Additional structure types can be envisaged that are compatible with the low temperature ESR pattern and cannot be rigorously eliminated.

Fluorescence detected magnetic resonance effects observed during the pulse radiolysis of anthracene-d_{10} in the presence of tetramethylethene portray an additional facet of the cyclobutane radical cation system [345, 346]. The spectra [eight (?) equivalent methyl groups, $a_d = 8.2$ G; approximately one half of the monomer splitting, $a_m = 17.1$ G] are compatible with a dimer cation. In analogy to the benzene dimer radical cation [347, 348] they were interpreted as evidence for a "'sandwich', one molecule above the other" [346].

Although this description fits a radical cation with two weakened C–C bonds (Scheme 10, type A), the experimental results do not rule out a structure of type B. Furthermore, since the dimer radical cation can be converted to a "trimer" [346], a species reacting (or interacting) readily with the monomer is implicated; the "bifunctional" intermediate of type C may be best suited for this reaction. Since the modest signal-to-noise ratios of these ODMR experiments may not clearly distinguish species with eight methyl groups from those with only four, structure type C cannot be ruled out. In this context we mention the ready fragmentation of tetraphenylcyclobutane upon pulse radiolysis [255]; this observation is compatible with an intermediate of type A, but does not prove its involvement.

Compared with the parent system and those with identical substitution in all four carbons, the structure of other derivatives should be affected by the substitution pattern and by the nature of the substituents. For 1,2-disubstituted derivatives, structure type C, in which the doubly substituted cyclobutane bond is weakened (and lengthened), or a related structure type in which the bond is cleaved, should be favored. This is born out by several observations mentioned earlier. For example, the geometric isomerization of 1,2-diaryloxycyclobutane (Sect. 4.1) can be rationalized by "one-bond rotation" in a type C radical ion. Similarly, the fragmentation of the *anti*-head-to-head dimer of dimethylindene (Sect. 4.4) may involve consecutive cleavage of two cyclobutane bonds in a type C radical ion. The (dialkylbenzene) substituents have a lower ionization potential (IP 9.25 eV) [349] than the cyclobutane moiety (IP 10.7 eV) [350]; hence, the primary ionization is expected to occur from one of the aryl groups.

In analogy to substituent effects discussed for the cyclopropane system, we point to the possibility that 1,1-disubstitution may stabilize cyclobutane radical cations of type D. To our knowledge, such a structure has not materialized to date.

Finally, we mention the radical cation of oxetane. Without any doubt, this is a π species produced by oxygen lone pair ionization [327, 329]. It has sizeable hfcs in the β and γ positions ($a_\beta = 65.5$ G; $a_\gamma = 10.8$ G) and is characterized by a highly anisotropic and positively shifted g-factor ($g'' = 2.0046$, $g = 2.0135$). The substantial difference between the β hfcs and g-factors of the oxetane ion and that derived from oxirane are an additional powerful argument for the ring-opened structure of the latter.

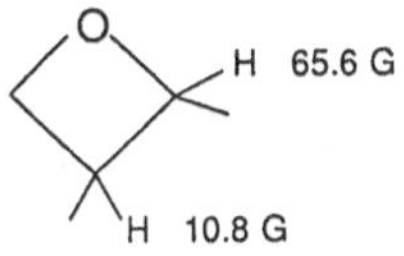

H 65.6 G

H 10.8 G

5.2.1 Radical Cations Derived From Cyclobutene Systems

The radical cations of cyclobutene and derivatives constitute the simplest system with the potential to undergo an electrocyclic reaction. Therefore, their reactions have attracted both theoretical and experimental scrutiny. The theoretical aspects have been treated in some detail, and a considerable body of experimental work has been carried out, utilizing a wide variety of techniques, particularly in the gas phase.

Two theoretical studies are of interest in connection with electrocyclic reactions of radical cations in general and with the ring opening of cyclobutene radical cation in particular. Haselbach and coworkers considered the role of orbital and state symmetries and predicted a barrier for the electrocyclic rearrangement, though considerably lower than for the corresponding neutral compound [351]. Similar conclusions were reached by Dunkin and Andrews, who focussed on the consequences of symmetry constraints in these reactions [352]. Both groups concluded that the electocyclic ring opening should have a barrier, though substantially lower than that for the parent compound.

Semiempirical calculations by Bauld and coworkers found substantial barriers to electrocyclic ring opening, the conrotatory opening being lower than the disrotatory mode. The lowest pathway located was a two-step process involving a cyclopropylcarbindiyl radical cation (**123**) as an intermediate [353]. It remains to be seen whether this interesting pathway can be confirmed by high-level ab initio calculations or proves to be an artefact of the semiempirical method employed.

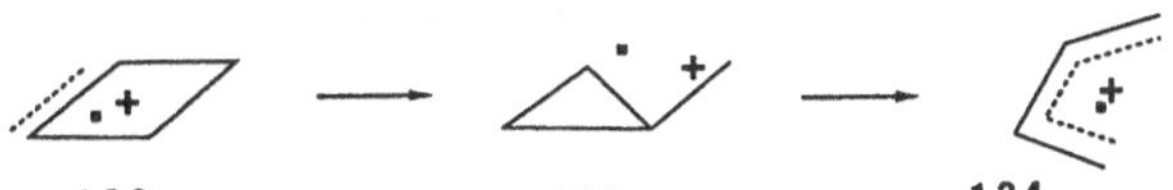

122 123 124

The barrier to ring opening is of special interest and has been examined for several systems, particularly in the gas phase. The barriers show a pronounced dependence upon the nature and position of substituents. A series of cyclobutene derivatives were studied by mass spectrometric and related gas-phase techniques. For example, recent studies by Gross and coworkers on the parent system and on several simple derivatives provided a coherent picture of the rearrangements of these ions [354]. Ion cyclotron resonance suggested that the initial ion generated from cyclobutene is a ring-closed species, with a barrier to ring-opening of <7 kcal mol^{-1} [354]. The technique of collisionally activated dissociation (CAD) was utilized to examine the ions of two phenylcyclobutenes. The rearrangement of the 1-phenyl derivative showed a barrier between 7 and 14 kcal mol^{-1}, whereas the barrier for the 3-phenyl derivative was too low to be detected [355]. Similar results were obtained for two methyl derivatives by Fourier transform mass spectrometry (FTMS). These experiments indicated barriers of <14 kcal mol^{-1} for the 1-methyl derivative and of <4 kcal mol^{-1} for the 3-methyl derivative [356].

The pronounced difference in the relative stability can be understood as a manifestation of molecular orbital interactions. The SOMO of the cyclobutene ion has high orbital coefficients on the "internal" sp^2 hybridized carbons (−CH=), whereas the butadiene SOMO has the highest coefficients in the terminal ($=CH_2$) carbons. Accordingly, substituents in the 1-(and 2-) position will stabilize the closed ion. On the other hand, substituents in the 3-(and 4-) position will favor the ring opened ion and lower the barrier to ring opening.

These considerations are borne out by results observed for 1,2- and 3,4-diphenyl derivatives in solution (Sect. 4.4). Schuster and coworkers detected the 1,2-diphenyl species **91** by optical spectroscopy at room temperature and found that it was thermally stable under these conditions [268]. In contrast, Miyashi and coworkers found that the 3,4-diphenyl derivative, under comparable conditions, suffered ring opening [266]. Remarkably, this reaction occurs with the same stereochemistry as the orbital symmetry controlled reaction of the neutral precursor.

The parent radical cation can be generated by radiolysis in frozen matrices at 77 K and studied by optical and ESR spectroscopy. Under these conditions, the optical spectrum of the cyclobutene radical cation remained unchanged up to 90 K [357]. However, irradiation with visible light converted the cyclobutene to a butadiene radical cation of unspecified conformation. More recently, the same radical cation was examined by ESR/ENDOR spectroscopy at temperatures as high as 130 K, suggesting a barrier of at least 10 kcal mol^{-1} [358, 359]. The coupling constants of the β protons (a_β = 28.0 G) far surpass those of the α protons (a_α = 11.1 G), indicating strong hyperconjugative stabilization of the radical cation. The rearranged species obtained as a result of visible irradiation is identified unambiguously by its ESR spectrum as the s-*trans* rotamer (**125**) of butadiene radical cation. The mechanism of this transformation poses an interesting mechanistic problem.

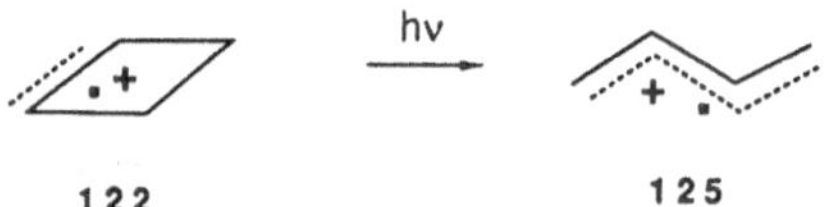

5.3 Strained Valence Isomers of Benzene

Among the radical cations derived from strained ring hydrocarbons those of composition $(CH)_6$, the valence isomers of benzene (**128**) (Scheme 11), have attracted considerable interest. In addition to the one-electron oxidation products of Dewar benzene (**126**) [360–368], bicylopropenyl (**129**) [369], benzvalene (**130**)

[370], and prismane (**127**) [364], the adduct between cyclopropenylium cation and cyclopropenyl radical [364, 371] have been investigated in some detail. The structures and interconversions of these species have been examined by a variety of spectroscopic methods, including steady-state optical spectroscopy in frozen glasses [360], nsec time resolved spectroscopy in fluid solution [362, 371], laser induced fluorescence spectroscopy [372], chemically induced magnetic polarization [361, 364, 369, 370], as well as electron spin resonance [363, 365–367] and optically detected magnetic resonance [368]. These experimental studies were supported by molecular orbital calculations [361, 364]. The elucidation of the unique structure of benzene [373] is generally recognized as a milestone in the development of organic chemistry. Similarly, we suggest that detailed insight into the radical cations of the $(CH)_6$ isomers and their intriguing potential energy surface, will contribute significantly to understanding the nature of radical cations in general.

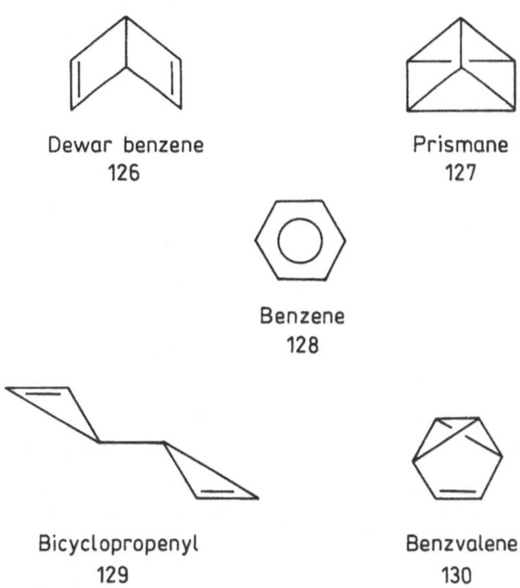

Scheme 11

5.3.1 Dewar Benzene Radical Cation

The first valence isomer to be investigated was the bicyclic (hexamethyl-, HM-) Dewar benzene. However, γ irradiation of this substrate produced the electronic spectrum of HM-**128** [360]. Similarly, nsec time-resolved laser spectroscopy failed to reveal evidence for the bicyclic radical cation [362]. The first indication for the existence of such a species as a discrete entity was provided by a CIDNP study [361]. These results are best discussed in connection with ab initio calculations carried out for the parent C_6H_6 system. At the 6-31 G level these calculations support the existence of two cationic states with the unpaired spin density either on the four olefinic carbons ($^2\mathbf{B_2}$) or on the two bridgehead carbons ($^2\mathbf{A_1}$). The $^2\mathbf{B_2}$ is the lower of the two states by $\sim$8 kcal mol^{-1} [361].

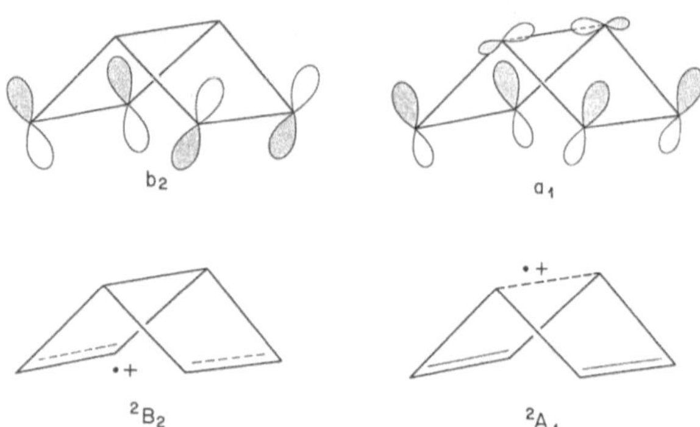

The spin density distribution of the two ions is expected to give rise to characteristic CIDNP effects: the $^2\mathbf{B_2}$ species would induce strong polarization for the allylic methyl groups; the $^2\mathbf{A_1}$ radical cation would cause the bridgehead methyl groups to be polarized. The fact that both signals showed sizeable CIDNP effects [361] was explained by the simultaneous involvement of both radical cations.

More recently, this system has been investigated by ESR spectroscopy as well as optically detected magnetic resonance. Five research groups on two continents and the British Isles exposed HM-**126** to ionizing radiation at 77 K in rigid matrices. In each laboratory, a thirteen-line ESR spectrum with a hyperfine splitting of 9.5 G was observed and ascribed to the $^2\mathbf{B_2}$ radical cation. However, this assignment, though unanimous, is not necessarily unambiguous, because a chemically reasonable alternative for the observed ESR pattern was neither considered nor eliminated.

The need for considering potential alternatives follows from ESR results obtained upon chemical oxidation of HM-**126** with $SbCl_5$ and/or $SbCl_5/SOCl_2$ [374]. Depending on sample preparation, we observed either the nineteen-line spectrum (a = 6.5 G) of HM-**128** radical cation [375, 376], or a thirteen-line feature (a = 9.8 G; Fig. 23), close to that reported for the radical cation of pentamethyl-(PM-) **128** [377, 378]. Indeed, essentially identical spectra are obtained upon oxidation of HM-**126**, HM-**128**, and PM-**128**. These results raise the question whether the thirteen-line spectrum ascribed to HM-**126**$^{\cdot+}$ ($^2\mathbf{B_2}$) [363, 365–368] can be distinguished from that of PM-**128**$^{\cdot+}$. The splitting of 9.5 G lies within 5% of the 10.05 G splitting reported for PM-**128**$^{\cdot+}$ in sulfuric acid at ambient temperatures. Small differences (5–10%) in hyperfine coupling constants (or zero-field splitting parameters, D, for triplet species) are not uncommon in two very different solvents or matrices.

Two structural changes are necessary for the conversion of HM-**126**$^{\cdot+}$ to PM-**128**$^{\cdot+}$, ring opening and demethylation; both reactions have precedent. The radiolysis of strained molecules in frozen matrices frequently results in rearrangements with ring opening. HM-**126**, quadricyclane, or dicyclopropylidene are but a few substrates undergoing such ring opening reactions. The proton induced

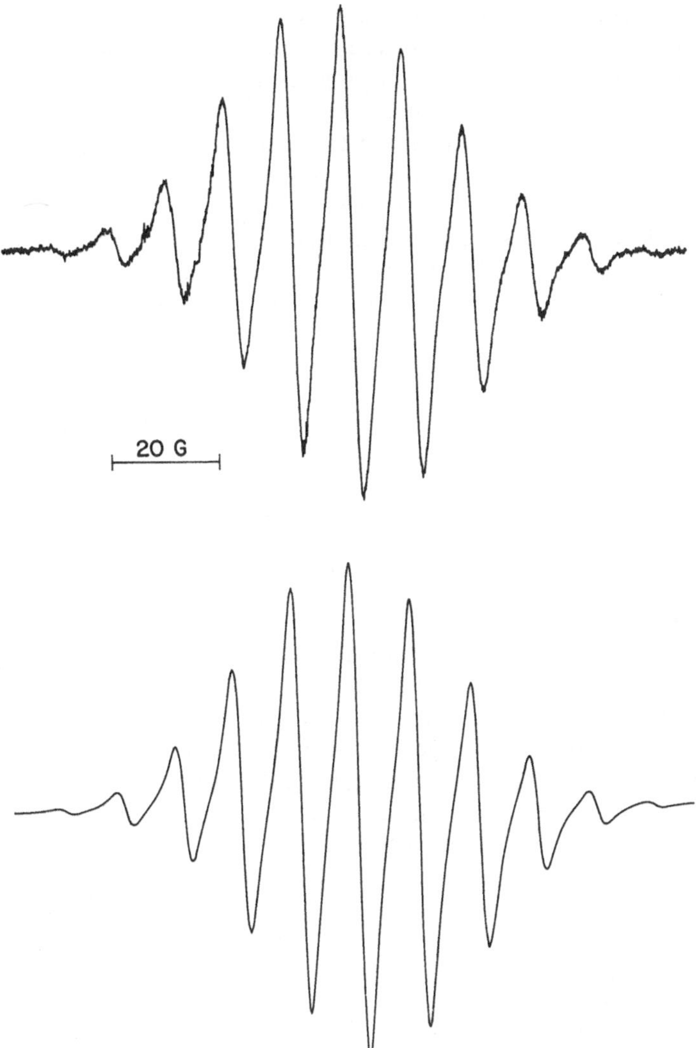

Fig. 23. ESR spectrum obtained upon oxidation of hexamethyl-(Dewar)-benzene with $SbCl_5$ at −78 °C *(top)* and simulation [374]

isomerization of strained rings is also well known. In fact, loss of a proton from HM-**126**$^{\cdot+}$ has been postulated [366, 367], and a bimolecular mechanism has been proposed [367].

The postulated generation of PM-**128**$^{\cdot+}$ from HM-**128** was observed as early as 1965 in sulfuric acid [378]. We view this reaction as a reverse Friedel-Crafts reaction, not unlike that complementing the formation of heptamethylbenzenonium ion from HM-**128** [379]. Although the formation of PM-**128**$^{\cdot+}$ requires several consecutive reactions, its observation under radiolysis-ESR conditions cannot be ruled out categorically, since the sample preparation requires several hours of irradiation. Therefore, the samples are essentially "annealed" (at liquid nitrogen temperature) by the time their ESR spectra are recorded. A plausible reaction sequence involves protonation of **PM-126** with ring opening, loss of methyl from the resulting hexamethylbenzenonium cation, followed by oxidation of **PM-128**.

One might argue that the results least subject to ambiguity are those with the shortest delay between the generation of the radical cation and its observation. In this respect, the time-resolved ODMR results of Trifunac and Qin (Fig. 24) [368] and time resolved CIDNP results observed in the author's laboratory (Fig. 25) [380], may provide the least distorted view of the species in question. Of course, neither of these experiments qualifies as the coveted "direct observation." Thus, the direct observation of the "elusive" hexamethyl-Dewar benzene radical cation must await further scrutiny.

HM-126$^{\cdot+}$ $\xrightarrow{-H^+}$ 131

In this context, we mention that the ESR spectrum claimed as evidence for the existence of HM-**126**$^{\cdot+}$ (^{2}**A**$_1$) [365] was later shown to be an erroneous assignment [366, 367]. The presumed seven-line spectrum was shown to be a iuxtaposition of an unknown single-line spectrum plus a six-line spectrum, a doublet of triplets, which was identified as a deprotonation product, **131**. Furthermore, the CIDNP evidence for the existence of the ^{2}**A**$_1$ radical cation [361] may also need reevaluation for the reason outlined in Sect. 3.7. Since the assignment was based on difference spectra, the possibility of an artefact is hard to eliminate.

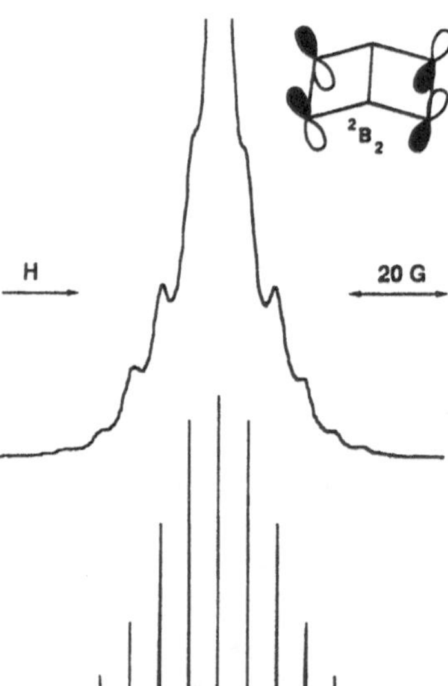

Fig. 24. ODMR spectrum obtained upon pulse radiolysis of hexamethyl-(Dewar)-benzene in solution [368]

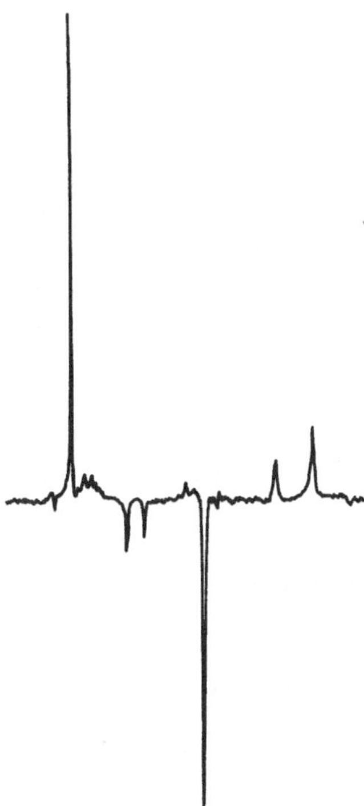

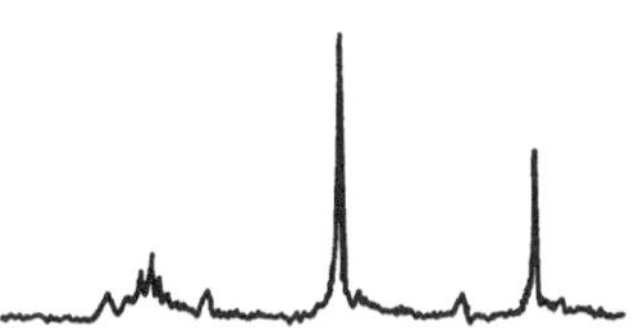

Fig. 25. Time resolved CIDNP spectrum observed upon photo-induced electron transfer oxidation of hexamethyl-(Dewar)-benzene (0.02 M) by chloranil (0.02 M). The *top* spectrum was observed 5 μs after excitation with the frequency-tripled output (355 nm) of a Nd/YAG laser. The *bottom* spectrum was recorded in the dark

5.3.2 Prismane Radical Cation

The most highly strained isomer of benzene is the quadricyclic derivative first considered by Ladenburg [381], now generally called prismane (**127**). The corresponding radical cation has proved very elusive. In fact, hexamethyl-**127** is rapidly rearranged by chloranil in polar solvents at room temperature. CIDNP experiments have provided tentative evidence for the fleeting existence of HM-$\mathbf{127}^{\cdot +}$ during the photo-reaction of HM-**127** with anthraquinone. The prismane polarization is very weak; apparently, only a small fraction of these ions reverts back to HM-**127**. The predominant reaction involves ring opening to HM-**126** [364]. In essence, **127** is related to **126** as **Q** is related to **N**. Because of the high symmetry of the substrate the CIDNP results for HM-**127** do not offer any clues as to the structure of the intermediate. However, the polarization observed for HM-**126** implicates a radical cation in which spin and charge are located in four equivalent positions and where the remaining centers are essentially without spin.

Ab initio calculations for the parent system support the existence of HM-**127**$^{\cdot+}$ and help to delineate its structure. The highest occupied molecular orbital in prismane is e″, which has significant bonding character in the cyclopropane-like bonds and antibonding character in the transannular bonds. The next lower orbital is e′, which has bonding character mainly in the three transannular bonds. Removal of an electron from the e″ orbital gives rise to a $^2E''$ state which on Jahn-Teller distortion yields a pair of states, $^2A_2$ and $^2B_1$ (in C_{2v} symmetry). These are expected to be the lowest electronic states of the prismane radical cation. Similarly, removal of an electron from the e′ orbital yields two states, $^2A_1$ and $^2B_2$, which are expected to be of higher energy.

Calculations using a split-valence 6-31 G basis set identified the $^2B_1$ state as the lowest in energy. Two of its cyclopropane bonds are lengthened to 1.73 Å whereas the transannular bonds are shortened to 1.48 Å (Scheme 12). These changes reflect the bonding pattern in the e″ orbital. The $^2A_2$ state lies close in energy to the $^2B_1$ state; four equivalent cyclopropane bonds are stretched to 1.63 Å and one transannular bond is shortened to 1.46 Å. This state was found not to be a minimum; it has an imaginary b_1 frequency that takes it to the $^2B_1$ state through C_s pathways. In essence, the $^2A_2$ state is a transition state in the pseudorotation interconverting two $^2B_1$ states. The $^2A_1$ and $^2B_2$ states lie much higher in energy; they are also connected on a pseudorotation hypersurface. In addition, the vibrational analysis of the $^2A_1$ state indicates a distortion to C_2 symmetry which connects it to the $^2A_2$ and, ultimately, to the $^2B_1$ structure.

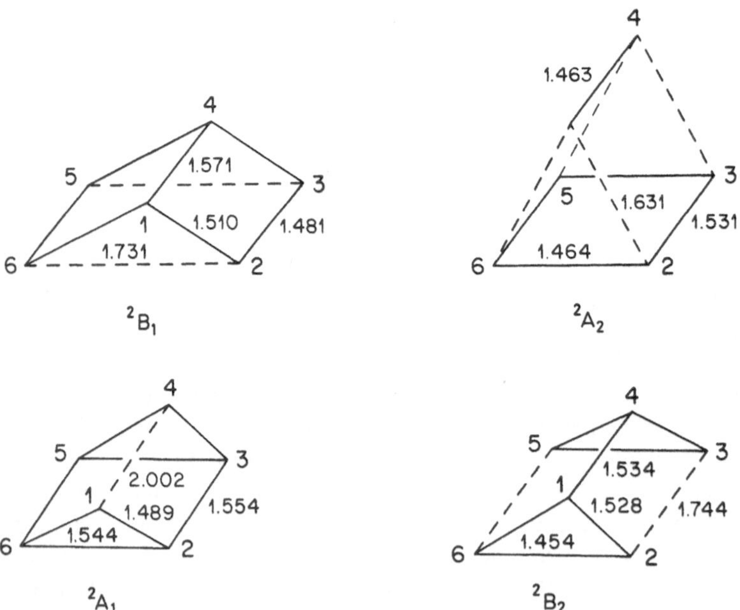

Scheme 12

The lowest state of prismane ($^2B_1$) cation lies 16 kcal mol^{-1} above the $^2B_2$ state of the Dewar benzene cation (at the MP2/6-31 G* level). This is considerably less than the corresponding energy difference of the neutral systems (37 kcal mol^{-1}). The ground electronic states of prismane and Dewar benzene ions do not correlate; their interconversion is forbidden from both state-symmetry and orbital-symmetry considerations. The CIDNP experiments indicate, however, that the actual barrier is quite small.

In summary, the prismane system exemplifies several characteristics of strained ring radical cations, particularly the low barrier to isomerization and the fact that its ion apparently has less strain energy than the parent molecule. Once again, bonding interactions with less than the full complement of electrons are accommodated with lesser strain energies. Further investigations of the (hexamethyl) prismane radical cation will be difficult in view of its kinetic instability and because of its high symmetry. One might expect a 13 line ESR spectrum showing an unexceptional coupling constant near 10 G, which hardly qualifies as unambiguous evidence.

5.3.3 Benzvalene and Bicyclopropenyl Radical Cations

Benzvalene (**130**) is a tricyclic benzene isomer containing a bicyclobutane ring system bridged by an ethene moiety. The radical cations of this system and its benzolog (naphthvalene) are accessible by photo-induced electron transfer. A CIDNP study indicated negative hfcs for the olefinic protons, strong positive hfcs for the (non-allylic) bridgehead protons, and negligible hfcs for the allylic bridgehead protons [370]. These results suggest that benzvalene radical cation has spin and charge essentially localized in the olefinic moiety. The strong positive hfcs of the non-allylic bridgehead protons is evidence for a strong hyperconjugative interaction between these protons and the *p*-orbitals bearing the unpaired spin. An analogous structure with an *o*-xylene type SOMO (B_2 symmetry) is indicated for the naphthvalene radical cation [370].

These assignments are consistent with radical cations corresponding to the HOMOs established by photoelectron spectroscopy and simple theoretical models [382–384]. They are also supported by theoretical calculations for the parent system [385]. Ab initio molecular orbital calculations were carried out for two low-lying radical cationic states of benzvalene radical cation, $^2B_1$ and $^2A_1$ in C_{2v} symmetry [385]. The calculated bond lengths and hyperfine coupling constants (Scheme 13) show characteristic differences.

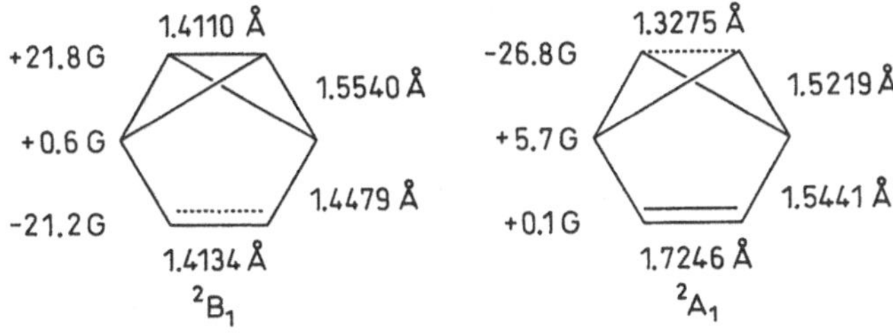

Scheme 13

The $^2B_1$ state was found to be the ground state and its predicted hfcs are fully compatible with the observed CIDNP results, whereas the hfcs calculated for the $^2A_1$ state show irreconcilable differences with the experimental findings for every type of proton.

The radical cation derived from bicyclopropenyl (**129**) is of interest a) because it is the only benzene isomer with any degree of conformational mobility, b) as a potential intermediate in the rearrangement of prismane radical cation (cf., the $^2B_2$ state, Scheme 12), and c) as a potential adduct between cyclopropenium cation and cyclopropenyl radical. It has proved practical to study a simple (3,3′-dimethyl-) derivative (**131**) of this species, because this precaution eliminates the prototropic rearrangement to the conjugated isomer.

CIDNP results indicate that the radical cation (**132**) formed from **131** is exceedingly short-lived and undergoes rearrangement to a benzvalene radical cation (**136**). This intermediate is identified by its unmistakable chemical shift and polarization patterns. The rearrangement poses an interesting mechanistic challenge. It may be initiated by a radical (or cationic) addition forming tricyclo[3.1.$0^{1,5}$.$0^{2,4}$]hexanediyl (**133**); an alkyl shift with cleavage of an internal cyclopropane bond (→ **134**), or a symmetry-allowed disrotatory ring opening (→ **135**) [253]. Since the CIDNP effect reflect the spin density distribution of **136**, any intermediate(s) preceding it cannot have lifetimes exceeding (fractions of) nanoseconds.

The pulse radiolysis of trimethyl-(TM-)cyclopropenylium cation, **TMCP⁺** examined by time resolved optical spectroscopy, provided evidence for the eventual formation of hexamethylbenzene cation, **HMB**$^{\cdot+}$. Apparently **TMCP⁺** and its one-

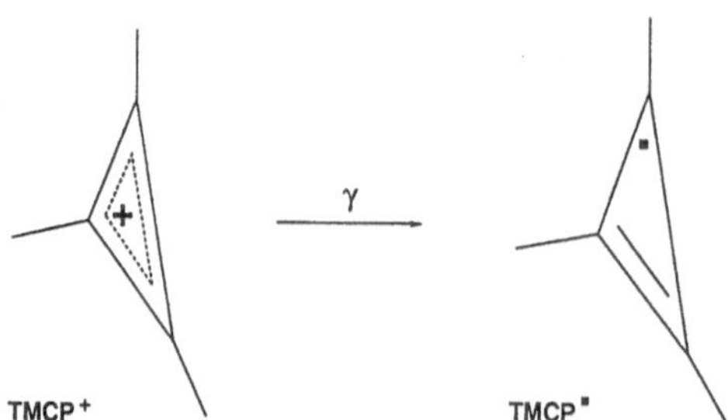

electron reduction product, TM-cyclopropenyl radical **TMCP˙**, undergo a coupling reaction. A short-lived transient preceding **HMB˙⁺** was tentatively identified as a sandwich dimer analogous to the benzene dimer cation [371]. However, ab initio calculations on this system (without methyl groups) do not support this assignment; rather, they support a species of lower symmetry [364].

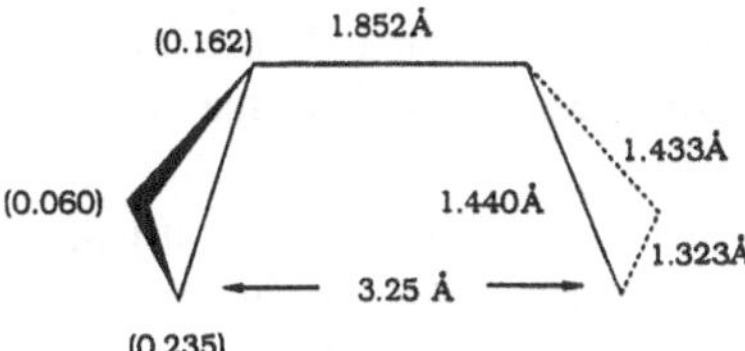

The optimized structure was found to be symmetric between the two interacting units and has overall C_2 symmetry. The spin density was essentially localized in a four-carbon plane with an interaction distance of 1.85 Å between the closest pair of carbon atoms. At the MP2/6-31 G* level of theory this structure was 23 kcal mol^{-1} higher in energy than the $^2B_1$ ground state of prismane. The calculated bond lengths and carbon spin densities of the adduct are shown above [364].

Although it is tempting to assign this calculated structure type to the transient documented by optical spectroscopy [371], there are several arguments against this assignment. In particular, the rapid rearrangement of bicyclopropenyl to benzvalene radical cation [369] might suggest an alternative identity for the transient. Clearly, additional information is required, before this issue can be resolved unambiguously.

In summary, the radical cations of the benzene valence isomers show several interesting structures. Although most of their structural features can be rationalized by considering the HOMOs of the precursor molecules, some of the species show substantial changes in individual bond lengths. Accordingly, species such as the $^2B_2$ and $^2A_1$ Dewar benzene radical cations, the $^2B_1$ and $^2A_1$ benzvalene radical cations, or the $^2B_1$ prismane radical cation cannot be expected to qualify as Koopmans radical cations. To date, most of the information available in this series is based on CIDNP results and ab initio calculations. It is safe to predict increasing involvement of ESR spectroscopy in this area.

5.4 Radical Cations Derived From Hexadiene Systems

Radical cations derived from a variety of hexadiene systems constitute an interesting family of intermediates, since they are related to the potential mechanistic extremes of the Cope rearrangement. The electrocyclic reaction of a hexadiene radical cation has three mechanistic extremes: a) addition precedes cleavage (associative mechanism); b) cleavage preceeds addition (dissociative mechanism); c) addition and cleavage occur in coordinated fashion (concerted mechanism). To date, radical cations corresponding to all three mechanistic extremes have been characterized. This illustrates remarkable differences between the potential surfaces of radical cations and neutral precursors. On the precursor potential surface, the states of intermediate geometry are saddle points (transition structures), whereas they are pronounced minima (e.g. Fig. 26) on the radical cation potential surface. In essence, the parent molecules undergo a concerted Cope rearrangement via a transition structure, whereas the radical cations undergo cycloaddition or cleavage reactions, which are "arrested" at intermediate geometries. A comparison between Figs. 14 or 22 on the one hand and Fig. 26 on the other illustrates most clearly the fundamental difference between a "distonic" and a non-vertical radical cation.

The first structure type to be established for a hexadiene radical cation was one in which cleavage is achieved without bonding, i.e. a representative of the dissociative mechanism. Dicyclopentadiene and several derivatives can be oxidized to radical cations (**137**) in which one of the bonds linking the monomer units is cleaved. The unique spin density distribution of **137** is reflected in an unmistakable polarization pattern [386–389].

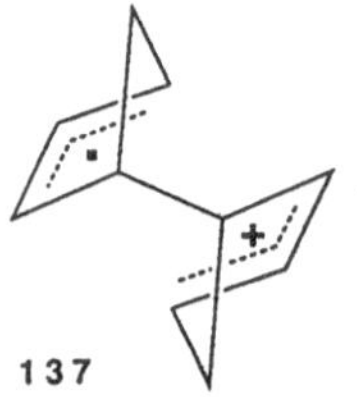

The resulting species can be viewed as a molecular assembly, in which two allyl moieties are linked to a four-carbon "spacer", each group attached to two adjacent carbon atoms. The resulting structure has three conformational minima; the

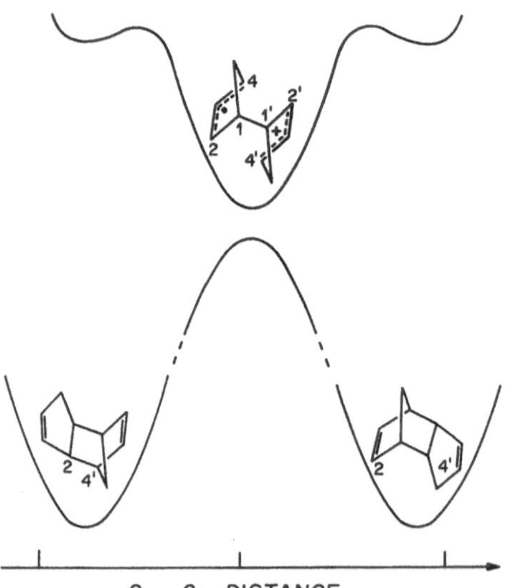

Fig. 26. Schematic cross section through the potential surfaces for dicyclopentadiene and its radical cation (**137**). The energy minimum on the radical cation surface corresponds to a saddle point or a shallow minimum on the potential surface of its precursor

termini are separated by 3–4 Å, clearly beyond a distance that would allow significant bonding interaction.

Subsequently, the radical cation structure **137** and that of its *(exo-)* stereoisomer were confirmed independently by an analysis of their electronic absorption spectra [390]. The large energy gap ($\Delta E = 1.67$ eV) observed for the *syn*-dimer radical cation is incompatible with the photoelectron spectrum of the parent molecule ($\Delta E = 0.15$ eV), but can be explained by ring-opened structures, such as **137** (Fig. 27).

Another structural possibility for a hexadiene radical cation arises, when the two allylic moieties are linked in pairwise fashion to two- or three-carbon spacers. This structure type can be approached by oxidation of molecules such as semibullvalene [391–393] or barbaralane [394]. In the resulting radical cations, the two allylic moieties are held in close proximity; model considerations suggest a "non-bonding" C–C distance of 2.2–2.3 Å, considerably closer than for the previously discussed structure type. At this distance, a moderately strong interaction of the twin moieties cannot be excluded. Accordingly, we assigned cyclic conjugated structures to radical cations derived from semibullvalene (→ **138**; cf.

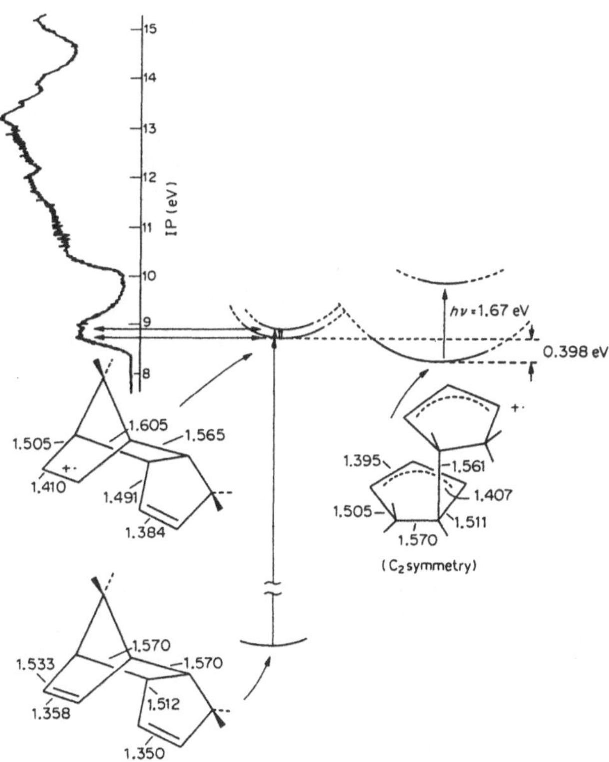

Fig. 27. Photoelectron spectrum of *endo*-dicyclopentadiene *(upper left)* and optimized geometries of the parent molecule and the ring-closed and ring-opened radical cations [390] (Reprinted by permission)

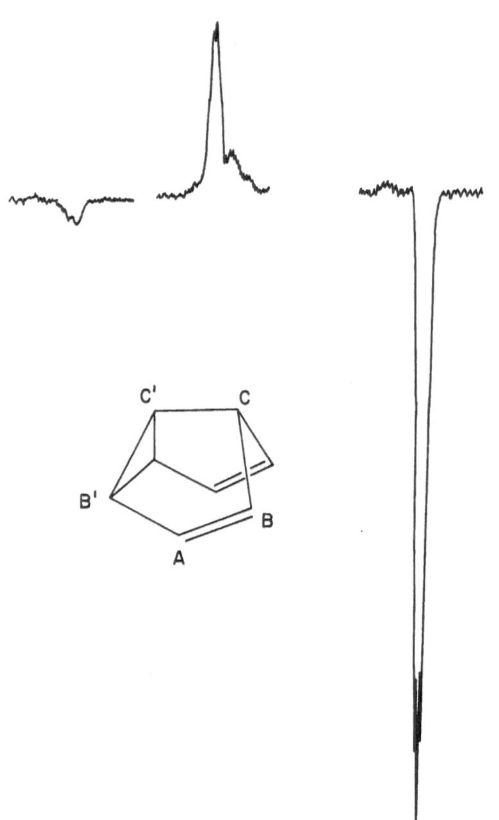

Fig. 28. [1]H CIDNP spectra (90 MHz) observed during the photoreaction of chloranil with semibullvalene (0.02 M each in acetonitrile-d_3) during UV irradiation [391]

Fig. 28) [391] and several barbaralane derivatives (→ **139**) [394]. In these systems, bond-making and bond-breaking processes are "frozen" at a stage permitting cyclic delocalization of five π-electrons.

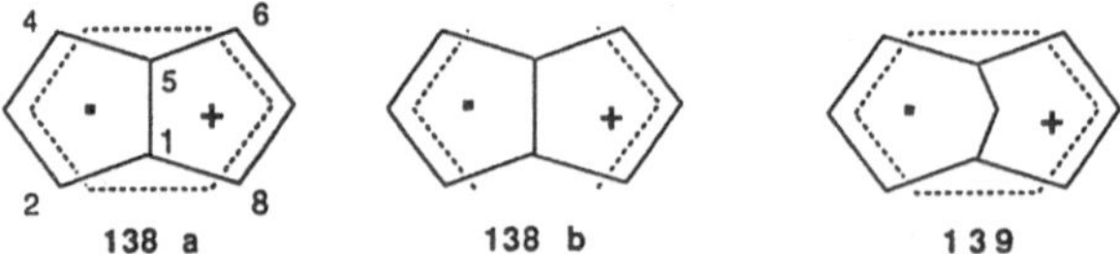

An ESR study of the semibullvalene radical cation [392] showed two strongly coupled protons (36.2 G) and four protons with intermediate hfc (7.7 G); the hfc of the "internal" olefinic protons was not resolved. The strong CIDNP effects ($a > 0$) for the bridgehead protons, the somewhat weaker effects ($a < 0$) for those alternating between olefinic and cyclopropane character and the very weak CIDNP effects for the internal olefinic protons (hfc 0.5–1.0 G) identify the signs of these interactions [391]. The strong hfc of the bridgehead protons provides an opportunity to differentiate between a cyclic conjugated structure (**138a**) and the time-average of two localized structures (**138b**). According to Whiffen [307], the hfc of H_1 and H_5 in the cyclic conjugated structure with spin densities, $\varrho = 0.25$, on four carbon atoms is given by:

$$a = B(0.25^{1/2} + 0.25^{1/2})^2 \cos^2 \Theta$$

the localized structure (spin densities, $\varrho = 0.5$, on two carbons) is

$$a = B(0.5^{1/2} + 0^{1/2})^2 \cos^2 \Theta$$

If one assumes a dihedral angle, $\Theta = 38°$, between the 2p axes at C_2 and C_8 (C_4 and C_6) and the C_1-H (C_5-H) bond, and a Heller-McConnell constant, $B = 46$ G, the measured hfc of 36.2 G clearly supports structure **138a** [392].

Interestingly, the radical cation **138** can be generated also by light-induced isomerization of cyclooctatetraene radical cation (**140**). The conversion of the red non-planar ion **140** (4n − 1 π electrons) upon irradiation with visible light had been observed previously [395], but the blue photo-product had not been recognized as the cyclic conjugated species **138** with 4n + 1 π electrons. This interconversion is one of only a few orbital symmetry allowed processes documented in radical cation chemistry [393].

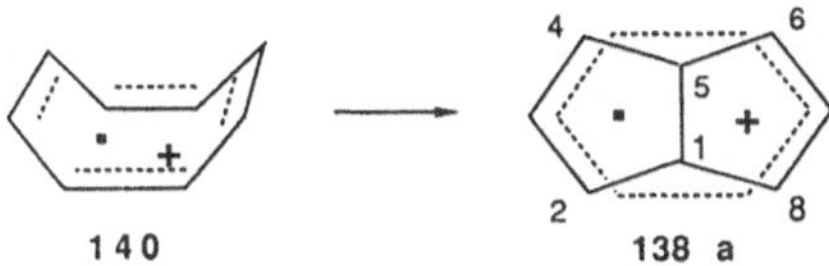

The third radical cation structure type for hexadiene systems is formed by radical cation addition without fragmentation. Two hexadiene derivatives were mentioned earlier in this review, allylcyclopropene (Sect. 4.4) [245] and dicyclopropenyl (Sect. 5.3) [369]. The products formed upon electron transfer from either substrate can be rationalized via an intramolecular cycloaddition reaction which is "arrested" after the first step (e.g. → **133**). Recent ESR observations on the parent hexadiene system indicated the formation of a cyclohexane-1,4-diyl radical cation (**141**). The spectrum shows six nuclei with identical couplings of 11.9 G, assigned to four axial β- and two α-protons (Fig. 29) [397–399]. The free electron spin is shared between two carbons, which may explain the blue color of the sample ("charge" resonance). At temperatures above 90 K, cyclohexane-1,4-diyl radical cation is converted to that of cyclohexene; thus, the ESR results do not support a radical cation Cope rearrangement.

Indirect (chemical) evidence for the cyclohexane-1,4-diyl structure type was provided by an elegant study of the electron transfer initiated rearrangement

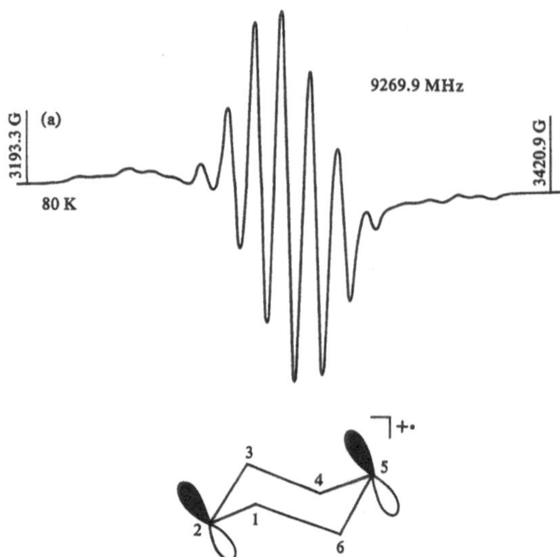

Fig. 29. ESR spectrum of a γ-irradiated CF_3CCl_3 solution of 1,5-hexadiene at 80 K (Reprinted by permission)

and oxygenation of 2,5-diphenylhexa-1,5-diene and 3,6-diphenylocta-2,6-diene (Scheme 14) [400]. The intermediate 1,4-cyclohexanediyls were intercepted by molecular oxygen; the stereochemistry of the products showed that the initial cycloaddition occurs in the same stereospecific manner established for the thermal rearrangement of the neutral parent [401].

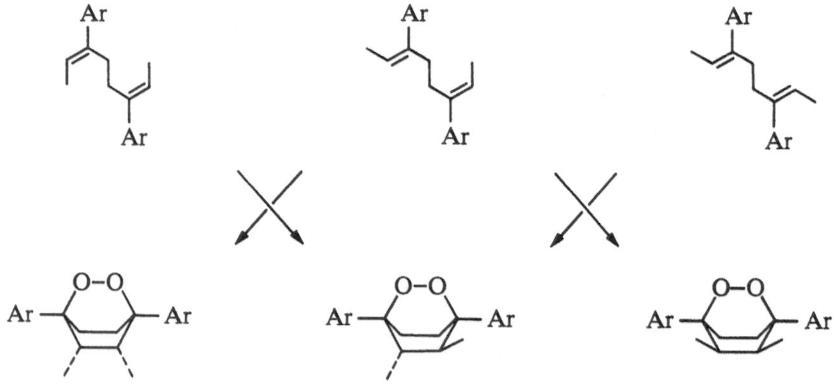

Scheme 14

One significant aspect of the cyclohexane-1,4-diyl radical cation has been a point of contention: the question whether it undergoes cleavage to the hexa-1,5-diene radical cation, i.e. whether it completes the Cope rearrangement. Results obtained in the laboratory of Miyashi, particularly the exchange of a deuterium label between the terminal olefinic and allylic positions, seem to suggest such a cycloreversion [400]. On the other hand, the hydrogen migration leading to cyclohexene radical cation [397] and the results of molecular orbital calculations [342] tend to argue against this possibility. While it may be necessary to carry the calculations to higher levels of theory, the ESR studies may reflect special matrix effects prevailing in the chlorofluorocarbon matrices employed in this work. In any event, this result can only place an upper limit on the rate of the putative ring opening. Similarly, the finding that Ce^{IV} oxidation of 1,4-diphenyl-2,3-diaza-bicyclo[2.2.2]oct-2-ene (**142**) gave rise to terphenyl [402], merely indicates that the oxidation of the 1,4-diyl radical cation is faster than its ring opening. In both cases, the observation of ring-closed products is a necessary, but not sufficient condition for the proposed interpretation.

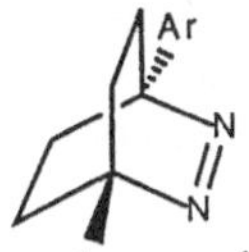

Ar Ar N N 142

The existence of a cyclohexanediyl boat conformer has also been considered. For example, the γ-irradiation of bicyclo[2.2.0]hexane was found to give rise to the ESR spectrum previously assigned to the chair conformer [399]. However, this experiment hardly qualifies as a subtle probe for an intermediate which may reside in a shallow potential well. More recently, chemical evidence for the existence of this structure type was reported by Tsuji and coworkers: the electron transfer induced reaction of the hexamethylbicyclo[2.2.0]hexane (**143**) led to the formation of *erythro*-(E,E)-diene (**145**) along with lesser amounts of **146** [403]. These results support an initial generation of 1,4-diyl (**144**) which is then partitioned between stereospecific ring opening (→ **145**) and hydrogen migration (→ **146**).

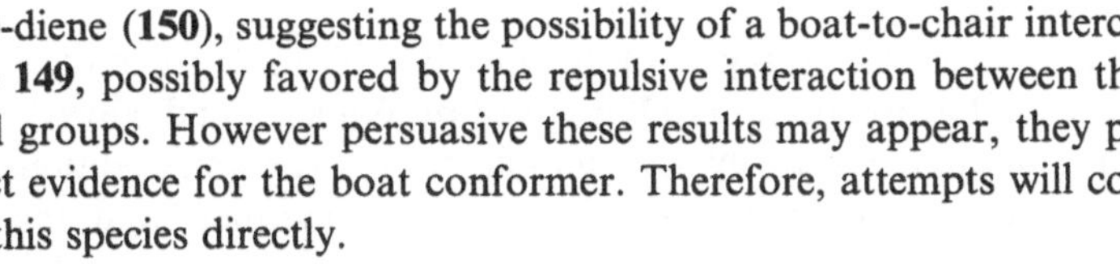

143 144 145 146

In contrast, the 2,3,5,6-tetrakis(*endo*-methyl)bicyclo[2.2.0]hexane (**147**) gave rise to the (E,Z)-diene (**150**), suggesting the possibility of a boat-to-chair interconversion, **148** → **149**, possibly favored by the repulsive interaction between the four *endo* methyl groups. However persuasive these results may appear, they provide only indirect evidence for the boat conformer. Therefore, attempts will continue to observe this species directly.

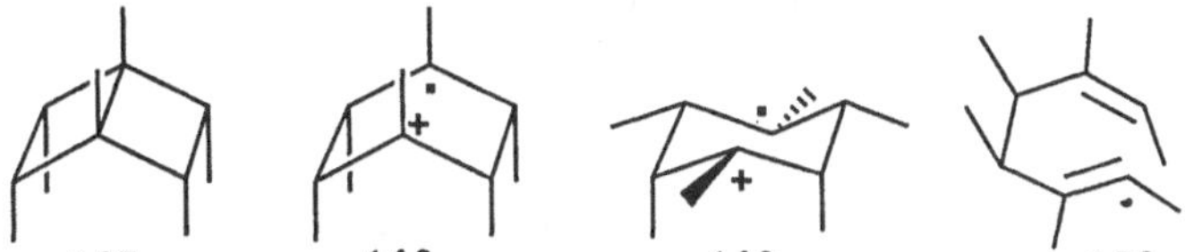

147 148 149 150

In summary, the existence of these three different structure types suggests a rather delicate balance between the different minima on this interesting potential energy surface. Molecular orbital calculations at the highest levels currently practical are certainly quite useful in elucidating the nature of these structures. This is also an area where one may safely predict continuing efforts and where we anticipate additional insights by both ESR and CIDNP spectroscopy.

5.5 Bifunctional (Doubly Allylic) Radical Cations

Two of the structure types established for hexadiene-derived radical cations, viz., **137** and **141**, feature two different sites, in which spin and charge are located (and stabilized). To date several structure types have been described that meet this general description; the following section contains a few brief comments on some representative examples.

We divide these structures into two categories: 1) species consisting of two identical or chemically similar moieties supporting spin and charge by some stabilizing interaction; and 2) species consisting of two different delocalizing (stabilizing) moieties, of which one supports the unpaired spin, whereas the other specifically stabilizes the charge. Of course, this subdivision is one of convenience and not one of principle. In fact, the preferred location of spin and charge in a given intermediate is subject to substituent effects in either of the fragments. This may result in an "umpolung" of a species in category 2 or in a conversion of intermediates from category 1 into a species of group 2. We will briefly discuss representatives of these families and delineate some characteristic features of these intermediates.

We have repeatedly referred to the involvement of symmetrical bifunctional radical cations. For example, they have been invoked as intermediates in electron transfer induced dimerizations of some mono-substituted olefins (Sect. 4.1) or, conversely, in the cleavage of appropriate dimers (Sect. 5.2). In most of these cases, the assignment of a bifunctional intermediate was based on chemical evidence, particularly on the regio- and stereochemistry of the reaction products. We have also encountered representatives of this family, whose existence was concluded on the basis of spectroscopic evidence (Sect. 5.1, 5.4).

Among the earliest examples of symmetrical bifunctional radical cations, the distonic trimethylene species (**103**) invoked by Williams and coworkers [293, 296, 297] are stabilized solely by hyperconjugation. The main rationale for their formation would be the relief of ring strain. On the other hand, the non-vertical radical cations **137** derived from cyclopentadiene dimers [386–389] are favored by two elements of allylic stabilization. This radical cation has three conformat-

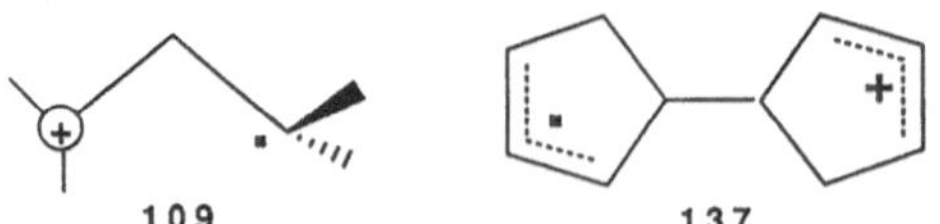

ional minima, though none allow substantial interaction between the allylic moieties. However, the possibility exists that the single unpaired electron may hop between the allyl functions. This should manifest itself in the hyperfine coupling pattern; no ESR study of this system has been published, as yet, that might elucidate this interesting question.

The relative stability of the delocalized, "non-vertical" radical cation relative to a localized, "vertical" isomer was demonstrated also in gas phase experiments [404]. The molecular ions of m/e 132 obtained by gas phase ionization of the [4 + 2] dimer exhibited a bimodal decay, a result which was interpreted as evidence for the presence of two isomeric ions with different structures. The possibility that the reactive ion is a species with excess internal energy was discounted, when equivalent decay curves were observed in experiments using 10 eV and 70 eV electron impact ionization energy. In dramatic contrast, the molecular ions derived from the [2 + 2] dimer fail to react; apparently the ion population resulting in this experiment is homogeneous [404].

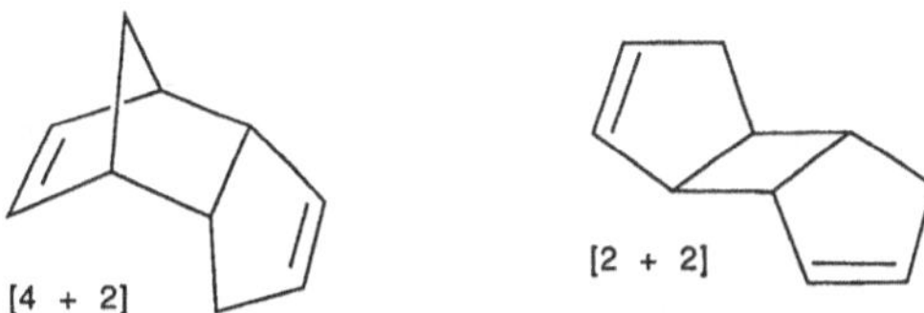

A recent example in this series, the tetramethyleneethane radical cation (**151**), was characterized by both ESR and ENDOR spectroscopy. This species was obtained upon γ-irradiation of dicyclopropenylidene, a rather thorough reorganization of the starting material. It contains two allyl functions, which are linked in the position (2,2′) of minimal spin and charge densities. The ESR spectrum documents the presence of two sets of four equivalent protons (Fig. 30) with isotropic coupling constants of 8.05 and 7.16 G for the *exo*- and *endo*-hydrogens, respectively. INDO calculations suggest that the singly occupied molecular orbital of **151** is a linear combination of two nonbonding allyl MOs. The two moieties are rotated by an angle of 90°, minimizing their steric interaction [405]. On the other hand, a three-dimensional conjugation analogous to spiroconjugation is unlikely because of the distance of ≧3.0 Å between the allyl termini.

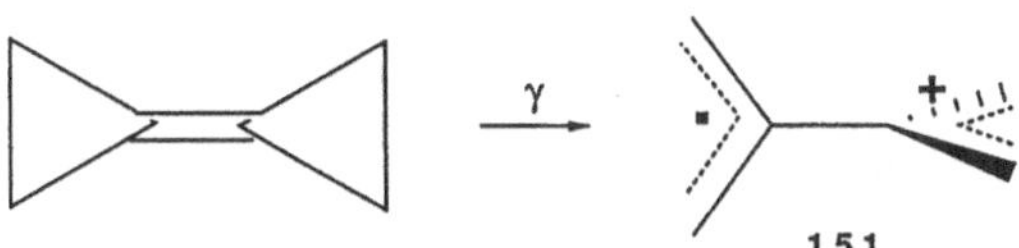

In the second group of radical cations to be discussed here, spin and charge are stabilized in two separate fragments of the molecule. In early examples of this family the charge was stabilized by diaryl or methoxy substitution. We mention the ring-opened trimethylenemethane radical cations (**152**), the various di-

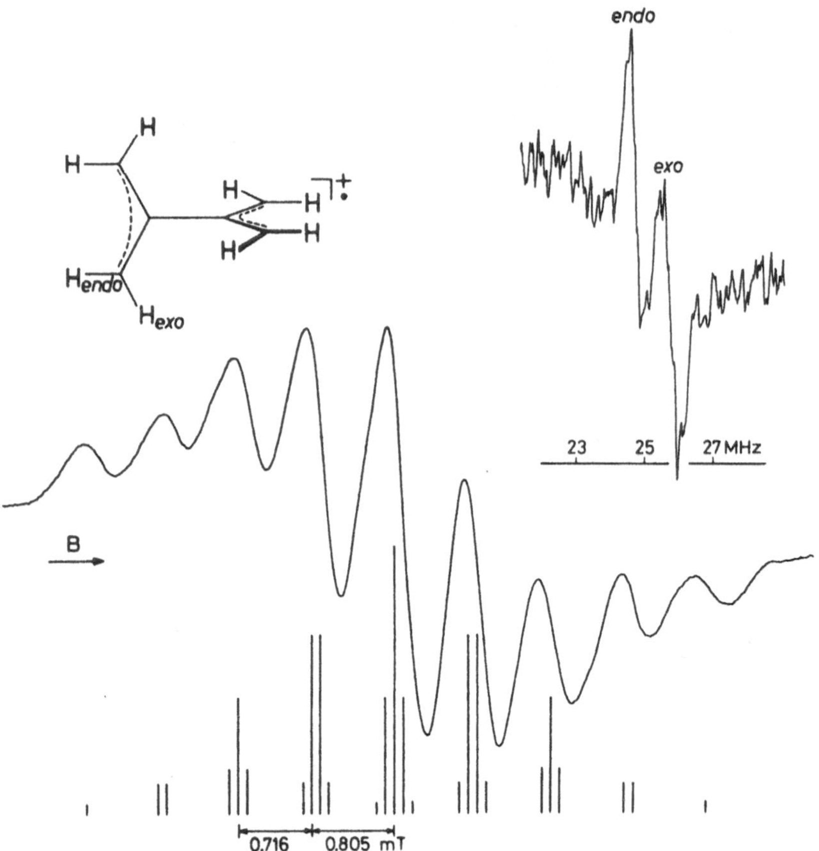

Fig. 30. ESR spectrum of tetramethyleneethane radical cation in a CF_3CCl_3 matrix at 140 K with ^1H ENDOR signals observed above the free proton frequency as an inset; a stick diagram of the ESR spectrum is shown on the bottom (Reprinted by permission)

arylfulvene radical cations (**154**), or the interesting species **155** derived from a 5-methylene-bicyclo[2.2.0]hex-2-ene system.

The photosensitized electron transfer reaction of 1,1-diaryl-2-methylene-cyclopropanes results in the rapid equilibration of the *exo*-methylene and the secondary cyclopropane carbons, whereas the third possible isomer is not formed (Sect. 4.4). These findings were interpreted as evidence for a ring-opened trimethylenemethane radical cation (**152**). The observed rearrangement can be explained if, upon electron return, the diarylmethylene group would couple with either of the allyl termini [406]. This assignment was confirmed by CIDNP results, which clearly indicate that spin and charge are localized in separate π-systems, the spin in an allyl group and the charge in a diarylmethylene moiety [407]. To explain the lack of any polarization and the absence of significant spin density in the aromatic rings, a perpendicular arrangement of the two groups was suggested. Structure **152** is clearly different from an isomeric structure **153**, derived from 1-(diphenylmethylene)-cyclopropane, which can be considered a "vertical" methylenecyclopropane radical cation [407].

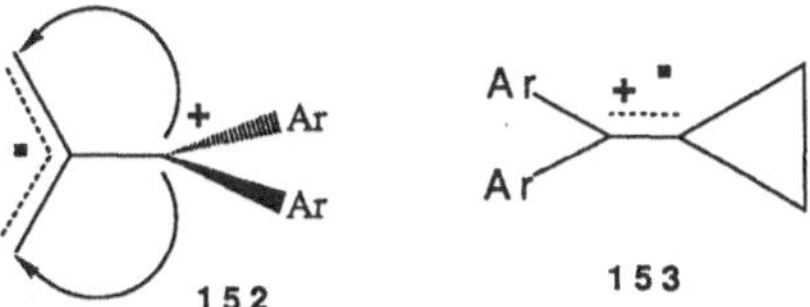

The radical cations of several fulvene derivatives, generated by electron transfer oxidation in solution, also appear to delocalize spin and charge into two separate functions. The respective cyclopentadienyl and methylene functions of structure **154** may be twisted relative to each other, but a perpendicular arrangement is ruled out by their "reluctant" interconversion [408]. The results leave little doubt that there is a clear 1:1 correspondence between the *Z*- and *E*-fulvenes and their radical cations, analogous to the relationship between other pairs of geometric or valence isomers and their radical cations (cf., Fig. 22). For comparison, we mention spectroscopic results which showed that the ethylene radical cation has some degree of twisting [409, 410], and more recent AM1/UHF calculations which suggest that increasing (alkyl-)substitution decreases the degree of twist in olefin radical cations [411].

154

Finally, oxidation of the 6-methoxy derivative (**94**) of 1,2,3,4,6-pentamethyl-5-methylenebicyclo[2.2.0]hex-2-ene gives rise to the unusual radical cation **155** (four methyl groups deleted for clarity) [274]. This species contains a strained cyclobutenyl radical coupled to a methoxyallyl cation; here, the charge is stabilized in the allylic moiety by strong resonance electron donation from the methoxy group. The CIDNP results observed in this system are remarkable, because the pairs of methyl groups in the 1,4- (1.5 ppm) and 2,3-positions (1.0 ppm) are magnetically almost indistinguishable in the parent molecule, but dramatically different in the radical cation (Fig. 31) [274].

155

In summary, a variety of interesting bifunctional radical cation structures have been established. At least one species (**152**) features an orthogonal arrangement of spin and charge, whereas another (**155**) clearly meets the definition of a distonic radical cation. The fulvene radical cations (**154**), on the other hand, are bifunctional and may deviate to some extent from planarity; however, they appear to be ordinary olefin radical cations rather than perpendicular ones.

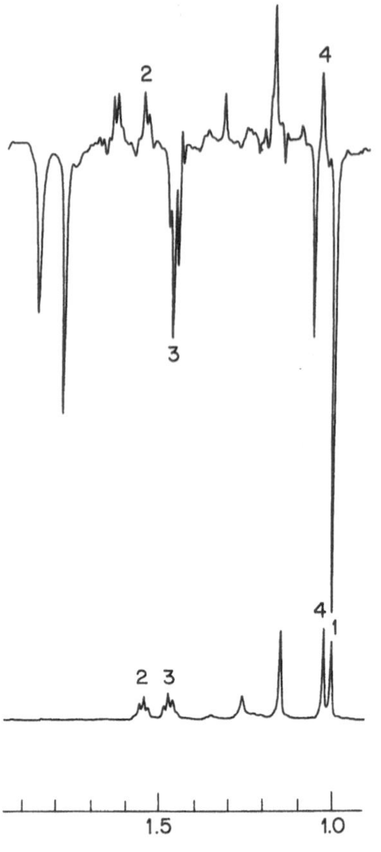

Fig. 31. ^{1}H CIDNP spectra (90 MHz) observed during the photoreaction of chloranil with the 6-methoxy-5-methylene-1,2,3,4,6-pentamethyl-bicyclo[2.2.0]hex-2-ene (**94**) (0.02 M each in acetone-d_6). The *bottom trace* is a dark spectrum

5.6 Cyclic Conjugation In Radical Cations – The Question of Homoaromaticity

In the previous section we encountered two factors favoring particular radical cation structures: relief of ring strain and stabilization by resonance, conjugation, or hyperconjugation. Upon one-electron oxidation, the ring strain inherent in neutral diamagnetic molecules may be relieved, partially or largely, by rupture of a strategic bond. The release of strain is recognized generally as a driving force for the reactions of ground states and intermediates alike. However, both the relief of ring strain and the extension of conjugated systems are subject to restrictions. For example, bifunctional structures (zwitterions, biradicals) are generally less stable on the ground state energy surface than their spin- or charge-paired alternatives. In contrast, several radical cations have bifunctional structures as energy minima. We suggest that a similar divergence is found between the relevance of cyclic homoconjugative (homoaromatic) stabilization in neutral diamagnetic systems and radical cations.

The concept of homoaromaticity was introduced by Winstein to account for the relative stability of molecules in which the cyclic conjugation of π orbitals is interrupted by an aliphatic fragment [412]. This stabilizing force is not considered important in neutral diamagnetic compounds [413], is being debated for anions [414], has been suggested to be significant in cations [415], and has been invoked expressedly for at least one radical anion (**156**), derived from bicyclo[6.1.0]nonatriene [416, 417]. In the following section we discuss radical cations for which we will consider homo- or bis-homoaromatic character.

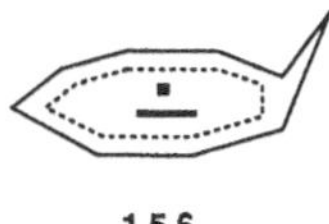

156

We are, of course, aware of the strict Hückel definition linking aromaticity to the presence of 4n + 2 π-electrons. On the other hand, we note that the radical cations discussed below have SOMOs, which have one electron less than the Hückel requirement and which show cyclic homoconjugation, even though the parent molecules assume alternative structures. In view of the precedence established for the radical anion (**156**) of bicyclo[6.1.0]nonatriene [416, 417], in order to emphasize the dramatic difference between the structures of these radical cations and their precursors, and for reasons of convenience, we will refer to these species as homo- or bis-homoaromatic. The paradigm of these remarkable species is found in the barbaralane system.

Barbaralane and its derivatives are systems with double-well potentials undergoing rapid degenerate Cope rearrangements [418]. In fact, the closely related semibullvalene has so low a barrier to interconversion that its rearrangement is fast even at −150 °C [419]. Yet, the cyclic conjugated structure with 6 π-electrons remains a transition structure. Theoretical considerations have been used to propose substitution patterns which may change semibullvalene to a (single-well) bis-homoaromatic 6 π-electron system [420–422]. Similarly, the barbaralyl cation has been studied as a potential single minimum system [423]. However, to date no conclusive evidence has been reported for a closed-shell bridged bicyclooctadienediyl.

157

In contrast, CIDNP results indicate that the radical cations of barbaralane (**157**, X = C=O) and semibullvalene (**157**, X = −) correspond to the elusive structure type with a single minimum [391, 424]. The spin density resides primarily on the termini (C-2,4,6,8) of the twin allyl moieties, whereas the remaining (internal) carbons of the 5 π-electron perimeter have negative spin density. This spin density distribution reflects the coefficients of orbital **158**, the HOMO of a bis-homoaromatic structure (Fig. 32) [424]. More recently, ESR results have confirmed this assignment [392, 393].

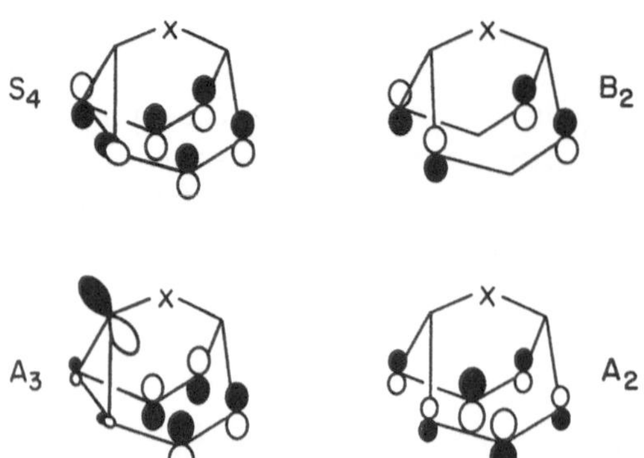

Fig. 32. High lying molecular orbitals of barbaralane and potential SOMOs of its radical cation

As mentioned earlier, appropriate substitution may change the ordering of high lying orbitals and, accordingly, the structure of the radical cations derived from a given ring system. For the bridged bicyclooctadienes, such a change in structure can be brought about by introducing an ethylidene bridge (**157**, X = $>C=CH_2$). The ethene FMO can interact favorably only with the antisymmetrical cyclic conjugated HOMO (**159**). Loss of an electron from this orbital gives rise to a radical cation similar to those of barbaralane and semibullvalene, except for a shift in spin (and charge) density from the terminal to the internal carbons (C-3,7) [425].

Another bis-homoaromatic radical cation is obtained upon one-electron oxidation of 9-methylenebicyclo[4.2.1]nonatriene (**160**). Because of the nature of its HOMO, it is reasonable to assume that this system initially forms a radical cation (**161**) with spin and charge restricted to the butadiene moiety. This intermediate undergoes rapid intramolecular cycloaddition to generate a bis-homoheptafulvene radical cation (**162**) formally derived from a highly strained tetracyclic structure. The existence of **162** is all the more remarkable as the parent hydrocarbon (**163**) had withstood all attempts of generation. Only at −50 °C can **163** be observed by NMR; at temperatures as low as −20 °C it rearranges within seconds to the bicyclic isomer (**160**). An activation energy ≤ 15 kcal/mol is suggested by the temperature dependence of the rearrangement (Fig. 33) [425].

We ascribe the increased stability of **162** to its bis-homoaromatic nature and to the fact that the increase in strain relative to **161** is minimized, as the two pivotal cyclopropane bonds are only partially formed. The cyclization of **161** must be fast even at −50 °C; the barrier is estimated to lie near 5 kcal/mol. The case of cycloaddition is surprising since it is formally of the [3+2]-type and, therefore, symmetry forbidden (vide supra). The driving force for the conversion **160** → **161** lies in the bis-homoaromatic stabilization of **161**, and the conversion must have an early transition state [425].

The final radical cation to be discussed in this section is derived from spirofluorenebicyclo[6.1.0]nonatriene (**164**, R−R = fluorenylidene). This system gives rise to an intermediate (**165**) in which a fluorenyl radical and a homocyclo-

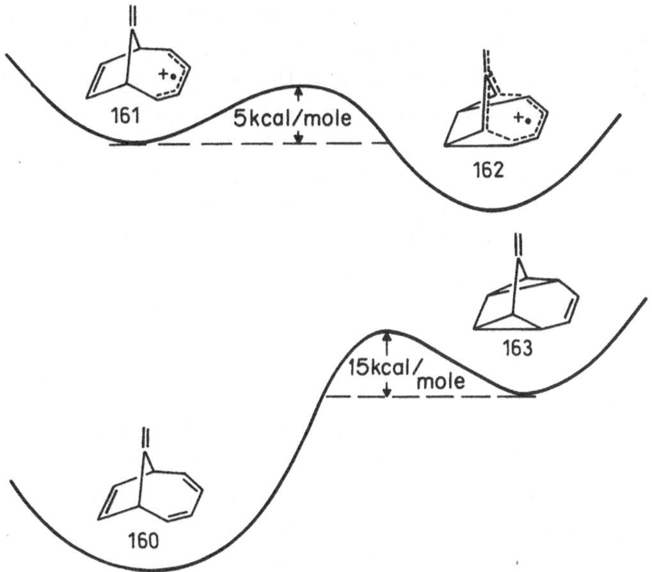

Fig. 33. Schematic comparison of relative energies between methylenebicyclo[4.2.1]-nonatriene and its quadricyclic valence isomer and the corresponding radical cations

heptatrienyl cation are connected with a single link. The limited delocalization of spin and charge is ascribed to the fact that a fluorenyl cation (4n π-electrons) would be antiaromatic whereas the cyclooctatrienyl cation is the paradigm of homoaromaticity. However, in spite of its perceived stabilization and in spite of an inversion barrier near 18 kcal mol^{-1}, the singly linked species is but a fleeting intermediate. Under the conditions of its formation it rapidly rearranges to the barbaralane radical cation, **166**. This conversion once again underscores the significant stabilization of this bishomobenzene radical cation type.

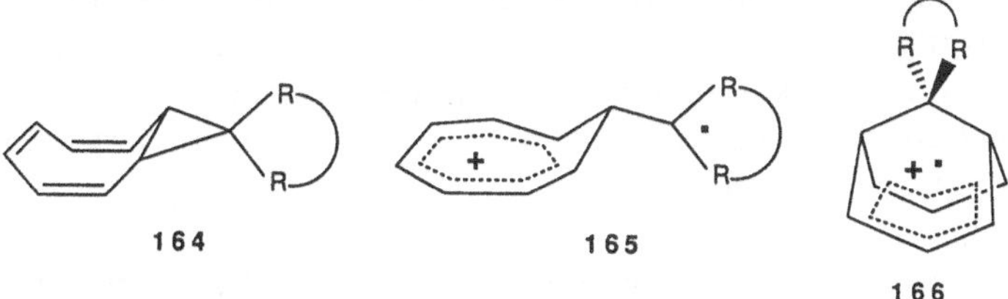

In summary, we emphasize that homoaromatic stabilization appears to be more important for radical cations than for their neutral diamagnetic precursors. There are several reasons why radical cations may assume cyclic conjugated structures with 4n + 1 π-electrons, one electron shy of the magic 4n + 2 π-electrons which achieve aromaticity, when the parent molecules opt for a less delocalized nonaromatic structure. The principal cause might lie in the strength of the carbon–carbon single bond, which generally disfavors biradical or zwitterionic structures. This major obstacle to cyclic conjugation is removed with one-electron oxidation. In addition, radical cations apparently have the ability to benefit from cyclic conjugation at slightly longer distances so that the strain energy is less of a factor. We note that similar to the radical anion derived from bicyclo[6.1.0]nonatriene all radical cations discussed above have $4n + 1$ π-electrons. To our knowledge, no system with $4n - 1$ π-electrons shows cyclic homoconjugation. It will be interesting to pursue the question whether systems with $4n + 1$ π-electrons indeed are different in principle from those having $4n - 1$ π-electrons. This problem may hold the key to the concept of homoaromaticity and needs to be pursued.

6 Perspective

The rich variety of radical cation reactions and the diversity of structure types portrayed in this article may lead the reader to the conclusion that this field has exhausted its growth potential. However, this impression would be premature; on the contrary, radical cation chemistry remains in a phase of rapid development and can be expected to retain a high level of attention for the foreseeable future.

It is noteworthy that the radical cation field owes the significant advances accomplished in the last decade mainly to experimental studies, whereas theoretical calculations have been largely limited to an auxiliary role. For example, most calculations of hyperfine coupling constants have utilized semiempirical methods with appropriate adhoc assumptions. Similarly, the unprecedented structure types, which have given new direction to the radical cation field, were recognized on the basis of experimental findings, and not as an outgrowth of theoretical predictions. One possible exception involves the tricyclo[3.1.0.0^{2,4}]hexane-3,6-diyl radical cation, which "materialized" in an attempt to evaluate an assumed sandwich dimer. Although theorists have been slow in examining any of the emerging novel structure types, it is safe to predict an increasing level of activity in the area of ab initio calculations along with continuing efforts to document novel structure types and delineate radical cation reactivities.

Considering the importance of radical cations and anions, it is surprising that many advanced organic chemistry texts essentially ignore these intermediates. It is the hope of this author that the advances portrayed in this article and the gains that are projected will contribute to a more general acceptance of radical ions as important intermediates.

Acknowledgements. It is a pleasure to acknowledge the contributions of many colleagues who helped shape my understanding of the material presented here. First and foremost, I thank my collaborators, especially M. L. Schilling, C. J. Abelt, and R. S. Hutton for their invaluable, sustained and diligent contributions. I am indebted also to a group of colleagues, including D. A. Arnold, T. Bally, J. Dinnocenzo, X.-M. Du, P. G. Gassman, R. C. Haddon, G. Jones, II, K. Krogh-Jespersen, D. Lemal, J. R. Miller, T. Miyashi, T. Mukai, W. A. Mulac, K. Raghavachari, C. C. Wamser, D. Whitten, and F. Williams, who collaborated on individual facets of my work or provided valuable samples for it, engaged in enlightening discussions, provided helpful criticism as well as stimulating disagreements, shared results prior to publication, or confirmed our assignments by alternative techniques. I apologize to those colleagues whose work I may have omitted, distorted, or represented poorly, and thank Enrico Radix-Ruber for useful soliloquy [426]. This paper is dedicated to Professor Albert Weller, a pioneer in donor-acceptor interactions and electron transfer [427], whose elegant work in the area of electron transfer helped stimulate my interest in radical ion reactions and structures.

7 References

1. Blankenship RE, Parson WW (1978) Ann Rev Biochem 47: 635
2. Kaiser ET, Kevan L (eds) (1968) Radical Ions, Interscience, New York
3. Forrester RA, Ishizu K, Kothe G, Nelson SF, Ohya-Nishiguchi H, Watanabe K, Wilker W (1980) Organic Cation Radicals and Polyradicals in Landolt Börnstein, Numerical Data and Functional Relationships in Science and Technology, Volume IX, Part d2, Springer Berlin, Heidelberg, New York
4. Shida T (1988) Electronic Absorption Spectra of Radical Ions, Elsevier, Amsterdam. In addition, approximately 1500 papers on the chemistry and spectroscopy of organic radical cations have appeared over the last ten years
5. Bard AJ, Ledwith A, Shine HJ (1976) Advances Phys Org Chem 13: 155
6. Hammerich O, Parker VD (1984) Advances Phys Org Chem 20: 55
7. Shida T, Haselbach E, Bally T (1984) Accounts Chem Res 17: 180
8. Nelsen SF (1987) Accounts Chem Res 20: 269
9. Roth HD (1987) Accounts Chem Res 20: 343
10. Bauld N, Bellville DJ, Hairirchian B, Lorenz KT, Pabon RA, Reynolds DW, Wirth DD, Chiou H-S, Marsh BK (1987) Accounts Chem Res 20: 371
11. Roth HD, Schilling ML, Wamser CC (1984) J Am Chem Soc 106: 5023
12. Green S (1981) Rev Phys Chem 32: 103
13. Wudl F (1984) Accounts Chem Res 17: 227
14. Maroulis AJ, Shigemitsu Y, Arnold DR (1978) J Am Chem Soc 100: 535
15. Kornblum N (1975) Angew Chem Int Ed Engl 17: 180
16. Laurent A (1835) Ann Chim 59: 367; (1836) Liebigs Ann Chem 17: 89
17. Kehrmann F, Speitel J, Grandmougin E (1914) Ber Dtsch Chem Ges 47: 2976
18. Kehrmann F, Diserens L (1915) Ber Dtsch Chem Ges 48: 318
19. Hirschon GM, Gardner DM, Fraenkel GK (1953) J Am Chem Soc 75: 4115
20. Hoijtink GJ, Weijland WP (1957) Rec Trav Chim 76: 836
21. Kon H, Blois MS (1958) J Chem Phys 28: 743
22. Meyer K (1910) Ber Dtsch Chem Ges 43: 161
23. Hilpert S, Wolf L (1913) Ber Dtsch Chem Ges 46: 2215
24. Kainer H, Hausser KH (1957) Chem Ber 86: 1563
25. Lewis IC, Singer LS (1965) J Chem Phys 43: 2712
26. Lewis IC, Singer LS (1966) J Chem Phys 44: 2082
27. Wurster C (1879) Ber Dtsch Chem Ges 12: 522
28. Wurster C, Schobig E (1879) Ber Dtsch Chem Ges 12: 1807
29. Wurster C (1879) Ber Dtsch Chem Ges 12: 2071
30. Wieland H (1907) Ber Dtsch Chem Ges 40: 4260
31. Wieland H, Wecker E (1910) Ber Dtsch Chem Ges 43: 699
32. Richardson TJ, Tanzella FL, Bartlett N (1986) J Am Chem Soc 108: 4937
33. Dinnocenzo JP, Bannash TE (1986) J Am Chem Soc 108: 6063
34. Corbin DR, Eaton DF, Ramamurthy V (1988) J Am Chem Soc 110: 4848
35. Ramamurthy V, Caspar JV, Corbin DR (1991) J Am Chem Soc 113: 594

36. Komatsu T, Lund A (1972) J Phys Chem 76: 1727
37. Knight LB (1986) Acc Chem Res 19: 313
38. Qin X-Z, Trifunac AD (1990) J Phys Chem 94: 4751
39. Yoshida K (1984) Electrooxidation in Organic Chemistry, Wiley, New York
40. Bard AJ, Faulkner LR (1980) Electrochemical Methods, Wiley, New York
41. Rao PS, Nash JR, Guarino JP, Ronayne MR, Hamill WH (1962) J Am Chem Soc 84: 500
42. Guarino JP, Ronayne MR, Hamill WH (1962) J Am Chem Soc 84: 4230
43. Guarino JP, Hamill WH (1962) Radiation Res 17: 379
44. Ledwith A (1972) Accounts Chem Res 5: 133
45. Mattes SL, Farid S (1983) Org Photochem 6: 233
46. Mattay J (1987) Ang Chem Int Ed Engl 26: 825
47. Knibbe H, Rehm D, Weller A (1969) Ber Bunsenges Phys Chem 73: 839
48. Saeva FD, Olin GR (1980) J Am Chem Soc 102: 299
49. Arnold DR, Maroulis AJ (1976) J Am Chem Soc 98: 5931
50. Park S-M, Caldwell RA (1977) J Electrochem Soc 124: 1859
51 McCullough JJ, Miller RC, Fung D, Wu W-S (1975) J Am Chem Soc 97: 5942
52. Mann CK, Barnes KK (1970) Electrochemical Reactions in Nonaqueous Systems
53. Kavarnos GJ, Turro NJ (1986) Chem Rev 86: 401
54. Gersdorf J, Mattay J, Görner H (1987) J Am Chem Soc 109: 1203
55. Howe I, Williams DH, Bowen RD (1981) Mass Spectrometry Principles and Applications, Wiley, New York
56. Lehman TA, Bursey MM (1976) Ion Cyclotron Resonance Spectrometry
57. Turner DW, Baker AD, Baker C, Brundle CR (1970) Molecular Photoelectron Spectroscopy, Wiley-Interscience, New York
58. Norrish RGW, Porter G (1949) Nature (London) 164: 658
59. Porter G (1950) Proc Roy Soc 200: 284
60. Novak JR, Windsor RW (1967) J Chem Phys 47: 3075
61. Porter G, Topp MR (1968) Nature 220: 1228
62. Miller A (1968) Z Naturforsch A 23: 946
63. Rentzepis PM (1968) Photochem Photobiol 8: 579
64. Rentzepis PM (1968) Chem Phys Lett 2: 117
65. Dutton PL, Kaufmann KJ, Chance B, Rentzepis PM (1975) FEBS Lett. 60: 275
66. Rockley MG, Windsor MW, Cogdell RJ, Parson WW (1975) Proc Natl Acad Sci USA 72: 2251
67. Shank CV, Fork RL, Yen RT (1982) Springer Ser Chem Phys 23: 2
68. Bondybey VE, Miller TA (1983) Molecular Ions, Spectroscopy, Structure, and Chemistry, North Holland, Amsterdam
69. Carrington A, McLachlan AD (1967) Introduction to Magnetic Resonance, Harper & Row, New York
70. Wertz JE, Bolton JR (1972) Electron Spin Resonance: Elementary Theory and Practical Applications, McGraw-Hill, New York
71. McLauchlan KA, Stevens DG (1987) Accounts Chem Res 21: 54
72. Fessenden RW, Schuler RH (1963) J Chem Phys 39: 2147
73. Buckley CD, McLauchlan KA (1985) Mol Phys 54: 1
74. McLauchlan KA (1985) Chem Britain 21: 825
75. Adrian FJ (1986) Rev Chem Intermed 7: 173
76. McLauchlan KA, Stevens DG (1986) Mol Phys 57: 223
77. Basu S, McLauchlan KA, Sealy GR (1983) J Phys E: Sci Instrum 16: 767
78. Smaller B, Remko JR, Avery EC (1968) J Chem Phys 48: 5147
79. Trifunac AD, Thurnauer MC, Norris JR (1978) Chem Phys Lett 57: 471
80. Trifunac AC, Norris JR, Lawler RG (1979) J Chem Phys 71: 4380
81. El Sayed MA (1972) in MTP Intern Rev Science, Spectrosc 119
82. Frankevich EL, Pristupa AI, Lesin VI (1977) Chem Phys Lett 47: 304
83. Anisimov OA, Grigoryants VM, Molchanov VK, Molin YuN (1979) Chem Phys Lett 66: 265
84. Trifunac AD, Smith JP (1980) Chem Phys Lett 73: 94

85. McConnel HM, Holm CH (1957) J Chem Phys. 27: 314; (1958) ibid. 27: 314
86. Eaton DR, Josey AD, Phillips WD, Benson RE (1962) Mol Phys 5: 407
87. de Boer E, van Willigen H (1967) Progr Nucl Magn Resonance Spectr 2: 111
88. Bargon J, Fischer H, Johnson U (1967) Z Naturforsch A 22: 1551
89. Ward HR, Lawler RG (1967) J Am Chem Soc 89: 5518
90. Roth HD, Lamola AA (1972) J Am Chem Soc 94: 1013
91. Roth HD, Lamola AA (1974) J Am Chem Soc 96: 6270
92. Lamola AA, Roth HD, Schilling MLM, Tollin G (1975) Proc Nat Acad Sci USA 72: 3265
93. Closs GL (1974) Adv Magn Reson 7: 157
94. Kaptein R (1975) Adv Free Radical Chem 5: 319
95. Adrian FJ (1979) Rev Chem Intermed 3: 3
96. Freed JH, Pedersen JB (1976) Adv Magn Reson 8: 2
97. Adrian FJ, Vyas HM, Wan JKS (1976) J Chem Phys 65: 1454
98. Adrian FJ (1977) in Chemically Induced Magnetic Polarization, Muus LT, Atkins PW, McLauchlan KA, Pedersen JB (eds) Reidel, Dordrecht, 36
99. Schilling MLM, Hutton RS, Roth HD (1977) J Am Chem Soc 99: 7792
100. Roth HD, Schilling MLM, Hutton RS (1979) J Chem Phys 71: 610
101. Closs GL, Czeropski MS (1978) J Am Chem Soc 99: 6127
102. Closs GL, Czeropski MS (1978) Chem Phys Lett 53: 321
103. Kaptein R (1978) Nature (London) 274: 293
104. Garrsen GJ, Kaptein R, Schoenmakers JGG, Hilbers CW (1978) Proc Natl Acad Sci USA 75: 5281
105. Toriyama K, Nunone K, Iwasaki M (1982) J Chem Phys 77: 5891
106. Shiotani M, Nagata Y, Sohma J (1984) J Phys Chem 88: 4078
107. Rhodes CJ (1988) J Am Chem Soc 110: 8567
108. Arnold A, Gerson F (1990) J Am Chem Soc 112: 2027
109. Williams F, Guo Q-X, Nelsen S (1990) J Am Chem Soc 112: 2028
110. Jones G, Becker WG, Chiang S-H (1982) J Photochem 19: 245
111. Evans TR, Wake RW, Sifain MN (1973) Tetrahedron Lett. 701
112. Okada K, Hisamitsu K, Miyashi T, Mukai T (1982) J Chem Soc, Chem Commun 974
113. Hasegawa E, Mukai T, Yanagi K unpublished results quoted by Lewis FD (1988) Photoinduced Electron Transfer Fox MA, Chanon M (eds) Part C, 1, Elsevier
114. Farid S, Hartman SE, Evans TR (1975) The Exiplex, Academic Press, New York
115. Evans TR, Wake RW, Jaenicke O (1975) The Exiplex, Academic Press, New York
116. Kuwata S, Shigemitsu Y, Odaira Y (1973) 21: 3803
117. Crellin RA, Ledwith A (1975) Macromolecules 8: 93
118. Yasuda M, Pac C, Sakuri H (1980) Bull Chem Soc Japan 53: 502
119. Kajima M, Sakuragi H, Tokumaru N (1981) Tetrahedron Lett 2889
120. Crellin RA, Lambert MC, Ledwith A (1970) J Chem Soc, Chem Comm 682
121. Roth HD, Hutton RS (1990) J Phys Org Chem 3: 119
122. Neunteufel RA, Arnold DR (1973) J Am Chem Soc 95: 4080
123. Mattes SL, Farid S (1980) unpublished results
124. Courtneidge JL, Davies AG (1987) Accounts Chem Res 20: 90
125. Courtneidge JL, Davies AG (1985) J Chem Soc, Chem Comm 1092
126. Shiotani M, Ohta K, Nagata Y, Sohma J (1985) J Am Chem Soc 107: 2562
127. Fox MA, Campbell KA, Hünig S, Berneth H, Maier G, Schneider KA, Malsch KD (1982) J Org Chem 47: 3408
128. Sheng D, Pappas RS, Chen G-F, Guo QX, Wang JT, Williams F (1989) J Am Chem Soc 111: 8759
129. Libman J (1976) J Chem Soc Chem Commun 361
130. Jones CR, Allman BJ, Mooring A, Spahic B (1983) J Am Chem Soc 105: 652
131. Pabon RA, Bellville DJ, Bauld NL (1983) J Am Chem Soc 105: 5158
132. Calhoun GC, Schuster GB (1984) J Am Chem Soc 106: 6870
133. Schutte R, Freeman GR (1969) J Am Chem Soc 91: 3715
134. Penner TL, Whitten DG, Hammond GS (1970) J Am Chem Soc 92: 2861
135. Roth HD, Schilling ML, Abelt CJ (1986) Tetrahedron 42: 6157

136. Roth HD, Schilling ML, Abelt CJ (1986) J Am Chem Soc 108: 6098
137. Gassman PG, Singleton DA (1984) J Am Chem Soc 106: 7993
138. Pabon RA, Bellville DJ, Bauld NL (1983) J Am Chem Soc 105: 5158
139. Reynolds DW, Lorenz KT, Chiou H-S, Bellville DJ, Pabon RA, Bauld NL (1987) J Am Chem Soc 109: 4960
140. Arnold DR, Wong PC, Maroulis AJ, Cameron TS (1980) Pure Appl Chem 52: 2609
141. Rentzepis PM, Steyert DW, Roth HD, Abelt CJ (1985) J Phys Chem 89: 3955
142. Klett M, Johnson RP (1985) J Am Chem Soc 107: 6615
143. Maroulis AJ, Shigemitsu Y, Arnold DR (1978) J Am Chem Soc 100: 535
144. Arnold DR, Snow MS (1988) Can J Chem 66: 3012
145. Eriksen J, Foote CS (1980) J Am Chem Soc 102: 6083
146. Schaap AP, Zaklika KA, Kaskar B, Fung LW-M (1980) J Am Chem Soc 102: 389
147. Mattes SL, Farid S (1980) J Chem Soc Chem Commun 457
148. Spada LT, Foote CS (1980) J Am Chem Soc 102: 393
149. Schaap AP, Siddiqui S, Prasad G, Magsudur-Rahman AFM, Oliver JP (1984) J Am Chem Soc 106: 6087
150. Miyashi T, Konno A, Takahashi Y (1988) J Am Chem Soc 110: 3676
151. Eriksen J, Foote CS, Parker TL (1977) J Am Chem Soc 99: 6455
152. Barton DHR, Leclerc G, Magnus PD, Menzies ID (1972) J Chem Soc Chem Commun 447
153. Nelson SF, Akaba R (1981) J Am Chem Soc 103: 1096
154. Clennan EL, Simmons W, Almgren CW (1981) J Am Chem Soc 103: 2098
155. Haynes RK (1978) Aust J Chem 31: 121
156. Haynes RK (1978) Aust J Chem 31: 131
157. Roth HD, Schilling MLM (1979) J Amer Chem Soc 101: 1898
158. Roth HD, Schilling MLM (1980) J Amer Chem Soc 102: 4303
159. Lewis FD, Petisce JR, Oxman JD, Nepras MJ (1985) J Amer Chem Soc 107: 203
160. Roth HD, Abelt CJ (1985) J Amer Chem Soc 107: 6814
161. Taylor GN, unpublished results, quoted in Roth HD (1973) Mol Photochem 5: 91
162. Hub W, Kluter U, Schneider S, Dörr F, Oxman JD, Lewis FD (1984) J Phys Chem 88: 2308
163. Hayashi H, Nagakura S (1978) Chem Phys Lett 53: 201
164. Arnold DR, Wong PC (1979) J Am Chem Soc 101: 1894; Wong PC, Arnold DR (1979) Can J Chem 57: 1037
165. Lewis FD, Simpson JT (1979) J Phys Chem 83: 2015
166. Adams BK, Cherry WR (1981) J Am Chem Soc 103: 6904
167. Goodman JL, Peters KS (1985) J Am Chem Soc 107: 1441
168. Goodman JL, Peters KS (1985) J Am Chem Soc 107: 5459
169. Ottolenghi M (1973) Accounts Chem Res 6: 153
170. Michel-Beyerle ME, Haberkorn R, Bube W, Steffens E, Schröder H, Neusser NJ, Schlag EW, Seidlitz H (1976) Chem Phys 17: 139
171. Schulten CK, Staerk H, Weller A, Werner HJ, Nickel B (1976) Z Phys Chem (Frankfurt am Main) 101: 371
172. Brown-Wensley KA, Mattes SL, Farid S (1978) J Am Chem Soc 100: 4162
173. Bally T, Nitsche S, Roth K, Haselbach E (1985) J Phys Chem 89: 2528
174. Shida T, Kato T, Nosaka Y (1977) J Phys Chem 81: 1095
175. Shida T, Egawa Y, Kubodera H, Kato T (1980) J Chem Phys 73: 5963
176. Cohen SG, Parola A, Parsons GH (1973) Chem Rev 73: 14
177. Roth HD, Manion ML (1975) J Amer Chem Soc 97: 6886
178. Roth HD (1977) in Chemically Induced Magnetic Polarization, Reidel, Dordrecht 39.
179. Peters KS, Freilich SC, Schaeffer CG (1980) J Am Chem Soc 102: 7566
180. Peters KS, Schaeffer CG (1980) J Am Chem Soc 201: 7566
181. Peters KS, Simon JD (1981) J Am Chem Soc 103: 6403
182. Schanze KS, Giannotti C, Whitten DG (1983) 105: 6326
183. Bellas M, Bryce-Smith D, Gilbert A (1967) J Chem Soc, Chem Comm 862
184. Bryce-Smith D, Gilbert A, Manning C (1974) Angew Chem Int Ed Engl 86: 350

185. Roth HD, Schilling MLM unpublished results
186. Ohashi M, Miyake K (1977) Chem Lett 615
187. Ohashi M, Miyake K, Tsujimoto K (1980) Bull Soc Chem Japan 53: 1683
188. Lewis FD, Ho T-I (1980) J Am Chem Soc 102: 1751
189. Lewis FD, Ho T-I (1977) J Am Chem Soc 99: 7991
190. Hub W, Schneider S, Dörr F, Simpson JT, Oxman JD, Lewis FD (1982) J Am Chem Soc 104: 2044
191. Lewis FD (1979) Acc Chem Res 12: 152
192. Lewis FD, Ho T-I, Simpson JT (1982) J Am Soc 104: 1924
193. Schilling MLM, Roth HD, Herndon WC (1980) J Am Chem Soc 102: 4271
194. Bellas M, Bryce-Smith D, Clarke MT, Gilbert A, Klunkin G, Krestonosich S, Manning C, Wilson S (1977) J Chem Soc Perkin Trans 1: 2571
195. Inbar S, Linschitz H, Cohen SG (1981) J Am Chem Soc 103: 1048
196. Monserrat K, Foreman TK, Graetzel M, Whitten DG (1981) J Am Chem Soc 103: 6667
197. Lewis FD, Zebrowski BE, Correa PE (1984) J Am Chem Soc 106: 187
198. Lee LYC, Ci X, Giannotti C, Whitten DG (1986) J Am Chem Soc 108: 175
199. Ci X, Lee LYC, Whitten DG (1987) J Am Chem Soc 109: 2536
200. Ci X, Whitten DG (1987) J Am Chem Soc 109: 7215
201. Kellett MA, Whitten DG (1989) J Am Chem Soc 111: 2314
202. Haugen CM, Whitten DG (1989) J Am Chem Soc 111: 7281
203. Ci X, Whitten DG (1989) J Am Chem Soc 111: 3459
204. Grob CA (1969) Angew Chem Int Ed Engl 8: 535
205. Maslak P, Asel SA (1988) J Am Chem Soc 110: 8260
206. Knight LB, Steadman J, Feller D, Davidson ER (1984) J Am Chem Soc 106: 3700
207. Iwasaki M, Toriyama K, Nunome K (1981) J Am Chem Soc 103: 3591–3592
208. Tabata M, Lund A (1983) Chem Phys 75: 379–388
209. Murov SL, Cole RS, Hammond GS (1968) J Am Chem Soc 90: 2957
210. Murov S, Hammond GS (1968) J Phys Chem 72: 3797
211. Solomon BS, Steel C, Weller AJ (1969) J Chem Soc Chem Comm 927
212. Taylor GN (1971) Chem Phys Lett 10: 355
213. Gassman PG, Yamaguchi R, Koser GF (1978) J Org Chem 43: 4393
214. Gassman PG, Yamaguchi RJ (1979) J Am Chem Soc 101: 1308
215. Rehm D, Weller A (1970) Israel J Chem 8: 259
216. Evans TR (1971) J Am Chem Soc 93: 2081
217. Gassman PG, Olson KD, Walter L, Yamaguchi R (1981) J Am Chem Soc 103: 4977
218. Wiberg KB, Connor HA (1976) J Am Chem Soc 98: 5411
219. Hogeveen H, Volger HC (1967) J Am Chem Soc 89: 2486
220. Gassman PG, Aue DH, Patton DS (1968) J Am Chem Soc 90: 7271
221. Gassman PG, Hershberger JW (1987) J Org Chem 52: 1337
222. Wong PC, Arnold DR (1979) Tetrahedron Lett 2101
223. Roth HD, Schilling MLM (1981) J Am Chem Soc 103: 7210
224. Gassman PG, Mlinarie-Majerski K (1986) J Org Chem 51: 2397
225. Dinnocenzo JP, Schmittel M. (1987) J Am Chem Soc 109: 1561
226. Dinnocenzo JP, Conlon DA (1988) J Am Chem Soc 110: 2324
227. Dinnocenzo JP, Conlon DA (1991) private communication
228. Arnold DR, Humphreys RWR (1979) J Am Chem Soc 101: 2743
229. Roth HD, Schilling MLM (1983) Can J Chem 61: 1027
230. Roth HD, Schilling MLM (1983) J Am Chem Soc 105: 6805
231. Rao VR, Hixson SS (1979) J Am Chem Soc 101: 6458
232. Mizuno K, Ogawa J. (1981) Chem Lett 741
233. Dinnocenzo JP, Todd WP, Simpson TR, Gould IR (1990) J Am Chem Soc 112: 2462
234. Mazzocchi PH, Somich C, Edwards M, Morgan T, Ammon HL (1986) J Am Chem Soc 108: 6828
235. Martini T, Kampmeier LA (1970) Angew Chem 82: 216; Angew Chem Int Ed Engl 9: 236
236. Mizuno K, Kamiyama N, Ichinose N, Otsuji Y (1985) Tetrahedron 41: 2207; Mizuno K, Kamiyama N, Otsuji Y (1983) Chem Lett 477

237. Schaap AP, Lopez L, Anderson SD, Gagnon SD (1982) Tetrahedron Lett 23: 5493
238. Schaap AP, Siddiqui S, Prasad G, Palomino E, Lopez L (1984) Photochem 25: 167
239. Miyashi T, Kamata M, Mukai T (1987) J Am Chem Soc 109: 2780
240. Miyashi T, Takahashi Y, Kameta M, Yokogawa, Ohaku H, Mukai T (1987) Studies Org Chem 31: 363
241. Takahashi Y, Mukai T, Miyashi T (1983) J Am Chem Soc 105: 6511
242. Miyashi T, Takahashi Y, Mukai T, Roth HD, Schilling MLM (1985) J Am Chem Soc 107: 1079
243. Roth HD, Schilling ML, Abelt CJ, Miyashi T, Takahashi Y, Konno A, Mukai T (1988) J Am Chem Soc 112: 5130
244. Miyashi T, Takahashi Y, Konno A, Mukai T, Roth HD, Schilling ML, Abelt CJ (1989) J Org Chem 54: 1445
245. Padwa A, Chou CS, Rieker WF (1980) J Org Chem 45: 455
246. Gassman PG, Carroll GT (1986) Tetrahedron 42: 6201
247. Gassman PG, Olson KD (1982) J Am Chem Soc 104: 3740
248. Abelt CJ, Roth HD, Schilling MLM (1985) J Am Chem Soc 107: 4148
249. Roth HD, Schilling MLM, Gassman PG, Smith JL (1984) J Am Chem Soc 106: 2711
250. Gassman PG, Hay BA (1985) J Am Chem Soc 107: 4075
251. Gassman PG, Hay BA (1986) J Am Chem Soc 108: 4227
252. Wiberg KB, Szeimies G (1968) Tet Letters 1235
253. Woodward RB, Hoffmann R (1971) The Conservation of Orbital Symmetry, Verlag Chemie, Weinheim
254. Gassman PG (1988) Photoinduced Electron Transfer, C: 70
255. Roth HD, Miller JA, Mulac WA (1991) unpublished results
256. Roth HD, Hutton RS (1990) J Phys Org Chem 3: 119
257. Arnold DR, Du X (1989) J Am Chem Soc 111: 7666
258. Rosenthal I, Elad D (1968) Biochem Biophys Res Commun 32: 599
259. Sutherland JC (1977) Photochem Photobiol 25: 435
260. Pac C, Ishitani O (1988) Photochem Photobiol 48: 767
261. Lamola AA (1972) Mol Photochem 4: 107
262. Kemmink J, Eker APM, Kaptein R (1986) Photochem Photobiol 44: 137
263. Young T, Nieman R, Rose SD (1990) Photochem Photobiol 52: 661
264. Van Camp JR, Young T, Hartman RF, Rose SD (1987) Photochem Photobiol 45: 365
265. Kim ST, Hartman RF, Rose SD (1990) Photochem Photobiol 52: 789
266. Miyashi T, Wakamatsu K, Akiya T, Kikuchi K, Mukai T (1987) J Am Chem Soc 109: 5270
267. Takahashi Y, Kochi JK (1988) Chem Ber 121: 253
268. Kawamura Y, Thurnauer MC, Schuster GB (1986) Tetrahedron 42: 6195
269. Brauer BE, Thurnauer MC (1987) Chem Phys Lett 133: 3
270. Aebischer JN, Bally T, Roth K, Haselbach E, Gerson F, Qin X-Z (1989) J Am Chem Soc 111: 5270
271. Bellville DJ, Chelsky R, Bauld NL (1982) J Comput Chem 3: 548
272. Evans TR, Wake RW, Sifain MM (1973) Tetrahedron Lett 701
273. Peacock NJ, Schuster GB (1983) J Amer Chem Soc 105: 3632
274. Roth HD, Schilling ML, Wamser CC (1984) J Am Chem Soc 106: 5023
275. Wamser CC, Ngo DD, Rodriguez MJ, Shama SA, Tran TL (1989) J Am Chem Soc 111: 2162
276. Shida T, Nosaka Y, Kato T (1978) J Phys Chem 82: 695
277. Bally T, Haselbach E, Lanyiova Z (1978) Helv Chim Acta 61: 2488
278. Forster P, Gschwind R, Haselbach E, Klemm U, Wirz J (1980) Nouv J Chim 4: 365
279. Haselbach E, Klemm U, Gschwind R, Bally T, Chassot L, Nitsche S (1982) Helv Chim Acta 65: 2464
280. Bally T, Neuhaus L, Nitsche S, Haselbach E, Janssen J, Lüttke W (1983) Helv Chim Acta 66: 1288
281. Nelsen SF, Buschek JM (1974) J Amer Chem Soc 96: 6424
282. Nelsen SF, Haselbach E, Gschwind R, Klemm U, Lanyiova Z (1978) J Amer Chem Soc 100: 4367

283. Yates BF, Bouma J, Radom L (1984) J Amer Chem Soc 106: 5805
284. Gleiter R (1980) Top Curr Chem 86: 197
285. Haselbach E (1970) Chem Phys Lett 7: 428
286. Rowland CG (1971) Chem Phys Lett 9: 169
287. Collins JR, Gallup GA (1982) J Am Chem Soc 104: 1530
288. Wayner DDM, Boyd RJ, Arnold DR (1983) Can J Chem 61: 2310
289. Wayner DDM, Boyd RJ, Arnold DR (1985) Can J Chem 63: 3283
290. Du P, Hrovat DA, Borden WT (1986) Chem Phys Lett 123: 337
291. Du P, Hrovat DA, Borden WT (1988) J Am Chem Soc 110: 3405
292. Krogh-Jespersen K, Roth HD (1991) J Am Chem Soc 113:
293. Qin X-Z, Snow LD, Williams F (1984) J Am Chem Soc 106: 7640
294. Iwasaki M, Toriyama K, Nunone K (1984) J Chem Soc Chem Comm 202
295. Shida T, Takemura Y (1983) Radiat Phys Chem 21: 157
296. Qin X-Z, Williams F (1984) Chem Phys Lett 112: 79
297. Qin X-Z, Williams F (1986) Tetrahedron 42: 6301
298. Dewar MJS, Dougherty RC (1975) The PMO Theory of Organic Chemistry, Plenum Press, NY
299. Haddon RC, Roth HD (1984) Croatica Chemica Acta 57: 1165
300. Roth HD, Schilling MLM, Schilling FC (1985) J Am Chem Soc 107: 4152
301. Truesdale EA, Hutton RS (1979) J Am Chem Soc 101: 6476
302. Roth HD, Schilling MLM, Hutton RS, Truesdale EA (1983) J Am Chem Soc 105: 153
303. Kawamura T, Tsumura M, Yokomichi Y, Yonegawa T (1977) J Am Chem Soc 99: 8251
304. Fessenden RW (1967) J Phys Chem 71: 74
305. Ohta K, Nakatsuji H, Hirao K, Yonezawa T (1980) J Chem Phys 73: 1770
306. Behrens G, Schulte-Frolinde D (1973) Angew Chem 85: 993
307. Whiffen DH (1963) Mol Phys 6: 223
308. Fessenden RW, Schuler RH (1963) J Chem Phys 38: 773
309. Symons MCR (1978) Chem Phys Lett 117: 381
310. Gross ML, McLafferty FW (1971) J Am Chem Soc 93: 1267
311. Sieck LW, Golden R Jr, Ausloos P (1972) J Am Chem Soc 94: 7157
312. van Velzen PNT, van der Hart W (1981) J Chem Phys 61: 335
313. Lias S, Buckley TJ (1984) Int J Mass Spectrom Ion Processes 56: 123
314. Sack TM, Miller DL, Gross ML (1985) J Am Chem Soc 107: 6795
315. Hoffmann R, Heilbronner E, Gleiter R (1970) J Am Chem Soc 92: 706
316. Hoffmann R (1971) Acc Chem Res 4: 1
317. Dewar MJS, Wasson JS (1970) J Am Chem Soc 92: 3506
318. Heilbronner E, Schmelzer A (1975) Helv Chim Acta 58: 936
319. Haselbach E, Bally T, Lanyiova Z, Baertschi P (1979) Helv Chim Acta 62: 583
320. Roth HD, Schilling MLM, Jones G (1981) J Am Chem Soc 103: 1246
321. Roth HD, Schilling MLM (1981) J Am Chem Soc 103: 7210
322. Raghavachari K, Haddon RC, Roth HD (1983) J Am Chem Soc 105: 3110
323. Toriyama K, Nunone K, Iwasaki M (1983) J Chem Soc, Chem Commun 1346
324. Gerson F, Qin X-Z (1989) Helv Chim Acta 72: 383
325. Roth HD, Abelt CJ unpublished results
326. Basch H, Robin MB, Kuebler NA, Baker C, Turner DW (1969) J Chem Phys 51: 52
327. Symons MCR, Wren BW (1983) Tetrahedron Lett 24: 2315
328. Rhodes CJ, Symons MCR (1987) Chem Phys Lett 140: 611
329. Snow LD, Wang JT, Williams F (1983) Chem Phys Lett 100: 193
330. Snow LD, Qin X-Z, Williams F (1983) J Am Chem Soc 107: 3366
331. Snow LD, Williams F (1983) Chem Phys Lett 143: 521
332. Feller D, Davidson ER, Borden WT (1983) J Am Chem Soc 105: 3347
333. Feller D, Davidson ER, Borden WT (1984) J Am Chem Soc 106: 2513
334. Clark T (1984) J Chem Soc Chem Commun 666
335. Bouma WJ, Poppinger D, Saebo S, MacLeod JK, Radom L (1984) Chem Phys Lett 104: 198
336. Nobes RH, Bouma WJ, MacLeod JK, Radom L (1987) Chem Phys Lett 135: 78

337. Bally T, Nitsche S, Haselbach E (1984) Helv Chim Acta 67: 86
338. Williams F, Dai S, Snow LD, Qin X-Z, Bally T, Nitsche S, Haselbach E, Nelsen S, Teasley MF (1987) J Am Chem Soc 109: 7526
339. Quinn CB, Wiseman JR, Calabrese JC (1973) J Am Chem Soc 95: 6121
340. Gerson F, Qin X-Z, Ess C, Kloster-Jensen E (1989) J Am Chem Soc 111: 6456
341. Raghavachari K, Roth HD (1989) unpublished results
342. Bauld NL, Bellville DJ, Pabon R, Chelsky R, Green G (1983) J Am Chem Soc 105: 2378
343. Bouma WJ, Poppinger D, Radom L (1983) Isr J Chem 23: 21
344. Ushida K, Shida T, Iwasaki M, Toriyama K, Numone K (1983) J Am Chem Soc 105: 5496
345. Saik VO, Anisimov OA, Lozovoy VV, Molin YuN (1985) Z Naturforsch 40a: 239
346. Desrosiers MF, Trifunac AD (1986) J Phys Chem 90: 1560
347. Badger B, Brocklehurst B (1970) Trans Faraday Soc 66: 2939
348. Howarth OW, Fraenkel GK (1970) J Chem Phys 52: 6258
349. Turner DW, Baker C, Baker AD, Brundle CR (1970) Molecular Photoelectron Spectroscopy
350. Bischoff P, Haselbach E, Heibronner (1970) Angew Chem 82: 952
351. Haselbach E, Bally T, Lanyiova Z (1979) Helv Chim Acta 62: 577
352. Dunkin JR, Andrews L (1985) Tetrahedron 41: 145
353. Bellville DJ, Chelsky R, Bauld NL (1982) J Comput Chem 3: 584
354. Gross ML, Russel DH (1979) J Am Chem Soc 101: 2082
355. Dass C, Gross ML (1983) J Am Chem Soc 105: 5724
356. Dass C, Sack TM, Gross ML (1984) J Am Chem Soc 106: 5780
357. Haselbach E, Bally T, Gschwind R, Hemm U, Lanyiova Z (1979) Chimia 33: 405
358. Aebischer JN, Bally T, Roth K, Haselbach E, Gerson F, Qin X-Z (1989) J Am Chem Soc 111: 7909
359. Gerson F, Qin X-Z, Bally T, Aebischer JN (1988) Helv Chim Acta 71: 1069
360. Haselbach E, Bally T, Gschwind R, Klemm U, Lanyiova Z (1978) Chimia 33: 405
361. Roth HD, Schilling MLM, Raghavachari K (1984) J Am Chem Soc 106: 253
362. Peacock NJ, Schuster GB (1983) J Am Chem Soc 105: 3632
363. Rhodes CJ (1988) J Am Chem Soc 110: 4446
364. Raghavachari K, Roth HD (1989) J Am Chem Soc 111: 2028
365. Rhodes CJ (1988) J Am Chem Soc 110: 8567
366. Arnold A, Gerson F (1990) J Am Chem Soc 112: 2027
367. Williams F, Guo Q-X, Nelsen S (1990) J Am Chem Soc 112: 2028
368. Qin X-Z, Werst DW, Trifunac A (1990) J Am Chem Soc 112: 2026
369. Abelt CJ, Roth HD (1985) J Am Chem Soc 107: 3840
370. Abelt CJ, Roth HD, Schilling MLM (1985) J Am Chem Soc 107: 4148
371. Closs GL, Gordon S, Mulac WA (1982) J Org Chem 47: 5415
372. Raghavachari K, Haddon RC, Miller TA, Bondybey VE (1983) J Chem Phys 79: 1387
373. Kekulé A (1865) Justus Liebigs Ann Chem 137: 129
374. Hutton RS, Roth HD unpublished results
375. Hulme R, Symons MCR (1965) J Chem Soc 1220
376. Carter MK, Vincow G (1967) J Chem Phys 47: 292
377. Dessau RM, Shih S, Heiba EI (1970) J Am Chem Soc 92: 243
378. Elson IH, Kochi JK (1973) J Am Chem Soc 95: 5060
379. Doering WvE, Saunders M, Boynton HG, Earhart HW, Wadley EF, Laber G (1958) Tetrahedron 4: 178
380. Abelt CJ, Schilling ML, Roth HD unpublished results
381. Ladenburg A (1869) Chem Ber 2: 140
382. Raghavachari K, Roth HD unpublished results
383. Bischof P, Gleiter R, Mueller E (1976) Tetrahedron 32: 2769
384. Harman PJ, Kent JE, Gan TH, Peel JB, Willett GD (1977) J Am Chem Soc 99: 943
385. Gleiter R, Gubernator K, Eckert-Maksic M, Spanget-Larsen J, Bianco B, Gandillion G, Burger U (1981) Helv Chim Acta 64: 1312
386. Roth HD, Schilling MLM (1982) Proc IUPAC Symp Photochem 9: 286

387. Roth HD, Schilling MLM (1985) J Am Chem Soc 107: 716
388. Roth HD, Schilling MLM, Abelt CJ (1986) Tetrahedron 42: 6157
389. Roth HD, Schilling MLM, Abelt CJ (1986) J Am Chem Soc 108: 6098
390. Momose T, Shida T, Kobayashi T (1986) Tetrahedron 42: 6337
391. Roth HD (1984) Proc IUPAC Symp. Photochem 10: 455
392. Dai S, Wang JT, Williams F (1990) J Am Chem Soc 112: 2835
393. Dai S, Wang JT, Williams F (1990) J Am Chem Soc 112: 2837
394. Roth HD, Abelt CJ (1986) J Am Chem Soc 108: 2013
395. Shida T, Iwata S (1973) J Am Chem Soc 95: 3473
396. Roth HD (1988) Pure Appl Chem 60: 933
397. Guo QX, Qin XZ, Wang JT, Williams F (1988) J Am Chem Soc 110: 1974
398. Williams F, Guo QX, Petillo PA, Nelsen SF (1988) J Am Chem Soc 110: 7887
399. Williams F, Guo QX, Bebout DC, Carpenter BK (1989) J Am Chem Soc 111: 4133
400. Miyashi T, Konno A, Takahashi Y (1988) J Am Chem Soc 110: 3676
401. Doering WvE, Roth WR (1962) Tetrahedron 18: 67
402. Adam W, Grabowski S, Miranda MA, Rübenacker M (1988) J Chem Soc Chem Commun 142
403. Tsuji T, Miura T, Sugiura K, Nishida S (1990) J Am Chem Soc 112: 1998
404. Reents WD Jr, Roth HD, Schilling MLM, Abelt CJ (1986) Int J Mass Spectr Ion Processes 72: 155
405. Gerson F, de Meijere, Qin X-Z (1989) J Am Chem Soc 111: 1135
406. Takahashi Y, Mukai T, Miyashi T (1983) J Am Chem Soc 105: 6511
407. Miyashi T, Takahashi Y, Mukai T, Roth HD, Schilling ML (1985) J Am Chem Soc 107: 1079
408. Roth HD, Abelt CJ (1985) J Am Chem Soc 107: 6814
409. Merer RJ, Schoonveld L (1969) Can J Phys 47: 1731
410. Koeppel H, Domcke W, Cederbaum LS, von Niessen W (1978) J Chem Phys 69: 4252
411. Clark T, Nelsen SF (1988) J Am Chem Soc 110: 868
412. Winstein SJ (1959) J Am Chem Soc 81: 6524
413. Doering WvE, Laber G, Vonderwahl R, Chamberlain NF, Williams RB (1956) J Am Chem Soc 78: 5448
414. Lee RE, Squires RR (1986) J Am Chem Soc 108: 5078
415. Childs RF (1984) Accounts Chem Res 17: 347
416. Rieke R, Ogliaruso M, McClung R, Winstein S (1966) J Am Chem Soc 88: 4152
417. Katz TJ, Talcott C (1966) J Am Chem Soc 88: 4732
418. Doering WvE, Ferrier BM, Fossel ET, Hartenstein JH, Jones M Jr, Klumpp G, Rubin M, Saunders M (1967) Tetrahedron 23: 3943
419. Cheng Ak, Anet FAL, Mioduski J, Meinwald J (1974) J Am Chem Soc 96: 2887
420. Hoffmann R, Stohrer W-D (1971) J Am Chem Soc 93: 6941
421. Dewar MJS, Lo DH (1971) J Am Chem Soc 93: 7201
422. Miller LS, Grohmann K, Dannenberg JJ (1983) J Am Chem Soc 105: 6862
423. Ahlberg P, Engdahl C, Jonsall G (1981) J Am Chem Soc 103: 1583
424. Abelt CJ, Roth HD (1986) J Am Chem Soc 108: 2013
425. Abelt CJ, Roth HD (1986) J Am Chem Soc 108: 6734
426. Silberszyc W (1963) J Polym Sci Polym Lett Ed 1: 577
427. Zachariasse K (1991) Albert-Weller Festschrift J Phys Chem 95: 1867

Author Index Volumes 151–163

Author Index Vols. 26–50 see Vol. 50
Author Index Vols. 50–100 see Vol. 100
Author Index Vols. 101–150 see Vol. 150

The volume numbers are printed in italics